RNA and the Regulation of Gene Expression

A hidden layer of complexity

Edited by Kevin V. Morris

Caister Academic Press

Contents

List of contributors   v

Preface   vii

1   The Hammerhead Ribozyme Revisited: New Biological Insights for the Development of Therapeutic Agents and for Reverse Genomics Applications   1
Justin Hean and Marc S. Weinberg

2   Epigenetic Regulation of Gene Expression   19
Kevin V. Morris

3   The Role of RNAi and Non-coding RNAs in Polycomb-mediated Control of Gene Expression and Genomic Programming   29
Manuela Portoso and Giacomo Cavalli

4   Heterochromatin Assembly and Transcriptional Gene Silencing under the Control of Nuclear RNAi: Lessons from Fission Yeast   45
Aurélia Vavasseur, Leila Touat-Todeschini and André Verdel

5   RNA-mediated Gene Regulation in *Drosophila*   59
Harsh H. Kavi, Harvey R. Fernandez, Weiwu Xie and James A. Birchler

6   MicroRNA-mediated Regulation of Gene Expression   73
Lena J. Chin and Frank J. Slack

7   Viral Infection-Related MicroRNAs in Viral and Host Genomic Evolution   91
Yoichi R. Fujii and Nitin K. Saksena

8   Regulation of Mammalian Mobile DNA by RNA-based Silencing Pathways   109
Harris Soifer

9   The Role of Non-coding RNAs in Controlling Mammalian RNA Polymerase II Transcription   133
Stacey D. Wagner, Jennifer F. Kugel and James A. Goodrich

10   Pyknons as Putative Novel and Organism-specific Regulatory Motifs   149
Isidore Rigoutsos

11   RNA-Mediated Recognition of Chromosomal DNA   165
      David R. Corey

12   RNA-mediated Transcriptional Gene Silencing: Mechanism and
      Implications in Writing the Histone Code   173
      Kevin V. Morris

13   Small RNA-mediated gene activation   189
      Long-Cheng Li

14   Therapeutic Potential of RNA-mediated Control of Gene Expression:
      Options and Designs   201
      L Scherer and JJ Rossi

      Index   227

Contributors

James A. Birchler
Division of Biological Sciences
University of Missouri
Columbia, MO
USA

BirchlerJ@missouri.edu

Giacomo Cavalli
Institute of Human Genetics
Centre National de la Recherche Scientifique
Montpellier
France

Giacomo.Cavalli@igh.cnrs.fr

Lena J. Chin
Department of Molecular, Cellular and
Developmental Biology
Yale University
New Haven, CT
USA

lena.chin@yale.edu

David R. Corey
Departments of Pharmacology and Biochemistry
University of Texas
Dallas, TX
USA

david.corey@utsouthwestern.edu

Harvey R. Fernandez
Division of Biological Sciences
University of Missouri
Columbia, MO
USA

Fernandez@Biology.Rutgers.Edu

Yoichi R. Fujii
Nagoya City University
Nagoya
Japan

fatfuji@hotmail.com

James A. Goodrich
Department of Chemistry and Biochemistry
University of Colorado at Boulder
Boulder, CO
USA

james.goodrich@colorado.edu

Justin Hean
Department of Molecular Medicine and
Haematology
University of the Witwatersrand
Parktown
South Africa

Justin.Hean@students.wits.ac.za

Harsh H. Kavi
Division of Biological Sciences
University of Missouri
Columbia, MO
USA

hhk966@mizzou.edu

Jennifer F. Kugel
Department of Chemistry and Biochemistry
University of Colorado at Boulder
Boulder, CO
USA

jennifer.kugel@colorado.edu

Long-Cheng Li
Department of Urology
University of California San Francisco
San Francisco, CA
USA

lilc@urology.ucsf.edu

Kevin V. Morris
Department of Molecular and Experimental
Medicine
The Scripps Research Institute
La Jolla, CA
USA

kmorris@scripps.edu

Manuela Portoso
Institute of Human Genetics
Centre National de la Recherche Scientifique
Montpellier
France

Manuela.PORTOSO@igh.cnrs.fr

Isidore Rigoutsos
Bioinformatics and Pattern Discovery Group
IBM Thomas J. Watson Research Center
Yorktown Heights, NY
USA

rigoutso@us.ibm.com

JJ Rossi
Division of Molecular Biology
Beckman Research Institute of the City of Hope
Duarte, CA
USA

jrossi@bricoh.edu

Nitin K. Saksena
Westmead Millennium Institute
Westmead Hospital
Sydney
Australia

nitin_saksena@wmi.usyd.edu.au

L Scherer
Division of Molecular Biology
Beckman Research Institute of the City of Hope
Duarte, CA
USA

lscherer@COH.org

Frank J. Slack
Department of Molecular, Cellular and
Developmental Biology
Yale University
New Haven, CT
USA

frank.slack@yale.edu

Harris S. Soifer
Division of Molecular Biology
Beckman Research Institute of the City of Hope
Duarte, CA
USA

hsoifer@coh.org

Leila Touat-Todeschini
Institut National de la Santé et de la Recherche
Médicale
Institut Albert Bonniot
Grenoble
France

leila.todeschini@ujf-grenoble.fr

Aurélia Vavasseur
Institut National de la Santé et de la Recherche
Médicale
Institut Albert Bonniot
Grenoble
France

aurelia.vavasseur@e.ujf-grenoble.fr

André Verdel
Institut National de la Santé et de la Recherche
Médicale
Institut Albert Bonniot
Grenoble
France

andre.verdel@ujf-grenoble.fr

Stacey D. Wagner
Department of Chemistry and Biochemistry
University of Colorado at Boulder
Boulder, CO
USA

stacey.wagner@colorado.edu

Marc S. Weinberg
Department of Molecular Medicine and
Haematology
University of the Witwatersrand
Parktown
South Africa

Marc.Weinberg@wits.ac.za

Weiwu Xie
Division of Biological Sciences
University of Missouri
Columbia, MO
USA

XieW@missouri.edu

Preface

Over the past few years it has become increasingly apparent that RNA is operative in the regulation of gene expression. Importantly, it appears that small RNAs are able to control gene expression by directed control of epigenetic modifications at particular genomic loci. The notion that RNA can regulate gene expression will no doubt prove to significantly change the current paradigm in molecular biology. This text has been compiled from several scientists working to some extent on understanding the role of RNA in the regulation of gene expression. When these writings are taken together it becomes overwhelmingly apparent that a 'hidden layer of complexity' is operative in the regulation of gene expression.

Kevin V. Morris

Other Books of Interest

Plasmids: Current Research and Future Trends	2008
Pasteurellaceae: Biology, Genomics and Molecular Aspects	2008
Vibrio cholerae: Genomics and Molecular Biology	2008
Pathogenic Fungi: Insights in Molecular Biology	2008
Helicobacter pylori: Molecular Genetics and Cellular Biology	2008
Corynebacteria: Genomics and Molecular Biology	2008
Staphylococcus: Molecular Genetics	2008
Leishmania: After The Genome	2008
Archaea: New Models for Prokaryotic Biology	2008
RNA and the Regulation of Gene Expression	2008
Legionella Molecular Microbiology	2008
Molecular Oral Microbiology	2008
Epigenetics	2008
Animal Viruses: Molecular Biology	2008
Segmented Double-Stranded RNA Viruses	2008
Acinetobacter Molecular Biology	2008
Pseudomonas: Genomics and Molecular Biology	2008
Microbial Biodegradation: Genomics and Molecular Biology	2008
The Cyanobacteria: Molecular Biology, Genomics and Evolution	2008
Coronaviruses: Molecular and Cellular Biology	2007
Real-Time PCR in Microbiology: From Diagnosis to Characterisation	2007
Bacteriophage: Genetics and Molecular Biology	2007
Candida: Comparative and Functional Genomics	2007
Bacillus: Cellular and Molecular Biology	2007
AIDS Vaccine Development: Challenges and Opportunities	2007
Alpha Herpesviruses: Molecular and Cellular Biology	2007
Pathogenic *Treponema*: Molecular and Cellular Biology	2007
PCR Troubleshooting: The Essential Guide	2006
Influenza Virology: Current Topics	2006
Microbial Subversion of Immunity: Current Topics	2006
Cytomegaloviruses: Molecular Biology and Immunology	2006
Papillomavirus Research: From Natural History To Vaccines and Beyond	2006
Epstein Barr Virus	2005
HIV Chemotherapy: A Critical Review	2005
Probiotics and Prebiotics: Scientific Aspects	2005
Foodborne Pathogens: Microbiology and Molecular Biology	2005

Caister Academic Press   www.caister.com

The Hammerhead Ribozyme Revisited: New Biological Insights for the Development of Therapeutic Agents and for Reverse Genomics Applications

Justin Hean and Marc S. Weinberg

Abstract

Hammerhead ribozymes are the smallest known naturally occurring ribozymes which are capable of catalysing the endonucleolytic *trans*-esterification of RNA. A recent re-examination of the catalytic properties of naturally derived hammerhead ribozymes has resulted in a better understanding of the catalytic efficiency of this enzyme *in vitro* and *in vivo*. The minimal *trans*-cleaving hammerhead ribozyme has been a ubiquitous tool in both genomics and therapeutics research over the last 20 years and these new insights into hammerhead ribozyme biochemistry may offer hope for the generation of improved *trans*-cleaving ribozymes which function effectively *in vivo*. Next-generation hammerhead ribozymes may play an important role as therapeutic agents, as enzymes which tailor defined RNA sequences, as biosensors, and for applications in functional genomics and gene discovery.

Catalytic RNAs

RNA has the dual task of being both an information carrier and a biological catalyst, the latter being a role previously thought to be the sole preserve of proteins. It was the discovery, by Cech, Altman and colleagues in the early 1980s, that RNA possesses independent catalytic activity (Guerrier-Takada *et al.*, 1983; Kruger *et al.*, 1982), which confirmed Crick, Orgel and Woese's earlier speculations that life may have emerged from a pre-cellular environment of RNA (Crick, 1968; Orgel, 1968; Woese, 1967). Gilbert later called this the 'RNA World' hypothesis (Gilbert, 1986). The name *ribozyme* was thus coined to denote all RNA sequences with enzyme-like functions. Yet few vestiges remain of this earlier world since the natural catalytic repertoire of ribozymes is limited mainly to RNA processing reactions (Canny *et al.*, 2007; Guerrier-Takada *et al.*, 1983; Kruger *et al.*, 1982; Peebles *et al.*, 1986; Winkler *et al.*, 2004), with peptidyl transfer reactions within the ribosome being the notable exception (Agmon *et al.*, 2005; Cech, 2000; reviewed by Rodnina *et al.*, 2007). The hammerhead ribozyme belongs to a class of extant ribozymes that catalyse the endonucleolytic *trans*-esterification of the phosphodiester bond backbone of RNA resulting in products with 2',3'-cyclic phosphate and 5'-hydroxyl termini (Prody *et al.*, 1986). These reactions usually require structural and/or catalytic divalent metal ions under physiological conditions. However the use of *in vitro* directed evolution techniques (SELEX) has greatly expanded the catalytic capabilities of both natural ribozymes and hammerhead ribozymes (reviewed by Fiammengo and Jaschke, 2005), underscoring the potential role of ribozyme-based reactions in a pre-DNA/protein world. Here we revisit the biological function and application of the hammerhead ribozyme, which has for over 20 years been the most well-studied ribozyme structure with a diverse and versatile set of distinct applications. New insight into the catalytic structure and function of the hammerhead ribozyme has given fresh life to old applications, and may spur renewed interest in this model biological molecule.

The smallest natural RNA catalyst

The hammerhead ribozyme is one of the smallest known catalytic RNAs and belongs to a group of small ribozymes: the hairpin ribozyme (Buzayan *et al.*, 1986; Dange *et al.*, 1990), hepatitis delta virus (HDV) ribozyme (Wu *et al.*, 1989) and the *Neurospora* mitochondrial Varkud satellite (VS) ribozyme (Saville and Collins, 1990). The hammerhead ribozyme was first identified as a self-cleaving motif in strands of both polarity within viroid and viroid-like satellite RNA sequences (Daros and Flores, 1995; Forster and Symons, 1987; Hutchins *et al.*, 1986) and later discovered within repetitive satellite DNA sequences of caudate amphibians (Epstein and Coats, 1991), *Dolichopoda* cave cricket species (Rojas *et al.*, 2000), schistosomes (Ferbeyre *et al.*, 1998) and most recently within a genomic region of *Arabidopsis thaliana* (Przybilski *et al.*, 2005). In contrast to the catalytic activity of hammerhead ribozymes in the newt, schistosome, cave cricket and *Arabidopsis* (which are mostly associated with transcribed repetitive DNA sequences), hammerhead ribozyme activity in the small plant pathogens is well defined and appears to be an integral component for genomic replication. The observed RNA processing involves the site-specific, self-cleavage (and possibly ligation) of linear RNA intermediates from multimeric RNA precursors. These pregenomic viral concatamers represent the precursor RNA multiples of monomeric plus and minus RNA template strands that undergo site-specific internal RNA editing (Bratty *et al.*, 1993; Symons, 1997). Spliced monomers then join head-to-tail to form a circularized single-stranded RNA genome (Buzayan *et al.*, 1986; Cote *et al.*, 2001; Cote and Perreault, 1997). This form of the rolling-circle replication mechanism is a feature shared by all ribozyme-containing RNA pathogens. Evidence from experiments using *in vitro* selection techniques suggests that the hammerhead ribozyme catalytic motif is ubiquitously conserved for the catalysis of phosphodiester bond hydrolysis, and that divergent organisms may thus have derived their hammerhead ribozyme function independently (Salehi-Ashtiani and Szostak, 2001). The biological role of hammerhead ribozymes in newt, schistosomes and cave crickets remains speculative. These ribozymes were found to be active *in vivo* and appear to impact RNA processing events at the riboprotein complex (Denti *et al.*, 2000; Luzi *et al.*, 1997). More specifically, with respect to the newt hammerhead ribozyme, dimeric and multimeric RNA transcripts, which are generated by all somatic tissues as well as in the testes, self-cleave into monomers at the hammerhead domain. Monomeric units contain intact hammerhead ribozyme sequences. These sequences associate with the newt ovary riboprotein complex with the help of a protein that specifically binds to the ovarian form of the newt ribozyme (Cremisi *et al.*, 1992; Denti *et al.*, 2000).

The hammerhead ribozyme as a catalyst: revisiting an old system

The hammerhead ribozyme catalytic core consists of eleven highly conserved nucleotides at the junction of three helices (Fig. 1.1). Structurally, the three helices of the hammerhead ribozyme form a three-dimensional γ or 'wishbone' shape, where helices II and III are aligned co-axially and where helix I lies adjacent to helix II (Fig. 1.1B). Two structural domains have been delineated (Hertel *et al.*, 1992). Domain I consists of nucleotides 5′ $C_3U_4G_5A_6$ 3′ and the scissile residue H_{17} (typically C_{17}). Domain II comprises nucleotides 5′ $G_{12}A_{13}A_{14}A_{15}$ 3′ and 5′ G_8A_9 3′ (outlined letters in Fig. 1.1). The conserved nucleotide U_{16} is on its own but forms part of a cleavage triplet 5′ $N_{15}U_{16}H_{17}$ 3′ (italicized letters in Fig. 1.1). Since the cleavage triplet forms part of the strand with the least conserved residues, it is a feature which has been heavily exploited in *trans*-cleaving versions of the hammerhead ribozyme and where cleavage occurs after H_{17} (Haseloff and Gerlach, 1989). The conserved elements of the hammerhead ribozyme were initially determined by removing nucleotides originally thought to be non-essential. The effects of non-conserved sequences peripheral to the catalytic core were largely ignored for over 15 years (Uhlenbeck, 2003). A kinetically well-behaved minimal hammerhead ribozyme was described as optimal with an observed cleavage rate constant (k_{obs}) of ~1 min^{-1} under millimolar concentrations Mg^{2+} (Hertel *et al.*, 1994). This

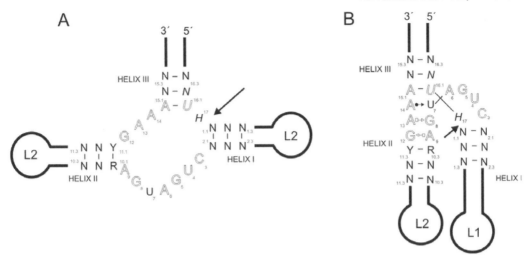

Figure 1.1 A two-dimensional representation of the hammerhead ribozyme with all three helical arms terminated by nucleotide loops. Conserved nucleotides are outlined and are labelled according to Hertel *et al.* (1992). The cleavage triplet is italicized. N represents any nucleotide; H represents any nucleotide except G. The dashes represent Watson-Crick base pairs. The catalytic core consists of domain I (C_3 to A_6) and domain II (U_7 to A_9 and G_{12} to A_{14}). The black arrow indicates the cleavage site. (A) A two dimensional representation of the hammerhead ribozyme structure is depicted, indicating the confluence of three helices surrounding a catalytic core of 11 conserved nucleotides. (B) A representation of the γ-shaped structure that results from both Watson-Crick and reversed-Hoogsteen base pairing, showing a co-axial alignment of helices II and III and a parallel alignment of helices I and II.

minimal hammerhead ribozyme is capable of catalysing the reverse ligation reaction but at a much lower rate. A recent re-examination of the kinetics of natural self-cleaving (*cis*-cleaving) hammerhead ribozymes, which have distinct helical-loop motifs not associated with the catalytic core, show much higher cleavage rates than was originally determined (De la Pena *et al.*, 2003; Khvorova *et al.*, 2003). In particular, the cleavage reaction is more pronounced at submillimolar concentrations of Mg^{2+}. Loops 1 and 2 of naturally derived *cis*-cleaving hammerhead ribozymes engage in non-canonical base pairing resulting in 'kissing loop' interactions that dramatically affect the cleavage rate (De la Pena *et al.*, 2003; Khvorova *et al.*, 2003). Similar fast *cis*-cleavage rates were also seen for the hammerhead ribozymes of *Schistosoma mansoni*, Smα1 (Canny *et al.*, 2004), and *Dolichopoda*, Pst3, where loop 2 interacts with an internal bulge in helix I (Yen *et al.*, 2004). In addition, a version of Smα1 targeted to cleave its native complementary sequence in *trans*, yielded rates that are up to 50-fold higher than a minimal *trans*-cleaving hammerhead ribozyme (Canny *et al.*, 2004; Penedo *et al.*, 2004). The natural hammerhead ribozyme is also capable of achieving *trans*-ligation rates which are over

2000 fold higher than the minimalist ribozyme, suggesting that the ligation reaction plays an equally important role in natural enzyme kinetics (Canny *et al.*, 2007). Recent crystallographic data indicate that loop–loop interactions cause a distortion in helix I, inducing partial unwinding, which likely facilitates the alignment and interaction of core-associated nucleotides (Martick and Scott, 2006) (Fig. 1.2). This rearrangement clearly assigns roles for catalytically important core nucleotides which were never seen within minimal ribozyme structures (Blount and Uhlenbeck, 2005). For example, canonical pairing between nucleotides G_8 and C_3 help to stabilize the catalytic core; other bases such as G_{12}, A_{13} and A_{14} allow for conformational locking of an active catalytic pocket. Nucleotide residues in the minimal hammerhead ribozyme core were often seen to be too far a part to interact. However, the conformational change that occurs in native or restored hammerhead ribozymes causes multiple hydrogen bond interactions between bases in a compact arrangement (Blount and Uhlenbeck, 2005; Martick and Scott, 2006). Structurally, G_5 and A_6 are wedged between C_{17} and $N_{1.1}$ ($C_{1.1}$), further facilitating cleavage (Martick and Scott, 2006) (Fig. 1.2).

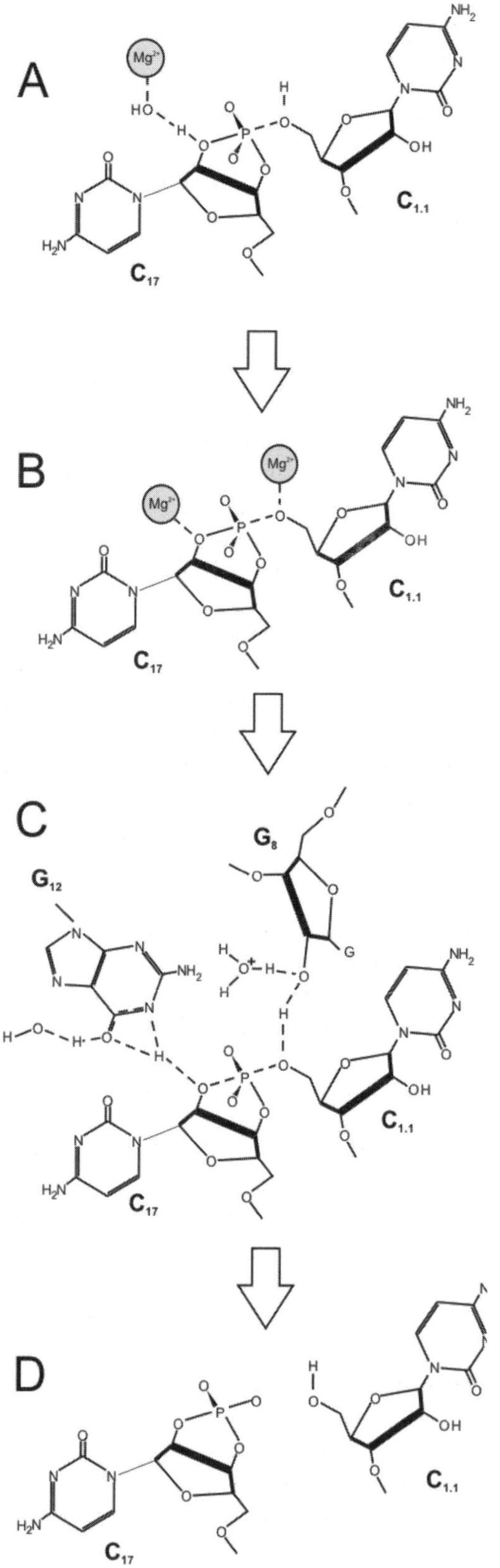

Figure 1.2 Theoretical mechanisms of cleavage within the native hammerhead ribozyme which are facilitated by the presence magnesium ions. (A) The single metal ion mechanism is shown whereby a hydroxylated Mg^{2+} acts as a general base by accepting the 2' H, facilitating the nucleophilic attack on the phosphate group (Mg^{2+} are represented as spheres). The double metal ion mechanism as seen in (B) involves the direct co-ordination of the Mg^{2+} with the 2' and the 5' oxygen atoms of C_{17} (arrows indicate the movement of electrons). This causes the $C_{1.1}$ phosphate to be more susceptible to nucleophilic attack by the C_{17} 2'-OH. (C) As adapted from Martick et al. (2006), demonstrating how G_{12} behaves as a general base, abstracting the H from the 2'-OH of C_{17}; G_8 acts as a general acid, donating a proton to the 5' leaving group of $C_{1.1}$, thus increasing the propensity of a nucleophilic attack on the $C_{1.1}$ phosphate. (D) C_{17} and $C_{1.1}$ are cleaved, leaving respective 2',5'-cyclic phosphate and 5'-OH termini.

Probably one of the most important findings in hammerhead ribozyme biology in the past few years is that divalent metal ions are not required for cleavage, but may enhance the tertiary structure formation (Martick and Scott, 2006). This goes against accepted dogma that hammerhead ribozymes are 'metalloenzymes'. The conformational adjustment of the core allows for another interesting change to occur: G_{12} behaves as a general base by allowing hydrogen abstraction to a 2'-OH, forming a penta-coordinated state (Han and Burke, 2005; Lambert et al., 2006; Martick and Scott, 2006). Conversely, G_8 behaves as a general acid, allowing for the stabilization of the 5'-O leaving group of $C_{1.1}$ (Martick and Scott, 2006) (Fig. 1.2).

Although it appears that the cleavage reaction of the hammerhead ribozyme has been largely elucidated, further characterization of the core may follow in the future. However one conundrum still remains: how was the minimal ribozyme able to cleave? There are a few ideas as to how this may happen. Firstly, the catalytic core of the minimal hammerhead ribozyme is dynamic in nature and is able to generate multiple conformational states, one of which may be the momentary active cleavage state. Another plausible reason for cleavage of the minimal system is that divalent metal ions such as Mg^{2+} do actually play a role, functioning in general acid-base catalysis as was initially suspected (Blount and Uhlenbeck, 2005; Kisseleva et al.,

2005). Magnesium, as a divalent cation may also bind to the negatively charged backbone of RNA causing further structural stability of the catalytic core (Penedo *et al.*, 2004). Ultimately, by simplifying the native hammerhead ribozyme to create a minimal system, a very different tertiary structure is formed. Yet although the minimal hammerhead ribozyme functions in a different manner to its native counterpart, the catalytic outcome for both is the same. Ironically, the hammerhead ribozyme structure was simplified in order to determine its catalytic behaviour in *trans*, yet it was only after restoration of the extra-core elements that natural catalytic mechanisms were elucidated. Since all *trans*-cleaving hammerhead ribozymes are derivatives of the minimal derivatives studied in the late 80s, a revised understanding of the catalytic properties of the natural *cis*-cleaving hammerhead ribozymes will probably stimulate the development of novel *trans*-cleaving versions. These hammerhead ribozymes are most likely to be exploited as novel therapeutic agents or as useful biotechnological tools.

Generating a *trans*-cleaving hammerhead ribozyme

The structural arrangement of hammerhead ribozyme helices and loops provide a natural classification for three different classes of hammerhead ribozymes (Fig. 1.3A). Type I and II hammerhead ribozymes have open 5' and 3' termini within loops 1 and 2 respectively. Loop–loop interactions are made possible by specific bulges on one of the helical arms. Although several examples of naturally occurring type I hammerhead ribozymes are known, thus far no type II ribozymes have been discovered. The vast majority of hammerhead ribozymes fall within the type III classification, where loops 1 and 2 are closed and helix III is open. Artificial *trans*-cleaving hammerhead ribozymes have been constructed in a number of ways: by closing helix III and opening helices I and II (Clouet-D'Orval and Uhlenbeck, 1996); with helix I closed and helices II and III open-ended (Jeffries and Symons, 1989; Uhlenbeck, 1987); and similarly by closing helix II and opening helices I and III (Haseloff and Gerlach, 1989). The hammerhead ribozyme described by Haseloff and Gerlach

(1989), where a nucleotide loop terminates helix II, is the most versatile of the *trans*-cleaving hammerhead ribozymes since almost all the conserved sequences necessary for the hammerhead ribozyme catalytic core formation can be positioned in one strand (Ruffner *et al.*, 1989) (Fig. 1.3B). The target or complementary RNA strand, which forms helices I and III upon hybridization, requires only the cleavage triplet (as described above) to generate an active site for effective catalysis. As a result, this conformation is the most suitable for applications that require the targeting of ribozymes to foreign RNA substrates, where the most common and effective cleavage triplet, 5'GUC3', is present every 64 bases (Shimayama *et al.*, 1995). However, the use of the Haseloff-Gerlach conformation may have contributed to our lack of understanding of native hammerhead ribozyme activity as loop 1 is absent in the context of an open helix I, thus preventing loop–loop interactions (Fig. 1.3B). There is nevertheless little doubt that the generation of *trans*-cleaving artificial hammerhead ribozymes resulted in a significant upsurge in interest in using this simple catalyst for biotechnological purposes, and in particular for therapeutic use.

The use of *trans*-cleaving hammerhead ribozymes as therapeutic agents

Trans-cleaving ribozymes can be distinguished from antisense RNAs in their ability to undergo multiple reactions. However, minimal hammerhead ribozyme catalytic efficiency is several orders of magnitude lower *in vivo* compared to *in vitro*, with single-turnover conditions prevailing (James and Gibson, 1998; Yen *et al.*, 2004). In fact, to date little is known about the true inhibitory effect of most *trans*-cleaving hammerhead ribozymes and it is seems that that these ribozymes behave mostly as antisense RNAs *in vivo*. Often post-transcriptional inhibition results obtained for catalytically inactive hammerhead ribozyme mirror those obtained for catalytically active hammerhead ribozymes. The therapeutic use of hammerhead ribozymes is thus curtailed by a limited understanding of their catalytic activity in an intracellular or *in vivo* environment. The inefficient *in vivo* inhibitory effects observed thus far for hammerhead ribozymes contrast

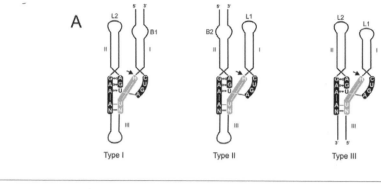

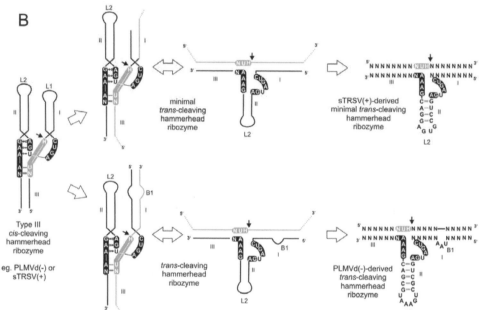

Figure 1.3 Naturally occurring *cis*-cleaving hammerhead ribozymes and the generation of *trans*-cleaving mimics. (A) Natural *cis*-cleaving hammerhead ribozymes fall into three types, type I, II and III, and each is identified by a specific helix which is open-ended. To date, no naturally occurring type II hammerhead ribozymes have been identified. (B) Adaptation of the type III *cis*-cleaving hammerhead ribozyme into *trans*-cleaving variants. The minimal hammerhead ribozyme derived from satellite RNA of tobacco ringspot virus, sTRSV(+) (Haseloff and Gerlach, 1989), is the most common ribozyme variant used for *trans*-cleavage. Recent adaptations of the hammerhead ribozyme derived from the negative strand of peach latent mosaic viroid, PLMVd(–), has resulted in a much more catalytically effective *trans*-cleaving hammerhead ribozymes, which functions under low Mg^{2+} concentrations (Saksmerprome *et al.*, 2004; Weinberg and Rossi, 2005). The conserved nucleotides within the core are shaded in black, and the cleavage triplet is shaded in grey. A black arrow indicates the cleavage site. Where possible, non-canonical base pair interactions are shown in the classification of (Leontis and Westhof, 2001).

markedly with the relatively efficient inhibitory activity of RNA Interference effector sequences such as short interfering RNAs (siRNAs) and short hairpin RNAs (shRNAs), which utilize an endogenous cellular pathway for post-transcriptional inhibition (reviewed in Kim and Rossi, 2007). Since the advent of RNAi, a lot of attention has shifted away from using ribozyme-based strategies for therapeutics. However,

recent new knowledge of natural hammerhead ribozyme catalytic activity and unique features associated with hammerhead ribozymes may be shifting the tide back. Ribozymes as therapeutic agents have distinct advantages over RNAi effectors. For example, siRNAs are less specific than ribozymes and are thus prone to off-target effects (Akashi *et al.*, 2005; Jackson *et al.*, 2003). In addition, double-stranded RNA is known to induce unwanted immunostimulation (Judge et al., 2005; Karpala et al., 2005; Sledz and Williams, 2004) and saturating the endogenous RNAi pathway has been shown to induce toxic harmful effects (An et al., 2006; Grimm et al., 2006). Unlike siRNA sequences, hammerhead ribozymes are capable of functioning independently and act on target RNA sequences located within all cellular compartments (Kato et al., 2001). Nevertheless, continued optimization of in vivo ribozyme inhibitory efficacy remains an important objective in order for hammerhead ribozymes to fulfill their therapeutic potential. Effective therapeutic ribozyme function hinges on a number of factors, which include: the identification of the optimal sequences which are accessible for targeting, improving the subcellular co-localization of ribozyme and target sequences, and determining the appropriate length of the hybridizing sequences to ensure specificity while maintaining enzymatic efficiency. Yet it is the intracellular hammerhead ribozyme catalytic activity which must be improved to truly distinguish ribozyme-mediated inhibition from antisense effects and for hammerhead ribozymes to rival the inhibitory effects of RNAi.

Improving hammerhead ribozyme trans-cleavage activity

Minimal trans-cleaving ribozymes require >5 mM Mg^{2+} for effective cleavage. These elevated Mg^{2+} ion concentrations are several fold higher than the 0.1–1 mM found for most intracellular Mg^{2+} levels (London, 1991) resulting in a 100-fold reduction in the cleavage rate of minimal hammerhead ribozymes. As mentioned earlier, natural cis-cleaving hammerhead ribozymes cleavage efficiently at submillimolar concentrations of Mg^{2+}, and are thus capable of significantly improved catalytic activity under physiologically conditions (Khvorova et al., 2003). Several attempts have been made to generate functionally useful trans-cleaving hammerhead ribozymes which adopt structural features from endogenous hammerhead ribozymes in order to improve their catalytic activity (Burke and Greathouse, 2005; Nelson et al., 2005; Saksmerprome et al., 2004; Weinberg and Rossi, 2005). We have shown that tertiary interactions

between internal loops found in diverse natural cis-cleaving hammerhead ribozymes can be successfully recapitulated in a trans-cleaving ribozyme format for faster cleavage under single-turnover conditions (Weinberg and Rossi, 2005). Interestingly, the most effective trans-cleaving activity was obtained from a mimic of the peach latent mosaic viroid (PLMVd) antigenomic hammerhead ribozyme. For the PLMVd(−) variant, only a 5′ UAA 3′ bulge in the arm of an open Helix I in the context of a natural PLMVd(−) loop 2 was necessary to dramatically improve hammerhead ribozyme trans-cleaving efficiency in vitro (Fig. 1.3B). These results were in agreement with Burke and colleagues, who used an in vitro selection and evolution (SELEX) approach to derive optimal trans-cleaving hammerhead ribozymes (Saksmerprome et al., 2004). Although these developments bode well for the generation of hammerhead ribozymes with improved intracellular activity, little is known at this stage whether modified fast-cleaving hammerhead ribozymes are truly more effective at inhibiting targeted RNA substrates in vivo than minimal hammerhead ribozymes.

Therapeutic targets for trans-cleaving hammerhead ribozyme

Some progress has been achieved in applying hammerhead ribozymes to target rogue genetic elements found within cellular targets or pregenomic and subgenomic RNAs of a wide variety of viruses (reviewed in Bartolome et al., 2004; Fanning and Symonds, 2006). However, the last 20 years have yielded inconsistent results and to date, the antiviral therapeutic use of hammerhead ribozymes does not yet match potent post-transcriptional effects of RNAi. There are some noteworthy exceptions where hammerhead ribozymes have been used creatively to suppress viral and cellular gene expression. HIV-1 remains an important target for hammerhead ribozymes and there have been many attempts at inhibiting this virus using synthetic RNAs or vector-delivered hammerhead ribozyme cassettes (Chang et al., 2005; Li and Rossi, 2005; Ramezani et al., 2006; Unwalla et al., 2006). For HIV-1, one particular strategy appears to be clinically relevant. Rossi and colleagues have previously described the expression and potent anti-HIV-1 inhibitory efficacy of a triple combination lentiviral construct comprising a U6 TAR RNA decoy with a U16 snoRNA for nucleolar localization, a U6 shRNA targeted to both *tat* and *rev* open reading frames, and a VA1 anti-CCR5 *trans*-cleaving hammerhead ribozyme (Li and Rossi, 2005). The triple construct efficiently transduced human progenitor CD34+ cells and was able to show improved suppression of HIV-1 over 42 days when compared to a single anti-*tat/rev* shRNA or double combinations of shRNA/ribozyme or decoy (Li and Rossi, 2005). Moreover, vector-transduced CD34+ cells were able to generate viral resistant T cells when injected into SCID-hu mice thy/liv grafts *in vivo* (Anderson et al., 2007). These results bode well for future *ex vivo* clinical trials.

Hammerhead ribozyme therapeutic strategies may play a significant role in treatment of cancer. One strategy relies on discriminating single nucleotide polymorphisms within the *ras* oncogenic transcript that creates a specific hammerhead ribozyme cleavage triplet (Irie et al., 1999; Kijima et al., 2004). Here, the therapeutic ribozyme would recognize both wild-type and mutant forms of the mRNA, but only cleave the oncogenic mRNA. This form of treatment has been used successfully *in vivo* against autosomal dominant retinitis pigmentosa (ARDP) (Gorbatyuk et al., 2007; Lewin et al., 1998; Sullivan et al., 2002). Unfortunately, the success of such treatment strategies is limited to the discovery of targets which contain discriminatory nucleotide mutations and is thus not generally applicable.

Expression and co-localization strategies

RNA polymerase III (Pol III) promoters naturally drive the expression of small RNAs such as tRNAs, 5S rRNA and most snRNAs. Transcript expression levels for Pol III promoters are significantly higher than for mRNAs generated from Pol II promoters. Unlike mRNAs, only short ancillary sequences are added to each transcript making them ideally suited for expression of hammerhead ribozymes (Cotten and Birnstiel, 1989). Hammerhead ribozyme inhibition of gene expression is greatly facilitated

by localizing both ribozymes and their target RNA to the same intracellular compartment (Arndt and Rank, 1997; Castanotto et al., 2000; Sullenger and Cech, 1993). Each ribozyme application is different, however, and the nature of the target mRNA is often carefully studied prior to establishing a co-localization strategy. Directing target mRNA and ribozymes to the same subcellular region is achieved by utilizing auxiliary sequences which are often promoter-derived. These sequences include: viral packaging signals (Pal et al., 1998; Sullenger and Cech, 1993), and either nuclear localization signals (Michienzi et al., 2000; Michienzi et al., 1998) or signals which allow co-localization of ribozymes and their pre-mRNA targets (Lee et al., 1999). Signals which direct the trafficking of target pre-mRNAs can be attached to ribozymes. This is best illustrated by the introduction, within the U1 snRNA coding region, of ribozymes that have 3'-end attached sequences which are part of the 3' untranslated regions (UTR) of certain mRNAs. Such ribozyme-encoding transcripts were channelled to 5' splice sites of targeted HIV-1 *rev* pre-mRNAs (Michienzi et al., 1996). The promoter itself encodes transcripts that can be exploited for their localization potential. For example, HIV-1 utilizes a tRNA^{Lys3} as a specific primer for reverse transcription. Ribozymes incorporated within the tRNA^{Lys3} coding region were able to generate a chimeric ribozyme-tRNA^{Lys3} transcript which co-packages along with HIV-1 into proviral particles and cleaves HIV RNA simultaneously with reverse transcriptase priming (Welch et al., 1997). Similarly, tRNA^{Val} (Yuyama et al., 1992), tRNA^{Meth} (Peter et al., 2007) and Adenoviral VAI (Prislei et al., 1997) Pol III promoters embed regulatory elements within their naturally transcribed regions. Ribozyme sequences can be inserted within stem–loop structures of these promoters systems to achieve high-level expression, ribozyme structural stability, and either nuclear or cytoplasmic localization (Kato et al., 2001; Kuwabara et al., 2001).

Multiple hammerhead ribozyme technologies

Viral RNA targets are prone to evolve mutations which resist the inhibitory effects of hybridizing nucleic acids such as hammerhead ribozymes. This is especially the case for mutations that occur within the hammerhead ribozyme cleavage triplet sequence, which may completely abrogate ribozyme-mediated cleavage. Mutations elsewhere may affect the accurate hybridization of ribozyme annealing arms with the target complementary sequence. Viruses like HIV-1 and HBV replicate using the error-prone reverse transcriptase, which lacks a proof-reading function (Preston *et al.*, 1988) and are thus especially prone to mutational escape. There have been reports of ribozyme (Bertrand and Rossi, 1996; Dropulic *et al.*, 1992) and more recently RNAi (Boden *et al.*, 2003; Das *et al.*, 2004; Sabariegos *et al.*, 2006) escape mutants generated for HIV-1 infections in cultured cells. Complicating matters further, HIV-1 has been shown to escape RNAi-mediated silencing by generating mutations in sequences which flank the targeted site, allowing for viral escape to occur by evolving alternative RNA secondary structures (Westerhout *et al.*, 2005). Thus, by targeting a single site for ribozyme-mediated cleavage, there exists the real possibility of evolving viral variants capable of evading ribozyme therapeutic action. To overcome the problem posed by the mutability of viruses such as HBV or HIV-1, several ribozymes can be applied to target simultaneously different sites on the viral sequence. A multiple targeting strategy for HIV-1 has been met with success when using RNAi effector sequences expressed from adjacent RNA Pol III promoters (ter Brake *et al.*, 2006). Several unique strategies have been devised for multiple hammerhead ribozymes. One approach is to use multiple ribozyme units that are joined together and expressed on the same transcript (Bai *et al.*, 2001; Chen *et al.*, 1992; Ramezani *et al.*, 1997b). Single transcripts harbouring different ribozymes bound head-to-tail on the same strand, have been constructed to target various sites on *BCR/ABL* mRNA (Leopold *et al.*, 1995), HIV-1 (Bai *et al.*, 2001; Chen *et al.*, 1992; Ohkawa *et al.*, 1993b; Ramezani *et al.*, 1997a) and HBV (Goila and Banerjea, 2004; Wands *et al.*, 1997). For these ribozymes, cleavage efficiency was shown to be directly proportional to the number of ribozyme units present on the transcript RNA.

However, there is a limit to the number of bound ribozymes which can exert maximal cleavage activity *in vitro*. No increase in cleavage efficiency was observed by adding more than three ribozyme units to the same transcript (Ohkawa *et al.*, 1993b), since this method lends itself to stearic hindrance between connected ribozymes. Here, hammerhead ribozymes bound together within a single transcript are catalytically constrained. Moreover, the kinetic mechanism of bound ribozyme units has not yet been adequately determined *in vivo*. A particularly useful strategy is the generation of multiple *trans*-cleaving hammerhead ribozymes within a single transcript such that they can be released from the parental chain through the action of flanking *cis*-cleaving hammerhead ribozymes present on both the 5′ and 3′ ends of a *trans*-cleaving ribozyme (Yuyama *et al.*, 1992). Released individual ribozyme monomers are capable of cleaving target RNA in *trans* (Inokuchi *et al.*, 1994; Ohkawa *et al.*, 1993a; Ohkawa *et al.*, 1993b; Yuyama *et al.*, 1992; Yuyama *et al.*, 1994). *Trans*-cleaving ribozymes processed from a single transcript are more efficient catalytically than the *trans*-cleaving action of ribozymes bound together on the same transcript (Inokuchi *et al.*, 1994). Later studies have used several different ribozymes that release themselves from a single transcript (Inokuchi *et al.*, 1994). In this system, 5′ and 3′ processed *trans*-cleaving multimeric ribozymes were capable of inhibiting target HIV RNA expression when expressed from retroviral vectors transduced into cultured cells (Xing *et al.*, 1995), but were no more effective than single-unit ribozymes in cell culture. A particularly interesting adaptation of this approach has been used to generate replication-competent vectors for HCV, whereby infectious viral RNA can be processed intracellularly by 5′ and 3′ flanking *cis*-cleaving hammerhead ribozymes (Heller *et al.*, 2005; Kato *et al.*, 2007). This technique has been subsequently used for the rescue of Encephalomyocarditis virus (EMCV) (Kobayashi *et al.*, 2007), Sleeping disease virus (SDV) (Moriette *et al.*, 2006) and Borna disease virus (BDV) (Yanai *et al.*, 2006). A processing system, where 5′ and 3′ *cis*-cleaving ribozyme motifs generate a *trans*-cleaving ribozyme was successfully developed for the inhibition of human papilloma virus 11 E6/E7 mRNA and HBV RNA *in vivo* (Pan *et al.*, 2004). Further modifications of the 'shotgun' multimeric ribozyme method (Inokuchi *et al.*, 1994) has resulted in a simplified system which includes multimeric units of *trans*-cleaving hammerhead ribozymes which also engage in *cis*-cleavage (Ruiz *et al.*, 1997) (Fig. 1.4). In this system each hammerhead ribozyme unit present on the transcript RNA includes a *cis*-cleaving ribozyme recognition sequence. *Cis*-cleaved individual monomeric ribozyme units are then capable of retaining their function to cleave a target RNA

in *trans*. The Ruiz model was applied successfully *in vitro* and *in vivo* to target simultaneously multiple *cis*- and *trans*-cleaving hammerhead ribozymes to different sites of the *HBx* transcript of HBV (Weinberg *et al.*, in press). Similar multimeric *cis*- and *trans*-cleaving hammerhead ribozymes targeted to the type 1 collagen gene, *Col1A1*, were shown to be far more effective than monomer ribozymes at inhibiting the target in cell culture experiments. Moreover, intracellular nuclear cleavage was observed and enhanced inhibitory effects correlated with an increased detection of monomeric hammerhead

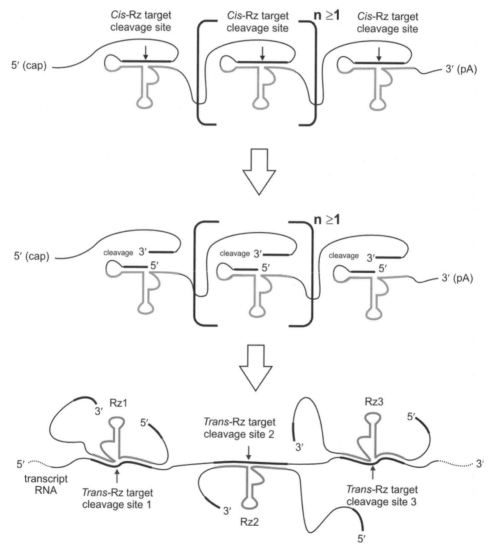

Figure 1.4 The principle of multimeric hammerhead ribozyme *cis*- and *trans*-cleavage action. A precursor transcript comprising several *cis*- and *trans*-cleaving hammerhead ribozyme releases single-unit ribozymes through internal *cis*-cleavage resulting in hammerhead ribozymes which retain their *trans*-cleaving function to cleave different target sites within a single transcript.

ribozymes (Peace *et al.*, 2005). There are some preliminary indications that intracellular *cis-* and *trans*-cleavage can be dramatically improved by using hammerhead ribozyme derivatives which include natural loop1–loop2 interactions (Hean, Weinberg, Arbuthnot, unpublished results).

Allosteric modulation of *trans*-cleaving hammerhead ribozymes

Hammerhead ribozymes have been adapted or modified in various ways to facilitate enhanced or controllable functioning. Ribozymes have a very useful trait in that they can be modified so that their activity is dependent on external components, such as an oligonucleotide or antibiotic. Typically helix II of the minimal *trans*-cleaving hammerhead ribozyme is adapted to provide binding sites for specific oligonucleotides which can inhibit or modulate ribozyme activity (Komatsu *et al.*, 2000). Such oligonucleotides can be used to create biological switches akin to those observed in electronic networks (Penchovsky and Breaker, 2005). The hammerhead ribozyme also lends itself to further modification by *in vitro* evolution technologies such as SELEX, resulting in allosteric ribozymes which are modulated or activated by small molecules like ATP (Tang and Breaker, 1997), flavin mononucleotide (Araki *et al.*, 1998), theophylin (Soukup and Breaker, 1999) and cAMP (Koizumi *et al.*, 1999). Morpholino oligonucleotides and the natural ribozyme inhibitor, toyocamycin, have been used to abolish the activity of natural *cis*-cleaving hammerhead ribozymes placed within a reporter gene construct, allowing for accurate tracking of transcripts *in vivo* (Yen *et al.*, 2004). Allosteric hammerhead ribozymes have also been adapted to detect structural conformational changes in proteins (Vaish *et al.*, 2002). In this example, hammerhead ribozymes were activated by the unphosphorylated or phosphorylated form of protein kinase ERK2, thus monitoring events during the post-translational modifications of proteins (Vaish *et al.*, 2002). Lastly, *trans*-cleaving hammerhead ribozymes can be modulated by the target RNA strand itself. A system known as targeted attenuated-ribozyme probe (TRAP) relies on a 3′ terminal 'attenuator' which inactivates the hammerhead ribozyme catalytic core (Burke *et* *al.*, 2002). The hammerhead ribozyme folds into the active conformation when bound to substrate RNA resulting in target strand cleavage.

Hammerhead ribozyme libraries

The therapeutic use of *trans*-cleaving hammerhead ribozymes has led to the development of other interesting applications, one of which includes the use of hammerhead ribozymes in functional genomics research. Hammerhead ribozymes appear to be especially useful as a gene discovery tool, for the identification of genes that are related to a specific phenotype (Fig. 1.5). A library of hammerhead ribozymes can be generated by introducing randomized nucleotides within the binding arms of helices I and III. These randomized ribozymes can be inserted into an appropriate expression system and introduced into cells where a specific phenotype can be selected through ribozyme-induced cleavage of a unique transcript. Subsequent amplification and sequencing of hammerhead ribozymes isolated from cells expressing specified phenotypic properties yield sequences which can be aligned to genome-wide databases, such as BLAST, to identify candidate targets (Kruger *et al.*, 2000; Li *et al.*, 2000; Welch *et al.*, 2000). Alternatively, candidate sites can be identified by detecting ribozyme cleavage products by cleavage-specific amplification of cDNA ends (C-SPACE) (Kruger *et al.*, 2001), or by using 5′ and 3′ rapid amplification of cDNA ends (RACE) (Kruger *et al.*, 2000). Target sequences can then be verified to produce the expected phenotype by designing new hammerhead ribozymes or RNAi effectors against the identified gene, thus targeting the same or different sequences within the same transcript (Fig. 1.5). Randomized hammerhead libraries have proven to be particularly useful at identifying functional genes in many diverse biological pathways. Notably, genes involved in cancer biology have been identified based on the selection of cells with discernable tumorigenic phenotypic properties (reviewed in Akashi *et al.*, 2005). Innovative selection strategies identify cells which display specific migratory patterns, abnormal proliferation and/or apoptotic signals (Suyama *et al.*, 2003; Suyama *et al.*, 2004; Wadhwa *et al.*, 2004). Randomized libraries have

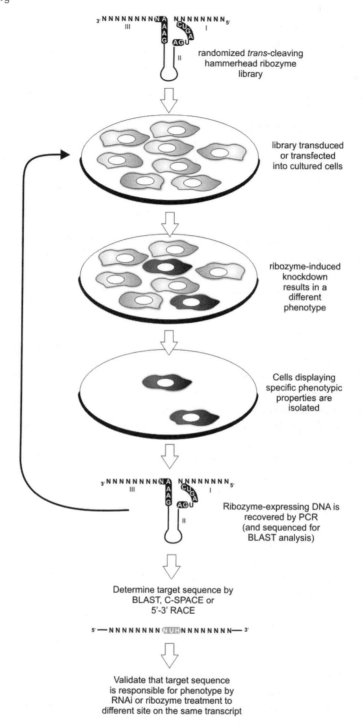

Figure 1.5 The use of a randomized *trans*-cleaving hammerhead ribozyme library for the identification of genes involved in selectable phenotypes. Libraries are introduced into cells transiently or permanently using viral vectors which express the library from ribozyme expression cassettes. Cells displaying selectable phenotypic properties are isolated, followed by the recovery of ribozymes genes and/or target cleavage products by PCR-based techniques. Recovered ribozymes are sequenced and aligned with genome-wide databases to identify candidate target genes. Genes putatively involved in generating the observed phenotype are verified by targeting alternative sites within the same mRNA using ribozyme or RNAi technologies.

been constructed from RNAi effector sequences such as shRNAs and siRNAs (Berns *et al.*, 2004; Paddison *et al.*, 2004). However, randomized hammerhead ribozyme libraries may be more specific than RNAi effectors since they hybridize and cleave 5′NUH3′ triplets, thus greatly simplifying target-site identification as many potential false positives can be removed from alignment searches which do not contain active cleavage triplets. However, as mentioned earlier, hammerhead ribozymes suffer from low intracellular efficacy and improved expression and cellular localization strategies are needed to produce meaningful results. It will be interesting to see whether *trans*-cleaving hammerhead ribozymes with enhanced *in vivo* catalytic activities can be applied to produce more effective randomized hammerhead ribozyme libraries.

Conclusion

Although 20 years have passed since the first hammerhead ribozyme was discovered, recent important discoveries continue to shed new light on the biology of this model enzyme. It remains surprising to observe the plethora of practical uses applied to this seemingly simple catalyst. Certainly in future we are likely to see further important biochemical and biotechnological discoveries now that the native structure and chemistry of the hammerhead ribozyme has been elucidated (Martick and Scott, 2006). Over the last few years, the therapeutic use of *trans*-cleaving hammerhead ribozymes has been severely hampered by its low-level activity *in vivo* and by the onset of RNAi technologies. Yet there are signs that the true catalytic potential of *trans*-cleaving hammerhead ribozymes may be recouped *in vivo* and that therapeutic derivatives will rightfully challenge or complement other nucleic acid hybridizing therapeutic strategies in future. Notwithstanding these new developments, already there are hammerhead ribozymes which march towards the clinic and undoubtedly there will be many more applications to come. The future outlook for hammerhead ribozyme research looks promising and hopefully the next twenty years will see many new advances and discoveries.

References

Agmon, I., Bashan, A., Zarivach, R., and Yonath, A. (2005). Symmetry at the active site of the ribosome: structural and functional implications. Biol. Chem. *386*, 833–844.

Akashi, H., Matsumoto, S., and Taira, K. (2005). Gene discovery by ribozyme and siRNA libraries. Nat. Rev. Mol. Cell. Biol. *6*, 413–422.

An, D.S., Qin, F.X., Auyeung, V.C., Mao, S.H., Kung, S.K., Baltimore, D., and Chen, I.S. (2006). Optimization and functional effects of stable short hairpin RNA expression in primary human lymphocytes via lentiviral vectors. Mol. Ther. *14*, 494–504.

Anderson, J., Li, M.J., Palmer, B., Remling, L., Li, S., Yam, P., Yee, J.K., Rossi, J., Zaia, J., and Akkina, R. (2007). Safety and Efficacy of a Lentiviral Vector Containing Three Anti-HIV Genes-CCR5 Ribozyme, Tat-rev siRNA, and TAR Decoy-in SCID-hu Mouse-Derived T Cells. Mol. Ther. *15*, 1182–8.

Araki, M., Okuno, Y., Hara, Y., and Sugiura, Y. (1998). Allosteric regulation of a ribozyme activity through ligand-induced conformational change. Nucleic Acids Res. *26*, 3379–3384.

Arndt, G.M., and Rank, G.H. (1997). Colocalization of antisense RNAs and ribozymes with their target mRNAs. Genome *40*, 785–797.

Bai, J., Rossi, J., and Akkina, R. (2001). Multivalent anti-CCR ribozymes for stem cell-based HIV type 1 gene therapy. AIDS Res. Hum. Retroviruses *17*, 385–399.

Bartolome, J., Castillo, I., and Carreno, V. (2004). Ribozymes as antiviral agents. Minerva Med. *95*, 11–24.

Berns, K., Hijmans, E.M., Mullenders, J., Brummelkamp, T.R., Velds, A., Heimerikx, M., Kerkhoven, R.M., Madiredjo, M., Nijkamp, W., Weigelt, B., *et al.* (2004). A large-scale RNAi screen in human cells identifies new components of the p53 pathway. Nature *428*, 431–437.

Bertrand, E., and Rossi, J.J. (1996). Anti-HIV therapeutic hammerhead ribozymes: targeting strategies and optimization of intracellular function, In Catalytic RNA, F. Eckstein, and D.M.J. Lilley, eds. (Berlin: Springer-Verlag), pp. 301–313.

Blount, K.F., and Uhlenbeck, O.C. (2005). The structure–function dilemma of the hammerhead ribozyme. Annu. Rev. Biophys. Biomol. Struct. *34*, 415–440.

Boden, D., Pusch, O., Lee, F., Tucker, L., and Ramratnam, B. (2003). Human immunodeficiency virus type 1 escape from RNA interference. J. Virol. *77*, 11531–11535.

Bratty, J., Chartrand, P., Ferbeyre, G., and Cedergren, R. (1993). The hammerhead RNA domain, a model ribozyme. Biochim. Biophys. Acta *1216*, 345–359.

Burke, D.H., and Greathouse, S.T. (2005). Low-magnesium, *trans*-cleavage activity by type III, tertiary stabilized hammerhead ribozymes with stem 1 discontinuities. BMC Biochem. *6*, 14.

Burke, D.H., Ozerova, N.D., and Nilsen-Hamilton, M. (2002). Allosteric hammerhead ribozyme TRAPs. Biochemistry *41*, 6588–6594.

Buzayan, J.M., Hampel, A., and Bruening, G. (1986). Nucleotide sequence and newly formed phosphodiester bond of spontaneously ligated satellite tobacco ringspot virus RNA. Nucleic Acids Res. *14*, 9729–9743.

Canny, M.D., Jucker, F.M., Kellogg, E., Khvorova, A., Jayasena, S.D., and Pardi, A. (2004). Fast cleavage kinetics of a natural hammerhead ribozyme. J. Am. Chem. Soc. *126*, 10848–10849.

Canny, M.D., Jucker, F.M., and Pardi, A. (2007). Efficient ligation of the schistosoma hammerhead ribozyme. Biochemistry *46*, 3826–3834.

Castanotto, D., Scherr, M., Lee, N.S., and Rossi, J. (2000). Targeting and Intracellular Expression Strategies for Therapeutic Ribozymes, In Ribozyme Biochemistry and Biotechnology, G. Krupp, and R.K. Gaur, eds. (Natick, MA: Eaton Publishing), pp. 421–440.

Cech, T.R. (2000). Structural biology. The ribosome is a ribozyme. Science *289*, 878–879.

Chang, L.J., Liu, X., and He, J. (2005). Lentiviral siRNAs targeting multiple highly conserved RNA sequences of human immunodeficiency virus type 1. Gene Ther. *12*, 1133–1144.

Chen, C.J., Banerjea, A.C., Harmison, G.G., Haglund, K., and Schubert, M. (1992). Multitarget-ribozyme directed to cleave at up to nine highly conserved HIV-1 env RNA regions inhibits HIV-1 replication – potential effectiveness against most presently sequenced HIV-1 isolates. Nucleic Acids Res. *20*, 4581–4589.

Clouet-D'Orval, B., and Uhlenbeck, O.C. (1996). Kinetic characterization of two I/II format hammerhead ribozymes. RNA *2*, 483–491.

Cote, F., Levesque, D., and Perreault, J.P. (2001). Natural 2′,5′-phosphodiester bonds found at the ligation sites of peach latent mosaic viroid. J. Virol. *75*, 19–25.

Cote, F., and Perreault, J.P. (1997). Peach latent mosaic viroid is locked by a 2′,5′-phosphodiester bond produced by *in vitro* self-ligation. J. Mol. Biol. *273*, 533–543.

Cotten, M., and Birnstiel, M.L. (1989). Ribozyme mediated destruction of RNA *in vivo*. EMBO J. *8*, 3861–3866.

Cremisi, F., Scarabino, D., Carluccio, M.A., Salvadori, P., and Barsacchi, G. (1992). A newt ribozyme: a catalytic activity in search of a function. Proc. Natl. Acad. Sci. USA *89*, 1651–1655.

Crick, F.H.C. (1968). The origin of the genetic code. J. Mol. Biol. *38*, 367–379.

Dange, V., Van Atta, R.B., and Hecht, S.M. (1990). A Mn2(+)-dependent ribozyme. Science *248*, 585–588.

Daros, J.A., and Flores, R. (1995). Identification of a retroviroid-like element from plants. Proc. Natl. Acad. Sci. USA *92*, 6856–6860.

Das, A.T., Brummelkamp, T.R., Westerhout, E.M., Vink, M., Madiredjo, M., Bernards, R., and Berkhout, B. (2004). Human immunodeficiency virus type 1 escapes from RNA interference-mediated inhibition. J. Virol. *78*, 2601–2605.

De la Pena, M., Gago, S., and Flores, R. (2003). Peripheral regions of natural hammerhead ribozymes greatly increase their self-cleavage activity. EMBO J. *22*, 5561–5570.

Denti, M.A., Martinez de Alba, A.E., Sagesser, R., Tsagris, M., and Tabler, M. (2000). A novel RNA-binding protein from Triturus carnifex identified by RNA-ligand screening with the newt hammerhead ribozyme. Nucleic Acids Res. *28*, 1045–1052.

Dropulic, B., Lin, N.H., Martin, M.A., and Jeang, K.T. (1992). Functional characterization of a U5 ribozyme: intracellular suppression of human immunodeficiency virus type 1 expression. J. Virol. *66*, 1432–1441.

Epstein, L.M., and Coats, S.R. (1991). Tissue-specific permutations of self-cleaving newt satellite-2 transcripts. Gene *107*, 213–218.

Fanning, G.C., and Symonds, G. (2006). Gene-expressed RNA as a therapeutic: issues to consider, using ribozymes and small hairpin RNA as specific examples. Handb Exp Pharmacol 289–303.

Ferbeyre, G., Smith, J.M., and Cedergren, R. (1998). Schistosome satellite DNA encodes active hammerhead ribozymes. Mol. Cell. Biol. *18*, 3880–3888.

Fiammengo, R., and Jaschke, A. (2005). Nucleic acid enzymes. Curr. Opin. Biotechnol. *16*, 614–621.

Forster, A.C., and Symons, R.H. (1987). Self-cleavage of plus and minus RNAs of a virusoid and a structural model for the active sites. Cell *49*, 211–220.

Gilbert, W. (1986). The RNA World. Nature *319*, 618.

Goila, R., and Banerjea, A.C. (2004). Sequence-specific cleavage of hepatitis X RNA in *cis* and *trans* by novel monotarget and multitarget hammerhead motif-containing ribozymes. Oligonucleotides *14*, 249–262.

Gorbatyuk, M., Justilien, V., Liu, J., Hauswirth, W.W., and Lewin, A.S. (2007). Preservation of photoreceptor morphology and function in P23H rats using an allele independent ribozyme. Exp. Eye Res. *84*, 44–52.

Grimm, D., Streetz, K.L., Jopling, C.L., Storm, T.A., Pandey, K., Davis, C.R., Marion, P., Salazar, F., and Kay, M.A. (2006). Fatality in mice due to oversaturation of cellular microRNA/short hairpin RNA pathways. Nature *441*, 537–541.

Guerrier-Takada, C., Gardiner, K., Marsh, T., Pace, N., and Altman, S. (1983). The RNA moiety of ribonuclease P is the catalytic subunit of the enzyme. Cell *35*, 849–857.

Han, J., and Burke, J.M. (2005). Model for general acid-base catalysis by the hammerhead ribozyme: pH-activity relationships of G8 and G12 variants at the putative active site. Biochemistry *44*, 7864–7870.

Haseloff, J., and Gerlach, W.L. (1989). Sequences required for self-catalysed cleavage of the satellite RNA of tobacco ringspot virus. Gene *82*, 43–52.

Heller, T., Saito, S., Auerbach, J., Williams, T., Moreen, T.R., Jazwinski, A., Cruz, B., Jeurkar, N., Sapp, R., Luo, G., and Liang, T.J. (2005). An *in vitro* model of hepatitis C virion production. Proc. Natl. Acad. Sci. USA *102*, 2579–2583.

Hertel, K.J., Herschlag, D., and Uhlenbeck, O.C. (1994). A kinetic and thermodynamic framework for the hammerhead ribozyme reaction. Biochemistry *33*, 3374–3385.

Hertel, K.J., Pardi, A., Uhlenbeck, O.C., Koizumi, M., Ohtsuka, E., Uesugi, S., Cedergren, R., Eckstein, F., Gerlach, W.L., Hodgson, R., and *et al.* (1992). Numbering system for the hammerhead. Nucleic Acids Res. *20*, 3252.

Hutchins, C.J., Rathjen, P.D., Forster, A.C., and Symons, R.H. (1986). Self-cleavage of plus and minus RNA transcripts of avocado sunblotch viroid. Nucleic Acids Res. *14*, 3627–3640.

Inokuchi, Y., Yuyama, N., Hirashima, A., Nishikawa, S., Ohkawa, J., and Taira, K. (1994). A hammerhead ribozyme inhibits the proliferation of an RNA coliphage SP in Escherichia coli. J. Biol. Chem. *269*, 11361–11366.

Irie, A., Anderegg, B., Kashi-Sabet, M., Ohkawa, T., Suzuki, T., Halks-Miller, M., Curiel, D.T., and Scanlon, K.J. (1999). Therapeutic efficacy of an adenovirus-mediated anti-H-ras ribozyme in experimental bladder cancer. Antisense Nucleic Acid Drug Development *9*, 341–349.

Jackson, A.L., Bartz, S.R., Schelter, J., Kobayashi, S.V., Burchard, J., Mao, M., Li, B., Cavet, G., and Linsley, P.S. (2003). Expression profiling reveals off-target gene regulation by RNAi. Nat. Biotechnol. *21*, 635–637.

James, H.A., and Gibson, I. (1998). The therapeutic potential of ribozymes. Blood *91*, 371–382.

Jeffries, A.C., and Symons, R.H. (1989). A catalytic 13-mer ribozyme. Nucleic Acids Res. *17*, 1371–1377.

Judge, A.D., Sood, V., Shaw, J.R., Fang, D., McClintock, K., and MacLachlan, I. (2005). Sequence-dependent stimulation of the mammalian innate immune response by synthetic siRNA. Nat. Biotechnol. *23*, 457–462.

Karpala, A.J., Doran, T.J., and Bean, A.G. (2005). Immune responses to dsRNA: implications for gene silencing technologies. Immunol Cell. Biol. *83*, 211–216.

Kato, T., Matsumura, T., Heller, T., Saito, S., Sapp, R.K., Murthy, K., Wakita, T., and Liang, T.J. (2007). Production of infectious hepatitis C virus of various genotypes in cell cultures. J. Virol. *81*, 4405–4411.

Kato, Y., Kuwabara, T., Warashina, M., Toda, H., and Taira, K. (2001). Relationships between the activities *in vitro* and *in vivo* of various kinds of ribozyme and their intracellular localization in mammalian cells. J. Biol. Chem. *276*, 15378–15385.

Khvorova, A., Lescoute, A., Westhof, E., and Jayasena, S.D. (2003). Sequence elements outside the hammerhead ribozyme catalytic core enable intracellular activity. Nat. Struct. Biol. *10*, 708–712.

Kijima, H., Yamakazi, H., Nakamura, M., Scanlon, K.J., Osamura, R.Y., and Ueyama, Y. (2004). Ribozyme against mutant K-ras mRNA suppresses tumor growth of pancreatic cancer. Int. J. Oncol. *24*, 559–564.

Kim, D.H., and Rossi, J.J. (2007). Strategies for silencing human disease using RNA interference. Nat. Rev. Genet. *8*, 173–184.

Kisseleva, N., Khvorova, A., Westhof, E., and Schiemann, O. (2005). Binding of manganese(II) to a tertiary stabilized hammerhead ribozyme as studied by electron paramagnetic resonance spectroscopy. RNA *11*, 1–6.

Kobayashi, T., Mikami, S., Yokoyama, S., and Imataka, H. (2007). An improved cell-free system for picornavirus synthesis. J. Virol. Methods.

Koizumi, M., Kerr, J.N.Q., Soukup, G.A., and Breaker, R.R. (1999). Allosteric ribozymes sensitive to the second messengers cAMP and cGMP. Nucleic Acids Symp. Ser. (Oxford) *42*, 275–276.

Komatsu, Y., Yamashita, S., Kazama, N., Nobuoka, K., and Ohtsuka, E. (2000). Construction of new ribozymes requiring short regulator oligonucleotides as a cofactor. J. Mol. Biol. *299*, 1231–1243.

Kruger, K., Grabowski, P.J., Zaug, A.J., Sands, J., Gottschling, D.E., and Cech, T.R. (1982). Self-splicing RNA: autoexcision and autocyclization of the ribosomal RNA intervening sequence of Tetrahymena. Cell *31*, 147–157.

Kruger, M., Beger, C., Li, Q.X., Welch, P.J., Tritz, R., Leavitt, M., Barber, J.R., and Wong-Staal, F. (2000). Identification of eIF2Bgamma and eIF2gamma as cofactors of hepatitis C virus internal ribosome entry site-mediated translation using a functional genomics approach. Proc. Natl. Acad. Sci. USA *97*, 8566–8571.

Kruger, M., Beger, C., Welch, P.J., Barber, J.R., and Wong-Staal, F. (2001). C-SPACE (cleavage-specific amplification of cDNA ends): a novel method of ribozyme-mediated gene identification. Nucleic Acids Res. *29*, E94.

Kuwabara, T., Warashina, M., Koseki, S., Sano, M., Ohkawa, J., Nakayama, K., and Taira, K. (2001). Significantly higher activity of a cytoplasmic hammerhead ribozyme than a corresponding nuclear counterpart: engineered tRNAs with an extended 3' end can be exported efficiently and specifically to the cytoplasm in mammalian cells. Nucleic Acids Res. *29*, 2780–2788.

Lambert, D., Heckman, J.E., and Burke, J.M. (2006). Three conserved guanosines approach the reaction site in native and minimal hammerhead ribozymes. Biochemistry *45*, 7140–7147.

Lee, N.S., Bertrand, E., and Rossi, J. (1999). mRNA localization signals can enhance the intracellular effectiveness of hammerhead ribozymes. RNA *5*, 1200–1209.

Leontis, N.B., and Westhof, E. (2001). Geometric nomenclature and classification of RNA base pairs. RNA *7*, 499–512.

Leopold, L.H., Shore, S.K., Newkirk, T.A., Reddy, R.M., and Reddy, E.P. (1995). Multi-unit ribozyme-mediated cleavage of bcr-abl mRNA in myeloid leukemias. Blood *85*, 2162–2170.

Lewin, A.S., Drenser, K.A., Hauswirth, W.W., Nishikawa, S., Yasumura, D., Flannery, J.G., and LaVail, M.M. (1998). Ribozyme rescue of photoreceptor cells in a transgenic rat model of autosomal dominant retinitis pigmentosa. Nat. Med. *4*, 967–971.

Li, M., and Rossi, J.J. (2005). Lentiviral vector delivery of siRNA and shRNA encoding genes into cultured and primary hematopoietic cells. Method. Mol. Biol. *309*, 261–272.

Li, Q.X., Robbins, J.M., Welch, P.J., Wong-Staal, F., and Barber, J.R. (2000). A novel functional genomics approach identifies mTERT as a suppressor of fibroblast transformation. Nucleic Acids Res. 28, 2605–2612.

London, R.E. (1991). Methods for measurement of intracellular magnesium: NMR and fluorescence. Annu. Rev. Physiol. 53, 241–258.

Luzi, E., Eckstein, F., and Barsacchi, G. (1997). The newt ribozyme is part of a riboprotein complex. Proc. Natl. Acad. Sci. USA 94, 9711–9716.

Martick, M., and Scott, W.G. (2006). Tertiary contacts distant from the active site prime a ribozyme for catalysis. Cell 126, 309–320.

Michienzi, A., Cagnon, L., Bahner, I., and Rossi, J.J. (2000). Ribozyme-mediated inhibition of HIV 1 suggests nucleolar trafficking of HIV-1 RNA. Proc. Natl. Acad. Sci. USA 97, 8955–8960.

Michienzi, A., Conti, L., Varano, B., Prislei, S., Gessani, S., and Bozzoni, I. (1998). Inhibition of human immunodeficiency virus type 1 replication by nuclear chimeric anti-HIV ribozymes in a human T lymphoblastoid cell line. Hum. Gene Ther. 9, 621–628.

Michienzi, A., Prislei, S., and Bozzoni, I. (1996). U1 small nuclear RNA chimeric ribozymes with substrate specificity for the Rev pre-mRNA of human immunodeficiency virus. Proc. Natl. Acad. Sci. USA 93, 7219–7224.

Moriette, C., Leberre, M., Lamoureux, A., Lai, T.L., and Bremont, M. (2006). Recovery of a recombinant salmonid alphavirus fully attenuated and protective for rainbow trout. J. Virol. 80, 4088–4098.

Nelson, J.A., Shepotinovskaya, I., and Uhlenbeck, O.C. (2005). Hammerheads derived from sTRSV show enhanced cleavage and ligation rate constants. Biochemistry 44, 14577–14585.

Ohkawa, J., Yuyama, N., Takebe, Y., Nishikawa, S., and Taira, K. (1993a). Importance of independence in ribozyme reactions: kinetic behavior of trimmed and of simply connected multiple ribozymes with potential activity against human immunodeficiency virus. Proc. Natl. Acad. Sci. USA 90, 11302–11306.

Ohkawa, J., Yuyama, N., Takebe, Y., Nisikawa, S., Homann, M., Sczakiel, G., and Taira, K. (1993b). Multiple site-specific cleavage of HIV RNA by transcribed ribozymes from shotgun-type trimming plasmid. Nucleic Acids Symp. Ser. 90, 121–122.

Orgel, L.E. (1968). Evolution of the genetic apparatus. J. Mol. Biol. 38, 381–393.

Paddison, P.J., Silva, J.M., Conklin, D.S., Schlabach, M., Li, M., Aruleba, S., Balija, V., O'Shaughnessy, A., Gnoj, L., Scobie, K., et al. (2004). A resource for large-scale RNA-interference-based screens in mammals. Nature 428, 427–431.

Pal, B.K., Scherer, L., Zelby, L., Bertrand, E., and Rossi, J.J. (1998). Monitoring retroviral RNA dimerization in vivo via hammerhead ribozyme cleavage. J. Virol. 72, 8349–8353.

Pan, W.H., Xin, P., Morrey, J.D., and Clawson, G.A. (2004). A self-processing ribozyme cassette: utility against human papillomavirus 11 E6/E7 mRNA and hepatitis B virus. Mol. Ther. 9, 596–606.

Peace, B.E., Florer, J.B., Witte, D., Smicun, Y., Toudjarska, I., Wu, G., Kilpatrick, M.W., Tsipouras, P., and Wenstrup, R.J. (2005). Endogenously expressed mul-timeric self-cleaving hammerhead ribozymes ablate mutant collagen in cellulo. Mol. Ther. 12, 128–136.

Peebles, C.L., Perlman, P.S., Macklenburg, K.L., Pertillo, M.L., Tabor, J.H., Jarrel, K.A., and Cheng, H.L. (1986). A self-splicing RNA excises an intron lariat. Cell 44, 212–223.

Penchovsky, R., and Breaker, R.R. (2005). Computational design and experimental validation of oligonucle-otide-sensing allosteric ribozymes. Nat. Biotechnol. 23, 1424–1433.

Penedo, J.C., Wilson, T.J., Jayasena, S.D., Khvorova, A., and Lilley, D.M. (2004). Folding of the natural hammerhead ribozyme is enhanced by interaction of auxiliary elements. RNA 10, 880–888.

Peter, J.U., Alenina, N., Bader, M., and Walther, D.J. (2007). Development of antithrombotic miniri-bozymes that target peripheral tryptophan hydroxy-lase. Mol. Cell Biochem. 295, 205–215.

Preston, B.D., Poiesz, B.J., and Loeb, L.A. (1988). Fidelity of HIV-1 reverse transcriptase. Science 242, 1168–1171.

Prislei, S., Buonomo, S.B., Michienzi, A., and Bozzoni, I. (1997). Use of adenoviral VAI small RNA as a carrier for cytoplasmic delivery of ribozymes. RNA 3, 677–687.

Prody, G.A., Bakos, J.T., Buzayan, J.M., Schneider, I.R., and Breuning, G. (1986). Autocatalytic processing of dimeric plant-virus satellite RNA. Science 231, 1577–1580.

Przybilski, R., Graf, S., Lescoute, A., Nellen, W., Westhof, E., Steger, G., and Hammann, C. (2005). Functional hammerhead ribozymes naturally encoded in the genome of Arabidopsis thaliana. Plant Cell 17, 1877–1885.

Ramezani, A., Ding, S.F., and Joshi, S. (1997a). Inhibition of HIV-1 replication by retroviral vectors expressing monomeric and multimeric hammerhead ribozymes. Gene Ther. 4, 861–867.

Ramezani, A., Ma, X.Z., Ameli, M., Arora, A., and Joshi, S. (2006). Assessment of an anti-HIV-1 combination gene therapy strategy using the antisense RNA and multimeric hammerhead ribozymes. Front Biosci. 11, 2940–2948.

Ramezani, A., Marhin, W., Weerasinghe, M., and Joshi, S. (1997b). A rapid and efficient system for screening HIV-1 Pol mRNA-specific ribozymes. Can. J. Microbiol. 43, 92–96.

Rodnina, M.V., Beringer, M., and Wintermeyer, W. (2007). How ribosomes make peptide bonds. Trends Biochem. Sci. 32, 20–26.

Rojas, A.A., Vazquez-Tello, A., Ferbeyre, G., Venanzetti, F., Bachmann, L., Paquin, B., Sbordoni, V., and Cedergren, R. (2000). Hammerhead-mediated processing of satellite pDo500 family transcripts from Dolichopoda cave crickets. Nucleic Acids Res. 28, 4037–4043.

Ruffner, D.E., Dahm, S.C., and Uhlenbeck, O.C. (1989). Studies on the hammerhead RNA self-cleaving domain. Gene 82, 31–41.

Ruiz, J., Wu, C.H., Ito, Y., and Wu, G.Y. (1997). Design and preparation of a multimeric self-cleaving hammerhead ribozyme. Biotechniques 22, 338–345.

Sabariegos, R., Gimenez-Barcons, M., Tapia, N., Clotet, B., and Martinez, M.A. (2006). Sequence homology required by human immunodeficiency virus type 1 to escape from short interfering RNAs. J. Virol. *80*, 571–577.

Saksmerprome, V., Roychowdhury-Saha, M., Jayasena, S., Khvorova, A., and Burke, D.H. (2004). Artificial tertiary motifs stabilize *trans*-cleaving hammerhead ribozymes under conditions of submillimolar divalent ions and high temperatures. RNA *10*, 1916–1924.

Salehi-Ashtiani, K., and Szostak, J.W. (2001). In vitro evolution suggests multiple origins for the hammerhead ribozyme. Nature *414*, 82–84.

Saville, B.J., and Collins, R.A. (1990). A site-specific self-cleavage reaction performed by a novel RNA in *Neurospora* mitochondria. Cell *61*, 685–696.

Shimayama, T., Nishikawa, S., and Taira, K. (1995). Generality of the NUX rule: kinetic analysis of the results of systematic mutations in the trinucleotide at the cleavage site of hammerhead ribozymes. Biochemistry *34*, 3649–3654.

Sledz, C.A., and Williams, B.R. (2004). RNA interference and double-stranded-RNA-activated pathways. Biochem. Soc. *Trans. 32*, 952–956.

Soukup, G.A., and Breaker, R.R. (1999). Design of allosteric hammerhead ribozymes activated by ligand-induced structure stabilization. Structure *7*, 783–791.

Sullenger, B.A., and Cech, T.R. (1993). Tethering ribozymes to a retroviral packaging signal for destruction of viral RNA. Science *262*, 1566–1569.

Sullivan, J.M., Pietras, K.M., Shin, B.J., and Misasi, J.N. (2002). Hammerhead ribozymes designed to cleave all human rod opsin mRNAs which cause autosomal dominant retinitis pigmentosa. Mol. Vis. *8*, 102–113.

Suyama, E., Kawasaki, H., Nakajima, M., and Taira, K. (2003). Identification of genes involved in cell invasion by using a library of randomized hybrid ribozymes. Proc. Natl. Acad. Sci. USA *100*, 5616–5621.

Suyama, E., Wadhwa, R., Kaur, K., Miyagishi, M., Kaul, S.C., Kawasaki, H., and Taira, K. (2004). Identification of metastasis-related genes in a mouse model using a library of randomized ribozymes. J. Biol. Chem. *279*, 38083–38086.

Symons, R.H. (1997). Plant pathogenic RNAs and RNA catalysis. Nucleic Acids Res. *25*, 2683–2689.

Tang, J., and Breaker, R.R. (1997). Examination of the catalytic fitness of the hammerhead ribozyme by *in vitro* selection. RNA *3*, 914–925.

ter Brake, O., Konstantinova, P., Ceylan, M., and Berkhout, B. (2006). Silencing of HIV-1 with RNA interference: a multiple shRNA approach. Mol. Ther. *14*, 883–892.

Uhlenbeck, O.C. (1987). A small catalytic oligoribonucleotide. Nature *328*, 596–600.

Uhlenbeck, O.C. (2003). Less isn't always more. RNA *9*, 1415–1417.

Unwalla, H., Chakraborti, S., Sood, V., Gupta, N., and Banerjea, A.C. (2006). Potent inhibition of HIV-1 gene expression and TAT-mediated apoptosis in human T cells by novel mono- and multitarget anti-TAT/Rev/Env ribozymes and a general purpose RNA-cleaving DNA-enzyme. Antiviral Res. *72*, 134–144.

Vaish, N.K., Dong, F., Andrews, L., Schweppe, R.E., Ahn, N.G., Blatt, L., and Seiwert, S.D. (2002). Monitoring post-translational modification of proteins with allosteric ribozymes. Nat. Biotechnol. *20*, 810–815.

Wadhwa, R., Yaguchi, T., Kaur, K., Suyama, E., Kawasaki, H., Taira, K., and Kaul, S.C. (2004). Use of a randomized hybrid ribozyme library for identification of genes involved in muscle differentiation. J. Biol. Chem. *279*, 51622–51629.

Wands, J.R., Geissler, M., Putlitz, J.Z., Blum, H., von Weizsacker, F., Mohr, L., Yoon, S.K., Melegari, M., and Scaglioni, P.P. (1997). Nucleic acid-based antiviral and gene therapy of chronic hepatitis B infection. J. Gastroenterol. Hepatol. *12*, S354–369.

Weinberg, M.S., Ely, A., Passman, M., Mufamadi, S., and Arbuthnot, P. (In press). Effective anti-HBV hammerhead ribozymes derived from multimeric precursors. Oligonucleotides *17*, 104–12.

Weinberg, M.S., and Rossi, J.J. (2005). Comparative single-turnover kinetic analyses of *trans*-cleaving hammerhead ribozymes with naturally derived non-conserved sequence motifs. FEBS Lett. *579*, 1619–1624.

Welch, P.J., Marcusson, E.G., Li, Q.X., Beger, C., Kruger, M., Zhou, C., Leavitt, M., Wong-Staal, F., and Barber, J.R. (2000). Identification and validation of a gene involved in anchorage-independent cell growth control using a library of randomized hairpin ribozymes. Genomics *66*, 274–283.

Welch, P.J., Tritz, R., Yei, S., Barber, J., and Yu, M. (1997). Intracellular application of hairpin ribozyme genes against hepatitis B virus. Gene Ther. *4*, 736–743.

Westerhout, E.M., Ooms, M., Vink, M., Das, A.T., and Berkhout, B. (2005). HIV-1 can escape from RNA interference by evolving an alternative structure in its RNA genome. Nucleic Acids Res. *33*, 796–804.

Winkler, W.C., Nahvi, A., Roth, A., Collins, J.A., and Breaker, R.R. (2004). Control of gene expression by a natural metabolite-responsive ribozyme. Nature *428*, 281–286.

Woese, C. (1967). The Genetic Code: The Molecular Basis for Genetic Expression (New York: Harper & Row).

Wu, H.N., Lin, Y.J., Lin, F.P., Makino, S., Chang, M.F., and Lai, M.M. (1989). Human hepatitis delta virus RNA subfragments contain an autocleavage activity. Proc. Natl. Acad. Sci. USA *86*, 1831–1835.

Xing, Z., Mahadeviah, S., and Whitton, J.L. (1995). Antiviral activity of RNA molecules containing self-releasing ribozymes targeted to lymphocytic choriomeningitis virus. Antisense Res. Dev. *5*, 203–212.

Yanai, H., Hayashi, Y., Watanabe, Y., Ohtaki, N., Kobayashi, T., Nozaki, Y., Ikuta, K., and Tomonaga, K. (2006). Development of a novel Borna disease virus reverse genetics system using RNA polymerase II promoter and SV40 nuclear import signal. Microbes Infect. *8*, 1522–1529.

Yen, L., Svendsen, J., Lee, J.S., Gray, J.T., Magnier, M., Baba, T., D'Amato, R.J., and Mulligan, R.C. (2004). Exogenous control of mammalian gene expression through modulation of RNA self-cleavage. Nature *431*, 471–476.

Yuyama, N., Ohkawa, J., Inokuchi, Y., Shirai, M., Sato, A., Nishikawa, S., and Taira, K. (1992). Construction of a tRNA-embedded-ribozyme trimming plasmid. Biochem. Biophys. Res. Commun. *186*, 1271–1279.

Yuyama, N., Ohkawa, J., Koguma, T., Shirai, M., and Taira, K. (1994). A multifunctional expression vector for an anti-HIV-1 ribozyme that produces a 5′- and 3′-trimmed *trans*-acting ribozyme, targeted against HIV-1 RNA, and *cis*-acting ribozymes that are designed to bind to and thereby sequester *trans*-activator proteins such as Tat and Rev. Nucleic Acids Res. *22*, 5060–5067.

Epigenetic Regulation of Gene Expression

2

Kevin V. Morris

Abstract

Epigenetics is the study of meiotically and mitotically heritable changes in gene expression which are not coded for in the DNA (Egger et al., 2004; Jablonka and Lamb, 2003). Three distinct mechanisms appear to be intricately related and implicated in initiating and/or sustaining epigenetic modifications; DNA methylation, RNA-associated silencing, and histone modifications (Egger et al., 2004). While chromatin remodelling and DNA methylation have been studied for several years now far less is known about how these epigenetic marks are directed to each particular gene. Recently, however the role of RNA in epigenetic gene regulation has begun to become apparent. In this chapter we will discuss the basic mechanisms of epigenetic regulation of gene expression and how epigenetics might be involved in the evolution of the cell.

Introduction

The quintessential essence of gene expression in eukaryotes is initiated in the cell nucleus. The nuclear compartment evolved in eukaryotes and essentially segregates DNA from the cytoplasm of the cell. Speculation as to why eukaryotes evolved a nuclear compartment whereas prokaryotes did not are to protect the cellular DNA from the sheer stresses induced upon the cell as a result of cell motility and/or as a result of the extensive editing of RNA molecules (Alberts B, 1994). The genome of each organism can be found in the chromosomes that make up each nucleated cell of that particular organism. The DNA of the organism is found within the nucleus packaged into chromosomes. The chromosomes are typically made up of nucleosomes which are essentially an octamer of histones (8 proteins) which the DNA is wrapped around (Fig. 2.1). Histones are small highly conserved proteins containing a large proportion of positively charged amino acids (lysines and arginines) which can function to maintain a charge-mediated interaction with the generally negatively charged nucleic acid (DNA) (Fig. 2.1). There are eight histones which make up the core histone that functions as the scaffolding for genomic DNA (~146 bp) to wrap around (Fig. 2.1) to produce the nucleosome. Traditionally chromatin, and in particular nucleosomes, were viewed simplistically as structural modalities which functioned mainly as a scaffolding for DNA to remain adhered. Interestingly, the N-terminal tails of histones H3 and H4 contain highly charged amino acids and as a result are generally tightly associated with DNA. Over the last decade it has become apparent that these highly charged histones (H3 and H4) tails can undergo several modifications such as acetylation, methylation, phosphorylation, ubiquitination, and sumoylation (Strahl and Allis, 2000; Strahl et al., 1999). These modifications have been referred to as the 'histone code' (Jenuwein and Allis, 2001). These modifications can have drastic effects on the local DNA, specifically those genes associated with the local chromatin and their subsequent transcriptional expression profile. Mechanistically the local histone modifications can regulate gene expression by altering the accessibility of the DNA (gene) to *trans*-

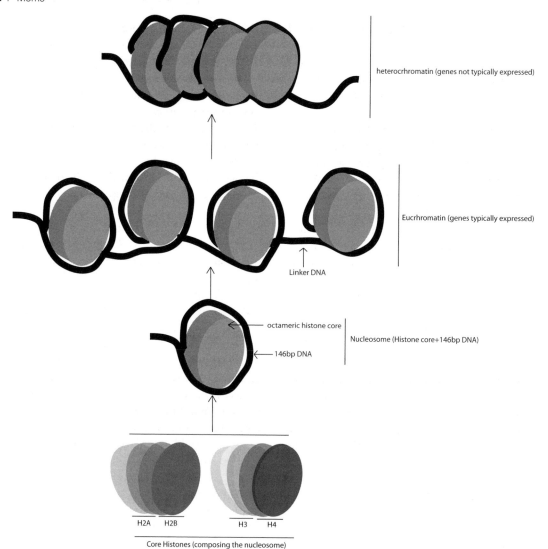

Figure 2.1 Eukaryotic chromatin. Chromatin is composed of both nucleic acids (DNA) and proteins (histones). An octameric histone core containing two of each core histone (H2A, H2B, H3, and H4) functions as the central scaffolding which ~146 bp of genomic DNA wraps around to compose the nucleosome. The nucleosomes essentially form the 'beads on a string' chromatin which can be either transcriptionally active (euchromatin) or transcriptionally silenced and condensed (heterochromatin).

acting factors or by directly recruiting proteins which recognize the subsequent histone modifications (Strahl and Allis, 2000). However, to better understand the suppressive or activating effects of the histone code on gene expression a brief study of transcription is required.

Gene expression (transcription)

There are currently three known RNA polymerases in human cells that function to transcribe DNA into RNA, RNA polymerase (RNAP) I, II, and III. Importantly, the RNA polymerases I and III transcribe a limited set of genes while RNAPII generally transcribes the messenger RNA (mRNA) for protein coding genes (Alberts B, 1994). RNAPI is generally localized to the nucleoli and transcribes the ribosomal RNA (rRNA) while RNAPIII transcribes the 5S rRNA and transfer RNAs (tRNA) (Paule and White, 2000). RNAPI and III transcription composes the majority of transcription (~80%) of the total RNA generated in the cell (Paule and White, 2000). Nonetheless, only RNAPII transcribes genes that are translated into proteins.

The transcription of a typical RNAPII expressed gene is a complex endeavour involving a plethora of factors and is a multi-stage process involving pre-initiation, initiation, and elongation. In order to initiate RNAPII-mediated transcription and in essence recognize the promoter region of the particular gene that is going to be transcribed the region must first be accessible to the basal transcription machinery. If for instance if the nucleosome is compacted with a silent state epigenetic mark or the addition of histone H1 it must first be modified or remodelled (Gerber and Shilatifard, 2003). There are predominantly two mechanisms by which epigenetically silent state chromatin can become remodelled and accessible to RNAPII-mediated transcription. Through ATP hydrolysis the nucleosomes can become shifted and moved along the DNA (Workman and Kingston, 1998) and/or evicted from the DNA (Boeger *et al.*, 2005; Schwabish and Struhl, 2004) or alternatively the histones making up the nucleosome can undergo covalent modifications (Strahl and Allis, 2000). Once the local genomic region corresponding to the gene promoter is accessible the RNAPII associated factors and transcription can take place. RNAPII requires the assistance of the TATA-binding protein (TBP), TFIIB and TFIIF (Holstege *et al.*, 1997). The TBP binds the TATA box located ~10–30 bp upstream of the typical transcriptional start site (Fig. 2.2) and is generally found in complex with several other proteins (Holstege *et al.*, 1997). The TFIIB and TFIIF are not necessarily required for promoter recognition as much as they are for the unwinding and opening up of the promoter for RNAPII to initiate transcription (Holstege *et al.*, 1997). Once the open complex is established and RNAPII binds the gene promoter it must undergo a series of events before productive elongation and in essence promoter clearance and transcription can occur (reviewed extensively in (Sims *et al.*, 2004). Once transcription is initiated and productive elongation established the RNAPII/complex continues transcribing the gene until transcriptional termination is established. Generally transcriptional termination and 3′ end processing is facilitated by the actions of RNAPII and in particular via interactions with the phosphorylated CTD-Ser-2 domain and several components of the 3′ end processing machinery (Hirose and Manley, 1998; Proudfoot and O'Sullivan, 2002). While it is clear that 3′ end processing involves the CTD-Ser-2 domain of RNAPII and 3′ end processing machinery there remain many unclear mechanistic details with regards to transcriptional termination. Thus, for active RNAPII-mediated transcription to occur the local chromatin for the particular genes promoter must be relatively relaxed. This requirement is in essence one important mechanistic details with regards to epigenetic modes of gene regulation.

Histone methylation and gene expression

Chromatin can exist inside the nucleus of the cell in either a less condensed and relatively transcriptionally active euchromatin or as the condensed relatively transcriptionally inactive heterochromatin (Fig. 2.1). Euchromatin is generally associated with acetylated histones (gener-

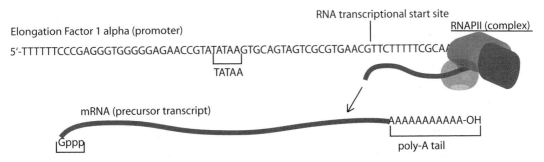

Figure 2.2 RNAPII-mediated transcription of the elongation factor 1 alpha gene. The EF1a promoter is shown with the TATA element ~30 bp upstream of the known mRNA transcriptional start site (Wakabayasi-Ito, 1994). Following RNAPII-mediated transcription of the EF1a gene a capped (Gppp) and polyadenylated mRNA is produced.

ally positively charged) and in particular with histone H3 di-methylation on lysine 4 (H3m-Lys-4), whereas the more silent heterochromatin is associated with histone H3 di-methylation on lysine 9 (H3mLys-9) (Lippman *et al.*, 2004). The acetylation of histone tails by histone acetyltransferase (HAT) ultimately results in relaxing of the chromatin and a disruption of histone-DNA interactions and gene activation (Fig. 2.3A). Conversely, the deacetylation of histones by histone deacetylases (HDACs) result in condensation of the chromatin and transcriptional repression (reviewed in Lusser, 2002) (Fig. 2.3B). Heterochromatin was initially discovered because it remained visibly condensed throughout

the cellular life cycle (Heitz, 1928). Upon further investigation it was determined that heterochromatin contained many repetitive elements and repetitive genomic regions, including past retroelements and transposons (Craig, 2005). There are now various forms of heterochromatin (constitutive and facultative). Of interest is the initiation of facultative heterochromatin, which can spend at least part of its overall lifecycle as euchromatin, but then can become permanently silenced through epigenetic mechanisms that are not fully understood (Craig, 2005). One example of facultative heterochromatin can be found in the mammalian X-chromosome which is permanently silenced following embryogenesis

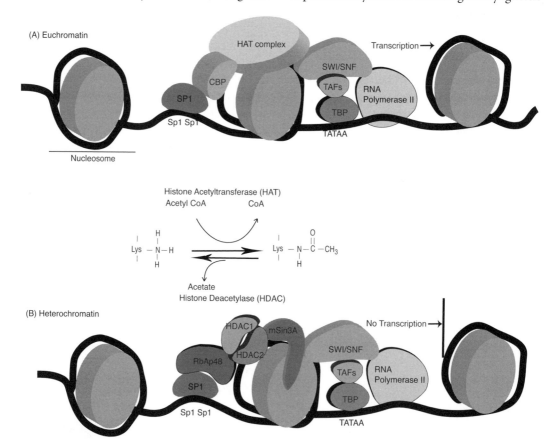

Figure 2.3 Euchromatin vs. heterochromatin. (A) Euchromatin (transcriptionally active) is shown with the Sp1 transcription factor bound to the Sp1 site (GGGCGG) which positively influences gene expression by recruiting the transcriptional co-activator cAMP response element binding protein (CBP). CBPcan intern associate with a histone acetyltransferase (HAT), which is functionally capable of acetylating histones (H2A, H2B, H3, and/or H4) resulting in the relaxation of the nucleosomes subsequently allowing for binding of the TATA binding protein (TBP), transcriptional activating factors (TAFs), SWI/SNF complex and eventual recruitment of RNA polymerase II. The result of this cascade is ultimately the transcription of the gene. (B) Heterochromatin (transcriptionally inactive) can also be found at the same nucleosomal region leading to transcriptional silencing and can be the result of the recruitment of one of 6 known chromatin remodelling complexes (reviewed in Dobosy and Selker, 2001) involving histone deacetylation by histone deacetylases (HDAC) resulting in nucleosomal compaction and possibly DNA methylation.

(Brockdorff, 2002). Importantly, it is becoming apparent that RNAs, small-interfering RNAs (siRNAs), antisense RNAs (asRNAs) and non-coding RNAs (ncRNAs) all appear to be involved in directing epigenetic silencing to corresponding genes with homology to the particular RNA (reviewed in Morris, 2005) possibly explaining the trigger as to how various regions of genomes become silenced. However, to date the most amount of information gathered on RNA-mediated regulation of organismal genomes has been sequestered in plants and yeast (Craig, 2005).

There are currently six well-defined histone deacetylase complexes (HDAC) which appear to be involved in the suppression of gene expression (reviewed by Dobosy and Selker, 2001). Indeed, HDACs are essential for the formation of heterochromatin and epigenetic gene silencing. HDACs are recruited to the particular gene to be silenced by various repressors and corepressors as well as proteins which bind to methylated regions of DNA. Generally, following DNA replication the newly formed nucleosomes contain histones that are predominantly acetylated and as such the gene is expressed (Ayer, 1999). HDACs can then be targeted to particular genes, by mechanisms that are not entirely clear; but may involve an RNA component, where they can initiate a cascade of deacetylation (Fig. 2.3). Of the six HDAC complexes involved in gene silencing 2 forms predominate; (1) the Sin3 HDAC complex which contains HDAC 1&2, Sin3A and MeCP2 (methyl-CpG-binding protein 2) (Dobosy and Selker, 2001) and (2) the MI2/NuRD complex, which is operative via histone methylation (H3 Lys-9) and has been shown to be involved in X inactivation in the mouse (Lee, 1999) as well as the recruitment of heterochromatin protein Swi6/HP1 (Hall, 2002; Volpe, 2002).

The HDAC-mediated modes of silencing ultimately result in the deacetylation of the histone tails, which is in essence the first step involved in writing a silent state histone mark, but ultimately relies on the methylation of the deacetylated histone tail to provide for the required charge ratio that will lead to targeted gene silencing. The methylation of histone 3 at the lysines residues is central to the localized HDAC-mediated silencing of gene expression. Histone methylation is carried out at the lysine residues specifically on deacetylated histone tails by the action of histone methyltransferases (Cao and Zhang, 2004; Sims et al., 2003). The most common of the three distinct methylation states is di-methylation followed by mono-methylation and then tri-methylation (Jenuwein, 2006). Histone methylation is not always directly associated with gene silencing and can, depending on the particular lysine methylated configuration, result in gene activation. Methylation at histone 3 residues lysines 9 and 27 (H3K9 and H3K27, respectively) have been shown to be associated with repression of gene expression (Sims et al., 2003) (Fig. 2.4). The methylation of lysines on the histone tails is the result of histone methyltransferases (HMTs). All the known HMTs associated with directing silent state histone methyl-marks on H3 contain SET domains (Craig, 2005). The most common silent state methyl mark is found on H3K9 (Fig. 2.4) and the result, specifically in human cells, of the HMT Suv39h1 or Suv39h2 (Craig, 2005). The

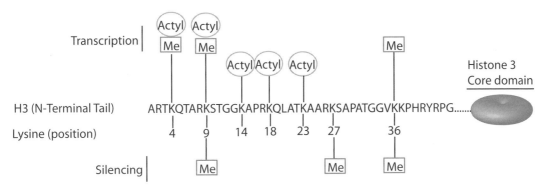

Figure 2.4 Some modifications found in lysines of histone 3. The amino acids in histone 3 tail and specifically 4–20 are involved in gene silencing. Various lysine methylation and acetylation sites are shown.

second most common methyl-mark for H3 is found on lysine 27 and is the result of the HMT enhancer of zeste homologue (EZH2) (Cao and Zhang, 2004) which can direct tri-methlyation at H3K27 (a methyl mark strongly associated with gene silencing, Fig. 2.4). While Suv39h1 or h2 as well as EZH2 are involved in the establishment of the histone methyl-mark, the maintenance and retention of these marks are probably the result of the HMT G9a (Cao and Zhang, 2004). Once the HMT has established the proper methyl-mark it is then possible, for example in the case of H3K9 methylation, for other gene repressive factors such as heterochromatin protein 1 (HP1) to bind the local chromatin specifically H3K9 via the HP1 chromodomain and modulate gene silencing and heterochromatinization (Lachner *et al.*, 2001; Stewart *et al.*, 2005).

DNA methylation and gene expression

While histone methylation is can be directly associated with gene silencing a second form of epigenetic control of gene expression has also been observed by the methylation of C^5 in CpG residues (Fig. 2.5) (Bird, 2002). The methylation of CpG residues is typically observed at intergenic regions and results in gene silencing (Razin and Riggs, 1980). DNA methylation is the result of DNA methyltransferases (DNMTs) which essentially transfer the methyl group from S-adenosylmethionine to the fifth position on the cytosine ring (Wu and Santi, 1985). The current dogma regarding DNA methylation is that it is established during development by DNMT3a and DNMT3b (Okano *et al.*, 1999) and then maintained in somatic cells by DNMT1 which can recognize the methylated parental strand and copy the methyl pattern to the unmethylated template (Razin and Szyf, 1984). Importantly, DNA methylation can be inherited to daughter cells and essentially functions to suppress gene expression by either hindering transcription factor binding or serving as a site for the binding of methylated DNA-binding domain containing proteins such as MBD1 (Sarraf and Stancheva, 2004) or MeCP2 (D'Alessio and Szyf, 2006). The binding of MeCP2 to the methylated CpG of a particular gene is not only directly with the methylcytosine but also with the core nucleosomal components (Chandler *et al.*, 1999). Furthermore, the activity of MeCP2 with regards to suppression of gene expression requires an HDAC complex containing HDAC1 and HDAC2 and as such supports the notion that chromatin and DNA epigenetic remodelling machinery are linked (Nan, 1998).

The role of epigenetic modifications in the evolution of the cell

With regards to human cells it is tempting to speculate that the methylation of CpG residues has acted as a naturally selective force in the

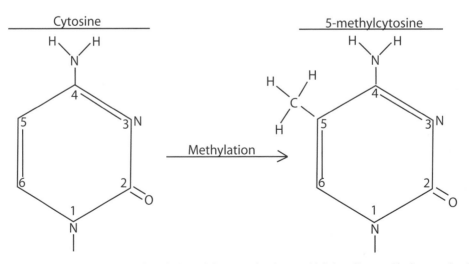

Figure 2.5 Cytosine methylation. Methylation of the cytosine base which is adjacent 5′ of a guanine in the DNA double helix results in CpG methylation that can transcriptionally silence the local gene.

evolution of the current state of the genome. Specifically the observation that in human cells methylated cytosines can, over a period of time, be eliminated due most likely to accidental deamination. The deamination of methylated cytosines subsequently gives rise to what appears to the cell as a thymine, which is not corrected by uracil DNA glycosylase and as such retained and recognized as a thymine during the next round of cellular and DNA replication. The result, upon cellular division, is that one of the daughter cells contains a thymine where a cytosine once resided and a genetic change that was induced by what was initially an epigenetic modification. In general, this paradigm makes sense if one is to consider the activity of the particular region of DNA. In promoters where consistent activity is required for cell fidelity the CpGs are maintained in a relatively demethylated state allowing for accidental deamination to be properly repaired, (i.e., an unmethylated cytosine is recognized as uracil by uracil DNA glycosylase and correctly excised and replaced with a cytosine). Whereas in non-promoter regions such as Alu repeated regions, an area typically littered with retrotransposons or mRNA coding regions actively targeted by lentiviruses such as HIV-1 (Wu *et al.*, 2003), one generally finds little in the form of CpG islands (Cross and Bird, 1995). To support this observation the vast majority of non-LTR retrotransposons (LINEs or L1s) in mammals are found to contain 5′ truncations, rearrangements, and non-sense mutations (Finnegan, 1997). Suggesting that a selective force was applied to inhibit the activity of these particular genomic regions.

Conclusion

The modification of histone tails, such as methylation, results in a 'histone code'. The 'histone code hypothesis' argues that the local histone environment (specifically in the nucleosomes) can have an effect on the expression profile of the corresponding local gene (Jenuwein and Allis, 2001). These 'marked' histone tails are then capable of dictating the recruitment of various specialized chromatin remodelling factors (Strahl *et al.*, 1999; Turner, 2000). To date the histone code is best exemplified by the sheer mulitiplic-

ity of modifications that can occur to histones (reviewed in Berger, 2002). Moreover, it has become apparent that chromatin and in particular epigenetic regulation of gene expression is linked between DNA CpG and histone methylation. Chromatin can affect DNA methylation and DNA methylation can affect chromatin. What is not clear is which comes first the DNA methylation or histone methylation. Some groups have observed histone methylation preceding DNA methylation (Bachman *et al.*, 2003) while others have observed DNA methylation preceding histone methylation (Curradi *et al.*, 2002). What is clear is that transgene silencing appears to precede the establishment of the respective silent state epigenetic mark (Mutskov and Felsenfeld, 2004).

Lastly, the fundamental underlying mechanism responsible for governing the histone code and epigenetic silencing is not yet understood. One potential mechanism to regulating the histone code could be mediated by the effect of small RNAs and will be discussed in greater detail in the following chapters. Interestingly, the recent discovery and characterization of a vast array of small (21- to 26-nt), non-coding RNAs suggests that there is an RNA component, possibly involved in epigenetic gene regulation which is weaved into the basic fabric of the cell and has been to date overlooked (Katayama *et al.*, 2005). These findings as well as the observation that small interfering RNAs can modulate transcriptional gene expression via epigenetic modulation of targeted promoters (refer to the following chapters) suggest that one day it may be possible to harness RNA to direct permanent epigenetic modifications resulting in superlative control of the human genome.

References

Alberts B, B. D., Lewis J, Raff M, Roberts K, and Watson JD. (1994). Molecular Biology of the Cell, 3rd edition (London: Garland Publishing, Inc.).

Ayer, D.E. (1999). Histone deacetylases: transcriptional repression with SINers and NuRDs. Trends Cell. Biol. 9, 193–198.

Bachman, K.E., Park, B.H., Rhee, I., Rajagopalan, H., Herman, J.G., Baylin, S.B., Kinzler, K.W., and Vogelstein, B. (2003). Histone modifications and silencing prior to DNA methylation of a tumor suppressor gene. Cancer Cell 3, 89–95.

Berger, S.L. (2002). Histone modifications in transcriptional regulation. Curr. Opin. Genet. Dev. *12*, 142–148.

Bird, A. (2002). DNA methylation patterns and epigenetic memory. Genes Dev. *16*, 6–21.

Boeger, H., Bushnell, D.A., Davis, R., Griesenbeck, J., Lorch, Y., Strattan, J.S., Westover, K.D., and Kornberg, R.D. (2005). Structural basis of eukaryotic gene transcription. FEBS Lett. *579*, 899–903.

Brockdorff, N. (2002). X-chromosome inactivation: closing in on proteins that bind Xist RNA. Trends Genet. *18*, 352–358.

Cao, R., and Zhang, Y. (2004). The functions of E(Z)/EZH2-mediated methylation of lysine 27 in histone H3. Curr. Opin. Genet. Dev. *14*, 155–164.

Chandler, S.P., Guschin, D., Landsberger, N., and Wolffe, A.P. (1999). The methyl-CpG binding transcriptional repressor MeCP2 stably associates with nucleosomal DNA. Biochemistry *38*, 7008–7018.

Craig, J.M. (2005). Heterochromatin – many flavours, common themes. Bioessays *27*, 17–28.

Cross, S.H., and Bird, A.P. (1995). CpG islands and genes. Curr. Opin. Genet. Dev. *5*, 309–314.

Curradi, M., Izzo, A., Badaracco, G., and Landsberger, N. (2002). Molecular mechanisms of gene silencing mediated by DNA methylation. Mol. Cell. Biol. *22*, 3157–3173.

D'Alessio, A.C., and Szyf, M. (2006). Epigenetic tete-a-tete: the bilateral relationship between chromatin modifications and DNA methylation. Biochem. Cell. Biol. *84*, 463–476.

Dobosy, J.R., and Selker, E.U. (2001). Emerging connections between DNA methylation and histone acetylation. Cell. Mol. Life Sci. *58*, 721–727.

Egger, G., Liang, G., Aparicio, A., and Jones, P.A. (2004). Epigenetics in human disease and prospects for epigenetic therapy. Nature *429*, 457–463.

Finnegan, D.J. (1997). Transposable elements: how non-LTR retrotransposons do it. Curr. Biol. *7*, R245–248.

Gerber, M., and Shilatifard, A. (2003). Transcriptional elongation by RNA polymerase II and histone methylation. J. Biol. Chem. *278*, 26303–26306.

Hall, I.M., G.D. Shankaranarayana, K. Noma, N. Ayoub, A. Cohen, S.I.S. Grewal. (2002). Establishment and Maintenance of a Heterochromatin Domain. Science *297*, 2232–2237.

Heitz, E. (1928). Das Heterochromatin der Moose. I. Jahrb. wiss. Bot. *69*, 726–818.

Hirose, Y., and Manley, J.L. (1998). RNA polymerase II is an essential mRNA polyadenylation factor. Nature *395*, 93–96.

Holstege, F.C., Fiedler, U., and Timmers, H.T. (1997). Three transitions in the RNA polymerase II transcription complex during initiation. EMBO J. *16*, 7468–7480.

Jablonka, E., and Lamb, M.J. (2003). [Epigenetic heredity in evolution]. Tsitologiia *45*, 1057–1072.

Jenuwein, T. (2006). The epigenetic magic of histone lysine methylation. FEBS J. *273*, 3121–3135.

Jenuwein, T., and Allis, C.D. (2001). Translating the histone code. Science *293*, 1074–1080.

Katayama, S., Tomaru, Y., Kasukawa, T., Waki, K., Nakanishi, M., Nakamura, M., Nishida, H., Yap,

C.C., Suzuki, M., Kawai, J., *et al.* (2005). Antisense transcription in the mammalian transcriptome. Science *309*, 1564–1566.

Lachner, M., O'Carroll, D., Rea, S., Mechtler, K., and Jenuwein, T. (2001). Methylation of histone H3 lysine 9 creates a binding site for HP1 proteins. Nature *410*, 116–120.

Lee, J. Lu N, Han Y. (1999). Genetic analysis of the mouse X inactivation center defines an 80-kb multifunction domain. Proc. Natl. Acad. Sci. USA *96*, 3836–3841.

Lippman, Z., Gendrel, A.V., Black, M., Vaughn, M.W., Dedhia, N., McCombie, W.R., Lavine, K., Mittal, V., May, B., Kasschau, K.D., *et al.* (2004). Role of transposable elements in heterochromatin and epigenetic control. Nature *430*, 471–476.

Lusser, A. (2002). Acetylated, methylated, remodeled: chromatin states for gene regulation. Curr. Opin. Plant Biol. *5*, 437–443.

Morris, K.V. (2005). siRNA-mediated transcriptional gene silencing: the potential mechanism and a possible role in the histone code. Cell. Mol. Life Sci. *62*, 3057–3066.

Mutskov, V., and Felsenfeld, G. (2004). Silencing of transgene transcription precedes methylation of promoter DNA and histone H3 lysine 9. EMBO J. *23*, 138–149.

Nan, X., H.H. Ng, C.A. Johnson, C.D. Laherty, B.M. Turner, R.N. Eisenman. (1998). Transcriptional repression by the methyl-CpG-binding protein MeCP2 involves a histone deacetylase complex. Nature *393*, 386–389.

Okano, M., Bell, D.W., Haber, D.A., and Li, E. (1999). DNA methyltransferases Dnmt3a and Dnmt3b are essential for *de novo* methylation and mammalian development. Cell *99*, 247–257.

Paule, M.R., and White, R.J. (2000). Survey and summary: transcription by RNA polymerases I and III. Nucleic Acids Res. *28*, 1283–1298.

Proudfoot, N., and O'Sullivan, J. (2002). Polyadenylation: a tail of two complexes. Curr. Biol. *12*, R855–857.

Razin, A., and Riggs, A.D. (1980). DNA methylation and gene function. Science *210*, 604–610.

Razin, A., and Szyf, M. (1984). DNA methylation patterns. Formation and function. Biochim. Biophys. Acta *782*, 331–342.

Sarraf, S.A., and Stancheva, I. (2004). Methyl-CpG binding protein MBD1 couples histone H3 methylation at lysine 9 by SETDB1 to DNA replication and chromatin assembly. Mol. Cell *15*, 595–605.

Schwabish, M.A., and Struhl, K. (2004). Evidence for eviction and rapid deposition of histones upon transcriptional elongation by RNA polymerase II. Mol. Cell. Biol. *24*, 10111–10117.

Sims, R.J., 3rd, Belotserkovskaya, R., and Reinberg, D. (2004). Elongation by RNA polymerase II: the short and long of it. Genes Dev. *18*, 2437–2468.

Sims, R.J., 3rd, Nishioka, K., and Reinberg, D. (2003). Histone lysine methylation: a signature for chromatin function. Trends Genet. *19*, 629–639.

Stewart, M.D., Li, J., and Wong, J. (2005). Relationship between histone H3 lysine 9 methylation, transcription repression, and heterochromatin protein 1 recruitment. Mol. Cell. Biol. 25, 2525–2538.

Strahl, B.D., and Allis, C.D. (2000). The language of covalent histone modifications. Nature 403, 41–45.

Strahl, B.D., Ohba, R., Cook, R.G., and Allis, C.D. (1999). Methylation of histone H3 at lysine 4 is highly conserved and correlates with transcriptionally active nuclei in tetrahymena [In Process Citation]. Proc. Natl. Acad. Sci. USA 96, 14967–14972.

Turner, B.M. (2000). Histone acetylation and an epigenetic code. Bioessays 22, 836–845.

Volpe, T.A., C. Kidner, I.M. Hall, G. Teng, S.I.S. Grewal, R.A. Martienssen. (2002). Regulation of Heterchromatic Silencing and Histone H3 Lysine-9 Methylation by RNAi. Science 297, 1833–1837.

Wakabayasi-Ito, N., and S. Nagata (1994). characterization of the regulatory elements in the promoter of the Human Elongation Factor-1 alpha gene. J. Biol. Chem. 269, 29831–29837.

Workman, J.L., and Kingston, R.E. (1998). Alteration of nucleosome structure as a mechanism of transcriptional regulation. Annu. Rev. Biochem. 67, 545–579.

Wu, J.C., and Santi, D.V. (1985). On the mechanism and inhibition of DNA cytosine methyltransferases. Prog. Clin. Biol. Res. 198, 119–129.

Wu, X., Li, Y., Crise, B., and Burgess, S.M. (2003). Transcription start regions in the human genome are favored targets for MLV integration. Science 300, 1749–1751.

The Role of RNAi and Non-coding RNAs in Polycomb-mediated Control of Gene Expression and Genomic Programming

3

Manuela Portoso and Giacomo Cavalli

Abstract

Regulation of gene expression is a complex, multi-layered process that is crucial to correctly drive and maintain cell identity during development and adult life. In this chapter, we discuss the functional and molecular links between two well-conserved gene silencing pathways, RNA interference (RNAi) and Polycomb. RNAi participates in post transcriptional as well as transcriptional gene silencing of natural genes as well as transposons and viruses. Polycomb group (PcG) proteins are well known for their role in silencing HOX genes through modulation of chromatin structure. However, both mechanisms were found to be involved in specific epigenetic processes like cosuppression in *Drosophila melanogaster* and the formation of *C. elegans mes* and SOP-2 complexes. Recent work has uncovered molecular links between RNAi components and Polycomb-mediated silencing in human cells and *Drosophila*. RNA polymerase II and Argonaute 1 interact to bring about chromatin modifications on endogenous Polycomb target gene promoters in human cells, while *Drosophila* RNAi components modulate the nuclear organization of PcG target DNA elements, thereby affecting the strength of PcG-mediated silencing. Finally, we discuss the findings of several microRNAs and non-coding RNAs in human and fly HOX gene loci, where they may regulate HOX gene expression both post-transcriptionally and co-transcriptionally.

Introduction

The regulation of gene expression by RNAi mechanisms occurs through different pathways including placement of epigenetic marks on specific chromatin domains or post-transcriptional mRNA degradation. RNAi is an evolutionary conserved eukaryotic mechanism that can induce both post-transcriptional gene silencing (PTGS) and transcriptional gene silencing (TGS). In PTGS, double-stranded RNA (dsRNA) molecules derived from endogenous and exogenous sources are processed into small interfering RNAs (siRNAs) of 21–25 nucleotides, which drive the degradation or the translation inhibition of the corresponding mRNA target through total or partial complementarity to the mRNA sequence, respectively. siRNAs are produced by the endonuclease III Dicer and are then incorporated in the RNA-induced silencing complex (RISC) that targets the homologous RNA for degradation or translational inhibition. Argonaute, a PAZ and PIWI domain protein, is the fundamental component of the effector RISC complex that binds the siRNAs and drives the PTGS response. In TGS, siRNAs can target the homologous DNA sequence leading to the deposition of repressive chromatin marks (histone modifications/DNA methylation).

Another well-conserved mechanism for gene silencing involves the Polycomb group of proteins (PcG). PcG proteins bind and maintain the chromatin structure of a large number of key genes, conferring cellular memory of their transcriptional states throughout development and contributing to the determination of cell identity as well as stem cell pluripotency. PcG proteins act in complexes that are recruited to specific regulatory elements of DNA and regulate their target genes via modulation of chromatin structure.

In this chapter, we present the evidence of an interplay between RNAi components and PcG proteins in different organisms and we discuss the possible role of RNAi in PcG-mediated control of gene expression and genome programming.

RNAi-mediated control of gene expression: exogenous and endogenous siRNAs

The introduction of double-stranded RNAs (dsRNA or hairpin) or siRNAs directed against a specific gene can cause silencing of the target gene in almost all the eukaryotes.

In 1998, Craig Mello and Andrew Fire discovered that double-stranded RNA is a very powerful system to induce gene silencing in *Caenorhabditis elegans* (Fire *et al.*, 1998). In tobacco and *Arabidopisis thaliana*, double-stranded hairpins were shown to cause both PTGS of the target mRNA or TGS when directed against a promoter by consequent methylation of the corresponding DNA promoter sequence (Mette *et al.*, 2000). In *Drosophila melanogaster*, different ds hairpin constructs have been found to induce PTGS of the target mRNA (Kennerdell *et al.*, 2000; Piccin *et al.*, 2001; Lee *et al.*, 2004) and siRNA injection has proven to be a very useful technique to generate gene knock down in cell culture. In *Schizosaccharomyces pombe*, a double-stranded hairpin construct targeted to green fluorescent protein (GFP) gene can cause PTGS of a GFP transgene (Sigova *et al.*, 2004). Tethering the effector complex RISC (RITS in fission yeast) to an endogenous gene like *ura4* causes TGS of this gene and an efficient RNAi response occurs on a second ura4 allele *in trans* only in the absence of the RNAi negative regulator gene *eri1* (Buhler *et al.*, 2006).

Also in cultured mammalian cells, the introduction of siRNAs and of ds hairpin constructs can drive an RNAi response. Like in plants, the RNAi response can produce PTGS of the target mRNA or TGS of the target promoter DNA sequence (Elbashir *et al.*, 2001; Caplen *et al.*, 2001; Svoboda *et al.*, 2004; Park *et al.*, 2004; Morris *et al.*, 2004; Castanotto *et al.*, 2005; Ting *et al.*, 2005).

Although RNAi is well conserved, many aspects of this phenomenon are different among eukaryotes. In plants and worms, silencing can spread from cell to cell and small RNAs can be amplified by the RNA-dependent RNA polymerase enzyme (RdRp). In flies and mammals, RdRp does not exist (Wassenegger and Krczal, 2006). In addition, in flies, plants and worms the introduction of multiple copies of a gene determines the phenomenon of cosuppression or downregulation of endogenous gene transcription. First observed in plants (Jorgensen *et al.*, 1990; Napoli *et al.*, 1990; van de Krol *et al.*, 1990a and 1990b), this phenomenon occurs also in *Drosophila* where it has been shown to be Polycomb dependent (Pal-Bhadra *et al.*, 1997).

Endogenous siRNAs between 19–30 nucleotides in length, have been cloned and described in different organisms. They function in different pathways. When they are homologous to genomic regions rich in repeats and transposable elements, their main task is to direct silent chromatin formation over these regions, that could otherwise become hotspots of non-homologous recombination. Other classes of endogenous siRNA have been recently discovered (reviewed in Kim 2005; Brodersen and Voinnet, 2006; Vazquez F., 2006). They are classified in natural siRNAs (nat-siRNAs), *trans*-acting siRNAs (tasiRNAs), repeat associated siRNAs (rasiRNAs), piwi-interacting (pi) RNAs and microRNAs (miRNAs). Natural siRNAs (Nat-siRNAs) in plants and endogenous siRNAs in nematodes are generated from dsRNAs derived from endogenous sense and antisense transcripts (Borsani *et al.*, 2005; Lee R.C., *et al.*, 2006). *trans*-acting siRNAs (tasiRNAs) in plants match intergenic regions and act on their mRNA targets in *trans* (Vazquez *et al.*, 2004; Peragine *et al.*, 2004). Similar to tasiRNA are the tiny non-coding RNAs (tncRNAs) found in nematodes (Ambros *et al.*, 2003). Repeat-associated siRNAs (rasiRNAs) match repetitive sequences and are involved in TGS in fission yeast, flies and plants (Volpe *et al.*, 2002; Reinhart and Bartel, 2002; Aravin *et al.*, 2001, 2003, 2004; Hamilton *et al.*, 2002; Llave *et al.*, 2002; Lippman *et al.*, 2004; Tran *et al.*; 2005). piwiRNAs/miliRNAs are important in mammalian gametogenesis and fly germline development (Aravin *et al.*, 2006; Girard *et al.*, 2006; Grivna *et al.*, 2006; Watanabe *et al.*, 2006;

Lau et al., 2006; Vagin et al., 2006). Micro RNAs (miRNAs) play important roles in development, proliferation, haematopoiesis and apoptosis in human, nematodes, flies and plants. They can act by direct mRNA cleavage or inhibition of mRNA translation (reviewed in Bartel, 2004). In Table 3.1 the RNAi components present in the different eukaryotes and their functions are represented.

In Tetrahymena, siRNA-like scan RNAs (scnRNAs) mark specific DNA sequences for deletion in the macronuclear genome (Mochizuki et al., 2002; Taverna et al., 2002; Mochizuki and Gorovsky, 2004; Lee and Collins, 2006). Worms defective in RNAi pathways lose silencing of transposons in the germline (Ketting et al., 1999; Tabara et al., 1999; Vastenhouw and Plasterk, 2004). In flies the RNAi machinery is required for silent chromatin formation and proper HP1 localization (Pal-Bhadra et al;, 2004). In mammals, the formation of higher order chromatin structure in pericentric heterochromatin requires specific histone modifications and an RNA component (Maison et al., 2002). Transcripts from the major satellite have been shown to arise from the centromeres (Saffery et al., 2003) and mutations in RNAi components induce loss of centromeric heterochromatin (Kanellopoulou et al., 2005; Fukagawa et al., 2004).

Functional links between RNAi and the chromatin associated protein Polycomb

The best characterized system in which RNAi induces TGS is fission yeast. In this organism, the RNAi component Argonaute has been found to be in a complex with a chromodomain protein, Chp1, that binds to H3K9me2, the methyl mark for heterochromatin in fission yeast centromeres (Verdel et al., 2004). These two proteins together with an uncharacterized third one, Tas3, constitute the effector complex, RITS (RNA-induced initiation of transcriptional gene silencing), necessary to establish heterochromatin. Moreover, RNA pol II takes part in this process and two different subunits of RNA pol II have been shown to be essential for RNAi-mediated heterochromatin assembly in this organism (Djupedal et al., 2005; Kato et al., 2005).

RNAi and TGS might be linked via direct interactions of RNAi components and chromatin associated proteins (reviewed in Bernstein and Allis, 2005). Argonaute proteins are common components of silencing complexes in different organisms and chromodomain proteins (CD) are linked to formation of silent chromatin domains. The ability of particular CD to bind both methylated histone tails and RNA (Bernstein et al, 2006) suggests a possible mechanism to determine specific chromatin structure through RNA moieties.

Polycomb could be a candidate CD protein interacting with RNA molecules to determine epigenetic silencing of HOX genes through the formation of localized facultative repressive chromatin structures. HOX genes encode homeodomain transcriptional factors that regulate genes required to specify the positional identities of cells along the anteroposterior axis of embryos of during development. HOX gene expression is tightly regulated during embryogenesis. A transcriptional cascade establishes the appropriate spatial pattern of expression of each HOX gene in early embryos. However, these early HOX regulators disappear at mid embryogenesis, and the PcG and trithorax group (trxG) proteins maintain the memory, respectively, of the repressive or active expression states throughout development. In flies, PcG proteins are recruited to their target loci by DNA binding proteins and they bind to sequence elements known as Polycomb response elements (PREs). In other organisms PREs have not been identified yet. In addition to DNA-binding proteins, a different mechanism possibly involving RNAi may contribute to target Polycomb proteins.

Two conserved PcG complexes have been identified, PRC2 contains Extra sex comb and enhancer of zeste and the PRC1 complex that contains Polycomb (PC), Posterior sex combs (PSC), Polyhomeotic (PH) and RING/Sce. The PRC2 complex has been shown to have histone deacetylase and histone methyltransferase activities that specifically methylate H3K9 and H3K27. PRC1 possesses an E3 ubiquitin ligase activity specific for H2A (reviewed in Bantignies and Cavalli, 2006; Schwartz and Pirrotta, 2007; Schuettengruber et al., 2007).

Table 3.1 Known components of the RNAi pathway include the highly conserved Dicer and Argonaute proteins and, except in human and flies, the RNA dependent RNA polymerase (RdRp) enzymes. Different enzymes belonging to the same family act in pathways to process exogenous (exo-siRNAs) or specific endogenous siRNAs. In the table various Dicer, Argonaute and RdRp homologues in human, C. elegans, D. melanogaster, S. pombe and A. thaliana, are indicated together with the pathway they are involved in (Sasaki et al., 2003; Okamura et al., 2004; Yigit et al., 2006; Henderson et al., 2006; Wassenegger and Krczal, 2006; Brennecke et al., 2007; Gunawardane et al., 2007). Question marks indicate that the protein function is not known so far.

	Dicer	Argonaute (PAZ/PIWI domain proteins)	RdRp
Human	Dicer (miRNA and exo-siRNA)	Ago-1 (miRNA and exo-siRNA) Ago-2 (miRNA and exo-siRNA) Ago-3 (miRNA and exo-siRNA) Ago-4 (miRNA and exo-siRNA) PIWIL1/HIWI (germline maintenance) PIWIL2/HILI (germline maintenance) PIWIL4/HIWI2 (germline maintenance) PIWIL3 (germline maintenance)	–
Drosophila melanogaster	Dicer-1 (miRNA) Dicer-2 (exo-siRNA)	Ago-1 (miRNA) Ago-2 (exo-siRNA) Ago-3 (rasiRNA) Piwi (rasiRNA and germline maintenance) Aubergine (rasiRNA)	–
Caenorhabditis elegans	Dicer (miRNA and siRNA)	Alg-1 and Alg-2 (miRNA) Rde-1 (exo -siRNA) Ergo-1 (tncRNAs) Crs-1 (chromosome segregation) Prg-1 (germline maintenance) Sago-1 and Sago-2 (secondary siRNAs) Ppw-1 (secondary siRNAs) 18 other Ago proteins with redundant function	RRf-1 (siRNA) RRf-2? RRf-3 (siRNA) Ego-1 (rasiRNA germline)
Schizosaccharomyces pombe	Dicer-1 (rasiRNA and exo-siRNA)	Ago-1 (rasiRNA and exo-siRNA)	Rdrp-1 (rasiRNA and exo-siRNA)
Arabidopsis thaliana	Dicer-like1 (miRNA, tasiRNA and natRNA) Dicer-like2 (exo-siRNA, tasiRNA and natRNA) Dicer-like3 (rasiRNA and tasiRNA) Dicer-like4 (tasiRNA and rasiRNA)	Ago-1(miRNA, tasi and exo-siRNA) Ago-2 Ago-3 Ago-4 (rasiRNA and miRNA) Ago-5, Ago-6? Ago-7 (tasiRNA) Ago-8, Ago-9? Ago10 (miRNA)	RDR-1 (exo-RNA) RDR-2 (rasiRNA) RDR-3/RDR3a? RDR-4/RDR3b? RDR-5/RDR3c? RDR-6 (tasiRNA, miRNA and natRNA)

In flies a link between RNAi and Polycomb comes from the study performed by Pal-Bhadra *et al.*, in which the introduction of multiple copies of a hybrid-transgene covering the regulatory region of the *white* eye colour gene and the *alcohol dehydrogenase* gene (*white-Adh*) results in repression of the transgene and of the endogenous *Adh* locus. This RNAi-related process is known as cosuppression (Pal-Bhadra *et al.*, 1997). In particular, the silencing mediated by the *white-Adh* transgene is relieved in *Polycomb*, *Polycomblike* and *piwi* mutants (Pal-Bhadra *et al.*, 2002). Polycomb is recruited to the transgene insertion sites as seen by polytene immunolocalization, but it is not required when full length *Adh* transgenes are introduced to induce a PTGS response. However Piwi was found to be necessary in the PTGS process (Pal-Bhadra *et al.*, 2002).

Evidence for a link between RNAi and Polycomb protein has also been shown in C. *elegans*. Among the genes required for RNAi, three of them are the *mes* genes (maternal effect sterile), namely, *mes-3*, *mes-4* and *mes-6* (Dudley *et al.*, 2002). These genes are required maternally for survival of larval germ cells. mes-3 is a novel protein and *mes-4* encodes a SET domain protein whose target is still unknown (Fong *et al.*, 2002). *mes-6* encodes for the orthologue of the *Drosophila* Extra Sex Combs (ESC) protein and it has been found in a complex with two other proteins, mes-2 and mes-3 (Xu *et al.*, 2001). mes-2 is the orthologue of *Drosophila* Enhancer of zeste E(Z) and it is responsible for H3K27me2 and H3K27me3 on the X chromosome that becomes inactivated in the germline (Bender *et al.*, 2004). *mes-3*, *mes-4* and *mes-6* but not *mes-2* are necessary for the induction of an RNAi response in presence of high concentration of ds RNA. In worms, a dsRNA amplification mechanism is provided by the RNA dependent RNA polymerase (RRF-3) and this amplification step might be sufficient to bypass the requirement of chromatin associated proteins like mes-3, mes-4 and mes-6 to set a stronger silencing response.

Another component of the Polycomb group of protein in C. *elegans*, SOP-2, the functional analogue of the fly and vertebrate PcG proteins PH and Sex Combs on Midleg (SCM), has been shown to have RNA binding activity (Zhang *et al.*, 2003). The very little similarity between SOP-2 and its counterparts is restricted to the protein-protein interaction domain SAM (Sterile Alpha Motif). SOP-2 interacts with the SUMO-conjugating enzyme UBC-9 and this interaction is important not only for the SOP-2 nuclear localization *in vivo* to nuclear bodies (SOP-2 bodies) similar to those formed in fly and vertebrate PH orthologues, but also for repression of the *egl-5* and *mad-5* HOX genes (Zhang *et al.*, 2004). In addition, SOP-2 directly interacts with SOR-1 a novel protein with an RNA binding domain. The RNA binding motifs present in both SOR-1 and SOP-2 proteins are necessary for their localization to SOP-2 bodies (Zhang *et al.*, 2006). Taken together, these results indicate a link between Polycomb, non-coding RNAs and the sumoylation pathways. The interplay among these three well-conserved pathways may be common to other organisms. For instance, it has been recently proposed that Polycomb bodies could act as centres for sumoylation. The human PcG protein Pc2 is a SUMO E3 ligase (Kagey et at., 2003).

An interaction between H3K27 methylation and RNAi has been very recently shown in *Tetrahymena*. Ezl1, a homologue of *Drosophila* E(z), methylates H3K27 in a RNAi-dependent manner. H3K27 methylation regulates also H3K9 methylation over specific loci, targeted by conjugation-specific siRNAs. Together, H3K27 and H3K9 trimethyl marks are recognized by the chromodomain protein Pdd1p that drives the 'programmed' DNA elimination through the formation of specialized heterochromatic structures (Liu *et al.*, 2007).

In flies and in mammals other pathways involve RNA molecules and chromodomain proteins to regulate chromatin. In flies, a dosage compensation process hyperactivates the genes located on the single male X chromosome in order to achieve an expression level equivalent to two X chromosomes in females. This process involves the action, among others, of the two CD proteins MOF and MSL3 that have been shown to bind RNA *in vitro* (Akhtar *et al.*, 2000). Two non-coding RNAs, *rox1* and *rox2*, are involved in targeting the MOF/MSL3 complex to its target chromatin (reviewed in Gilfillan *et al.*, 2004; Bernstein and Allis, 2005).

In mammals, dosage compensation involves inactivation of one of the two female X chromosomes, leaving the other to express its genes to similar levels as the single male X chromosome. In female X inactivation, the Xist gene is transcribed in both sense and antisense orientation to produce *Xist/Tsix* non-coding RNAs. The Xist RNA coats the inactive X chromosome *in cis*. Subsequent maintenance of the inactive state is ensured by a combination of epigenetic marks like DNA (CpG) methylation, incorporation of macro H2A, H3K9me2, H4K20me1 and H3K27me3. Xist recruits directly or indirectly the Polycomb group complex Ezh2/Eed (PRC2) to catalyse H3K27me3 (Plath *et al.*, 2003; Silva *et al.*, 2003; Kohlmaier *et al.*, 2004) and also members of PRC1 like the chromodomain protein Cbx2, Polyhomeotic 1 (Phc1), Polyhomeotic 2 (Phc2) and Bmi1 (Plath *et al.*, 2004). Another chromodomain protein, Cbx7, has been found to associate to the inactive X chromosome and this association has been found to depend partially on RNA (Bernstein *et al.*, 2006). The molecular mechanisms of this process are still unknown and no siRNAs complementary to *Xist* have been cloned so far.

Molecular links between RNAi and Polycomb silent chromatin assembly

A molecular link between RNAi and Polycomb-mediated silent chromatin assembly comes from studies performed in human cells. siRNAs directed against the elongation factor 1 alpha (EF1A) promoter were shown to induce TGS of the GFP reporter gene and the endogenous EF1A gene when they were transfected in human 293FT cultured cells bearing an integrated EF1A promoter driven GFP reporter. Silencing was associated with DNA methylation of the targeted sequence, detected by a methyl-sensitive restriction enzyme in a PCR-based assay, and it was reversed by treatment with trichostatin, an inhibitor of histone deacetylases, and 5-azacytidine, an inhibitor of DNA methyltransferases. In addition, active transport of siRNAs or permeabilization of the nuclear envelope by lentiviral transduction was necessary for transcriptional silencing to occur. However transfection with siRNAs targeted to a GFP exon caused PTGS with no change in the transcriptional rate, as confirmed by transcriptional run-on experiments (Morris *et al.*, 2004).

In a subsequent study using the same system, Weinberg and colleagues (2006) demonstrated in more details how siRNAs specifically delivered to the nucleus mediate TGS in human cells by establishing histone marks characteristic of facultative heterochromatin (like H3K9me2 and H3K27me3) at the targeted EF1A promoter, the requirement of active transcription, and how only antisense specific siRNAs can induce specific TGS. The requirement of transcription for siRNA-mediated TGS was established adding α-amanatin, an inhibitor of RNA pol II, 24h after the transfection with the specific EF1A siRNAs (EF52), and assaying the H3K9 methylation state on the target sequence. In an siRNA pull-down assay, they found that the antisense siEF52 RNA binds the DNA methyltransferase DNMT3A as efficiently as double-strand siEF52 RNA. In experiments targeting antisense, sense or combined dsRNA to two regions of the U3 in the HIV-1 LTR promoter by cotransfection of 1G5 cells, they reproduced similar results, leading to the conclusion that antisense siRNAs are sufficient to induce histone modifications on a target promoter in a RNA pol ll-dependent manner. In addition, in a triple pull down experiment they showed that the antisense siRNA bound to DNMT3A is accompanied by H3K27 methylation on the target DNA sequence suggesting the existence of a complex formed by antisense siRNAs bound to DNMT3A. The siRNAs would then target DNMT3A by RNA homology to the promoter where the nascent transcript has been made, and direct H3K9 and H3K27 methylation (Weinberg *et al.*, 2006).

A further piece of information on the mechanism of siRNAs directed TGS was obtained from the subsequent study by Kim *et al.* (2006) which showed that Ago1 directs siRNA-TGS in human cells at the promoters of the genes coding for the human immunodeficiency virus-1 coreceptor CCR5 and the tumour suppressor RASSF1A. Ago1 associates with the targeted DNA region, together with RNA pol II, and is required to set H3K9 methylation at the targeted region, as its deletion by siRNA transfection results in loss of this histone mark. This work also showed that the double-stranded RNA binding protein TRBP2 that associates with Ago2 in human cells and is required for the formation of the RISC complex (Chendrimada *et al.*, 2005), is necessary for Ago1-directed TGS. Interestingly, the Polycomb protein EZH2 was recruited at the siRNA-targeted RASSF1A promoter and ChIP experiments showed that EZH2, H3K27me3, TRBP2 and Ago1 are associated to endogenous PcG target promoters, like MYT1 and CCR5. Taken together, these data implicate that RNAi may be involved in endogenous mechanisms of Polycomb silencing and establish a connection between RNAi components, the RNA pol II machinery and PcG-mediated control of gene expression (Kim *et al.*, 2006) (Fig. 3.1). A model suggested by Han *et al.* (2007) proposes that a variant species of mRNA, a promoter-associated mRNA, containing an extended 5′ UTR, may be recognized by the antisense siRNAs or endogenous antisense RNAs produced by RNA pol II transcription of the RNA-targeted promoter. These antisense siRNAs may then drive a transcriptional silencing complex composed of histone-modifying enzyme, EZH2, DNA methyltransferase DNMT3A and Ago1 to the

targeted promoter where gene silencing takes place. Another study performed in mammalian cells, established the involvement of both Ago1 and Ago2 in transcriptional silencing. In that case, silencing was shown to require high activity of the targeted promoter and association of Ago1 and Ago2 with it. No change in H3K4me2 or H3K9me2 status was shown at the targeted silent region in that study (Janowski *et al.*, 2006).

miRNAs across HOX genes

MicroRNAs (miRNAs) are endogenous 21–24 nt RNAs that play important regulatory roles in animal and plant development by targeting the 3′ untranslated region (UTR) of mRNAs for translational repression or direct mRNA cleavage (Lai *et al.*, 2003; Bartel 2004). Hox clusters are groups of related transcription factor genes important for development in animals. In mammals, there are four HOX clusters (HOX A to D) and it has been shown that the miR-196 miRNA has complementarity to sequences in the 3′UTRs of HOX genes in each cluster, namely, HOXA7, HOXB8, HOXC8 and HOXD8. Although miR-196 exhibits perfect complementarity to HOXB8 and drives HOXB8 mRNA

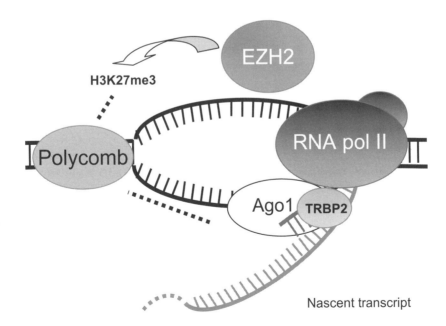

Figure 3.1 AGO1 together with a dsRNA binding protein TRBP2 are associated at the endogenous Polycomb target promoter gene CCR5 in human cells. This locus is marked by H3K27me3 catalysed by EZH2, association with Polycomb and unphosphorylated RNA pol II (Kim *et al.*,2006). A non-coding nascent transcript from CCR5 promoter may be the target of the Ago1 protein and initiate the spreading of Polycomb-mediated gene silencing.

cleavage, the 3'UTRs of HOXA7, HOXC8 and HOXD8 have multiple conserved matches to residues 2 to 8 of miR-196, that are known as 'seed matches' and their regulation through miR-196 occurs through translational inhibition (Yekta et al., 2004). In a more detailed study on the mechanism of action of miR-196, it has been shown that miR-196 acts upstream of HOXB8 and Sonic hedgehog (Shh) in vivo during limb development (Hornstein et al., 2005). Shh is a key signal for anteroposterior polarity in the fore and hindlimb bud and it becomes actively transcribed following retinoic acid (RA) exposure by HOXB8-mediated activation. RA is a morphogen that can induce time- and spatially regulated activation not only of HOX genes but also of non-coding RNAs. This study supports the existence of a 'fail-safe' mechanism of control of gene expression in the HOX clusters. First, HOX genes are transcriptionally regulated by PcG of proteins and, if they ever escape this mode of silencing, the ensuing transcription produces this miRNA which can then post-transcriptionally repress the HOX genes. This prompts to hypothesize the existence of a tightly time regulated expression of miRNAs, in such a way that they are expressed slightly later than their HOX targets, in order to ensure the correct pattern of expression along the anteroposterior axis.

An unrelated miRNA, miR-iab-4, maps to the corresponding region in the Bithorax complex of fly HOX genes and it targets Ubx, the insect counterpart of the HOX 6 to 8 paralogous groups (Ronshaugen et al., 2004). Ectopic expression of miR-iab-4 determines decrease of Ubx protein levels with the appearance of a homeotic transformation of halteres into wings. However, it cannot be excluded that iab-4 might have other targets apart from Ubx in the BX-C complex. homothorax has been identified by computational analysis as another possible target of iab-4 (Grun et al., 2005). The mechanism of action of this miRNA is similar to miR-196 in vertebrate and follows the rule of 'posterior prevalence', in which posterior HOX genes repress the expression of more anterior HOX genes, i.e. the posteriorly expressed miRNAs repress anterior HOX genes at a post-transcriptional level (Chopra and Mishra, 2006). Polycomb on the other end controls transcription of miR-iab-4 that occurs in the iab4 PRE region in both senses. The existence of a feedback loop mechanism acting at transcriptional level through Polycomb and at post-transcriptional level through RNAi may exist to regulate the expression of miR-iab-4 that then modulates Ubx protein accumulation.

Another miRNA, miR-10, is located in the other fly HOX cluster, named the Antennapedia gene complex (ANT-C). miR-10 locates between the HOX genes Deformed and Sex combs reduced (Lagos-Quintana et al., 2001). Sex combs reduced has been proposed to be its target by computational analysis (Brennecke et al., 2005). Most probably, other miRNA genes exist within the HOX clusters and wait to be characterized for their homeotic activities.

Intergenic transcripts among human HOX clusters and the fly BX-C complex – are they really independent from the RNAi pathway?

Recent evidences have shown that a large part of the regulatory regions and intergenic sequences in the human HOX cluster and in the fly BX-C complex are transcribed and their transcription occurs colinearly with the HOX gene expression (Cheng et al., 2005; Sessa et al., 2007; Rinn et al., 2007; Bae et al., 2002). Rinn et al., identified 30 kb new non-coding RNAs across the four human HOX loci and showed that they define active and silent chromatin domain within the HOX loci. They are differentially expressed in different cell types following their position along the body axis. In addition, they characterized the non-coding HOTAIR RNA from the HOXC locus and found that it represses transcription in trans, of the HOXD locus. HOTAIR interacts with Suz12, and EZH2, two components of PRC2 by pull-down experiments and it is necessary for H3K27me3 presence over the HOXD locus in vivo. These results pinpoint the existence of a mechanism where non-coding transcripts dictate transcriptional silencing of distant chromosomal domains.

In Drosophila, intergenic transcription was suggested to play an important role in epigenetic control of the BX-C transcription pattern, counteracting Polycomb group-mediated silencing

(Rank *et al.*, 2002; Schmitt *et al.*, 2005; Sanchez-Elsner *et al.*, 2006). However, a recent paper by Mazo and coworkers, (Petruk *et al.*, 2006) has shown by a detailed single cell analysis that transcriptional interference occurs among the BX-C complex between the *Ubx* gene and a non-coding RNA produced in the upstream regulatory region of this gene, called *bithoraxoid* (*bxd*). This phenomenon is similar to what happens in yeast cells for the *ser3* gene. When cells are growing in rich medium, Ser3 protein is not required and its repression is achieved by transcription of an intergenic promoter (*srg1*) located upstream of it. Once RNA pol II starts the transcription of *srg1*, it runs over the *ser3* promoter and prevents the binding of the *ser3* transcriptional activators causing *ser3* repression (Martens *et al.*, 2004). In this case, like in the *Drosophila bxd/Ubx* gene, it is the process of transcription itself that causes repression, and it is likely that RNA pol II plays a central role in this mechanism. Thus, specific RNA pol II interactions with key transcription factors or repressors which can influence its processivity, or with chromatin remodelling factors that help opening up chromatin structure, must be fine-tuned to ensure the correct shut down of specific genes. In particular, in flies, the *bxd* RNA does not seem to act through an RNAi based mechanism (siRNA or miRNA) to control *Ubx* transcription. It is the trithorax protein complex TAC1, comprising the histone H3K4 methyltransferase Trithorax (Trx), Sbf1 and the histone acetyltransferase dCBP, that associates with *bxd* or *Ubx* to facilitate the alternative transcription of *bxd* and *Ubx* (Petruk *et al.*, 2006). Thus, TAC1 and elongation factors may favour RNA pol II read-through from *bxd* to the *Ubx* promoter, preventing *Ubx* transcription. Although the non-coding *bxd* controls *Ubx* transcription in the posterior part of the *Drosophila* embryo, it is possible that multiple mechanisms employed at different developmental stages occur, as *bxd* RNAs are not present at later embryonic stages and in imaginal discs (Petruk *et al.*, 2006).

Even if the RNAi pathway and siRNA production do not seem to be implicated directly in the regulation of the non-coding *bxd* transcript, the question whether specific RNAi components are involved in this process remains open. It is now clear that RNA pol II directly interacts with Ago1 at Polycomb target promoters in human cells (Kim *et al.*, 2006). In addition, it has been shown that intergenic transcripts in the human β-globin locus are regulated by Dicer. In *dicer* mutants, the intergenic β-globin transcripts and unspliced globin transcripts are upregulated. This may be due to read-through intergenic transcription into the β-globin gene (Haussecker and Proudfoot, 2005). The authors suggest that the intergenic transcripts together with the unspliced globin transcripts may become target of an RNAi-based turnover mechanism instead of a typical mRNA processing mechanism. They also observed by Chip experiments a robust presence of RNA pol II over the actively transcribed globin genes but almost no signal for intergenic regions. A similar mechanism might also be involved in regulating non-coding transcripts that are located along the BX-C complex and control the expression of homeotic genes.

A role for RNAi in Polycomb-directed nuclear organization

Recently it is becoming clear that the spatial organization of chromatin in the nucleus is correlated with the regulation of transcription (Lanctot *et al.*, 2007). Studies performed in flies have shown that RNAi impinges on the nuclear organization of Polycomb target genes. In *Drosophila*, PcG proteins are recruited at PREs that are generally located in the neighbourhood of the promoter of the genes that they repress. A well-defined *cis*-regulatory element involved in the regulation of the homeotic gene *Abdominal-B* (*Abd-B*) at the BX-C locus is *Fab-7*. This is a multipartite regulatory element that contains a PRE and a region defined as chromatin boundary, since it was shown to be able to block enhancer-promoter interactions (Hagstrom *et al.*, 1996; Zhou *et al.*, 1996). When *Fab-7* was cloned upstream a reporter gene and inserted in the X chromosome at the *scalloped* (*sd*) locus, it was able to ectopically recruit PcG proteins and induce PcG dependent silencing of the transgene. Moreover, the *Fab-7* containing transgene was shown to induce silencing of the endogenous *sd* gene, whose promoter is located 18.4 kb downstream of the *Fab-7* PRE (Bantignies *et al.*, 2003). Silencing of *sd* is strongly enhanced by long-range pairing interactions (gene kissing) between the transgenic *Fab-7* element and the endogenous *Fab-7*, which resides in a different chromosome (Bantignies *et al.*, 2003). More recent work has analysed the role of the RNAi machinery in this phenomenon. Mutations in *dicer2* and *piwi* genes determine loss of silencing of both the transgene and the endogenous *sd* gene, concomitant with the loss of gene kissing. Thus, RNAi components play a role in higher-order PcG complex interactions. Although Dicer2 and Piwi colocalize to Pc bodies in nuclei of larvae, they are dispensable for Pc recruitment to their endogenous target sequences. Furthermore, the RNAi machinery regulates the nuclear organization of endogenous PcG target genes, such as HOX genes. Indeed, the same study showed that the BX-C and the *Antennapedia* HOX genes kiss in tissues where they are silenced and gene kissing is reduced in RNAi mutants. As for the cosuppression phenomenon, siRNAs matching the boundary of *Fab-7* were detected when the transgene is present in multiple copies and they were found to decrease in *dicer2* and *piwi* mutants (Grimaud *et al.*, 2006). Since the boundary region is the source of siRNAs and the recruitment of Polycomb to *Fab-7* PRE is not affected by RNAi mutants, a dual and distinct role of these two sequences in the *Fab-7* element may exist. Only boundary sequences may have properties that make them direct targets of the RNAi machinery or RNAi factor binding sites.

PRE elements exhibit very low homology and siRNAs homologous to *Fab-7* have been found only in the transgenic line, thus further analyses are required to determine whether siRNAs are directly involved in bringing about gene kissing, or rather specific protein-protein interactions between RNAi components and PcG proteins mediate clustering of PREs at nuclear compartments enriched in PcG proteins (Polycomb bodies) (Fig. 3.2). Another insulator element in the BX-C sequence, *Mcp*, has been shown to mediate long-range chromosome-chromosome interactions in *Drosophila* imaginal disc nuclei, through an *in vivo* microscopy assay based on Lac Repressor/operator recognition (Vazquez *et al.*, 2006). The effect of RNAi mutants on *Mcp* driven long distance interactions has not been tested so far.

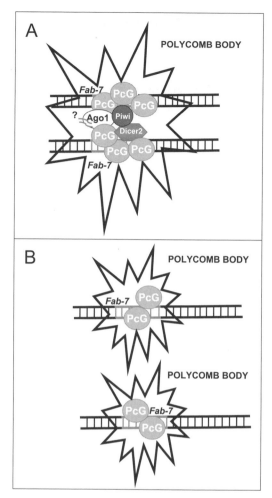

Figure 3.2 (A) RNAi-mediated long distance PRE nuclear interactions: two Fab-7 sequences (the transgene and the endogenous one) associate together in a single Polycomb body. Pairing sensitive silencing occurs and RNAi components interacting with PcG proteins or siRNAs reinforce this association. (B) Long distance interactions among PRE are lost in RNAi mutants: two separated Polycomb bodies form and Polycomb-mediated silencing becomes loose (Grimaud *et al.*, 2006).

RNAi components modulate nuclear architecture, not only of HOX loci but also of chromatin insulators in *Drosophila*. The *gypsy* insulator determines the formation of distinct transcriptional domains organized in a higher order structure, termed insulator bodies. The DNA binding protein Suppressor of Hairy wing (Su(Hw)) together with the DNA binding Centrosomal protein 190 (CP190) and Modifier of mdg4 2.2 (Mod(mdg4)2.2) forms a complex that builds insulator bodies. By immunoaffinity purification, CP190 has been found to associate with the DEAD-box putative RNA helicase Rm62 required in dsRNA-mediated silencing, heterochromatin formation and transposon silencing. *Rm62* mutations enhanced insulator function. This suggested that wild type Rm62 inhibits insulator activity *in vivo*. However *gypsy* insulator activity decreased in two other RNAi mutants: *piwi* and *aubergine* (*aub*). Polytene immunostaining of insulator proteins (Su(Hw), Mod(mdg4)2.2 and CP190 remained unaltered in *piwi* or *aub* mutants. This suggested that RNAi components are not necessary for targeting insulator proteins to their genomic binding sites as also seen for Polycomb recruitment at PRE sequences. However, RNAi components seem to play a role in the higher-order organization of insulator complexes, since double *mod(mdg4)*^{u1};*piwi*^{1}/*piwi*^{2} and *mod(mdg4)*^{u1};*aub*$^{ΔP-3a}$/*aub*^{QC42} mutants exhibited more diffuse nuclear CP190 staining than *mod(mdg4)*^{u1} single mutants (Lei and Corces, 2006). The molecular role of RNAi components in the regulation of insulator bodies is unknown and remains to be explored in future studies.

RNAi, together with specific histone modifications, takes part in the formation of nucleolar architecture and rDNA organization in *Drosophila*. In an analysis performed on imaginal disc tissues and polytene larval salivary glands, multiple nucleoli were found in homozygous mutants for the Su(var)3–9 histone H3K9 methyltransferase and *dicer2*. This phenotype was reversed by mutation in the Ligase IV, an essential regulator of nonhomologous end joining pathway, (NHEJ) that is required for extrachromosomal circular (ecc) DNA formation in mammals. Indeed, ecc repeated DNA increased in *Su(var)3–9* and RNAi mutants. This implies that H3K9 methylation, together with the RNAi pathway, regulates nucleolar architecture by influencing the chromatin structure and repressing formation of rDNA and ecc DNAs. This pathway does not involve the two PcG genes tested (*Pc* and *ph*), nor *dsir2* (*Drosophila* NAD-dependent histone deacetylase) (Peng and Karpen, 2007).

Conclusions and perspectives

Undoubtedly RNAi factors play a role in Polycomb-mediated gene expression and ge-

nomic reprogramming. However, the molecular mechanisms by which this interplay occurs are not clear and they may differ in mammals and flies. Despite the current role attributed to RNA moieties as scaffold or mediator or sequence-specific recruitment of chromatin factors to target sequences, recent results imply that RNAi components play roles independent from their usual 'way of action' that is beyond the mere processing of dsRNA into siRNAs and RNA degradation. Most probably it is the interaction between Dicer and Argonaute proteins with the RNA pol II machinery that somehow regulates gene transcription.

Thus, different models could explain how RNAi factors influence Polycomb-mediated gene expression. A direct interaction between RNAi components and Polycomb may occur through RNA pol II. An RNAi component (such as Ago1) interacting with Pol II would recruit *in cis* chromatin modifying enzymes to set facultative silent marks (H3K9me2 and H3K27me3) on target genes that become then associated with Polycomb. This scenario is likely to exist in human cells with Ago1 being recruited at siRNA targeted promoter and at endogenous Polycomb target genes together with E(Z) and H3K27me3 (Kim *et al.*, 2006) (Fig. 3.1). It remains to be shown whether the same applies in *Drosophila* and other species. Moreover, although human Ago1 associated with Polycomb target genes, no endogenous siRNAs complementary to these Polycomb target genes have been detected so far.

As mentioned before, several non-coding transcripts are present at the homeotic gene clusters of mammals and flies. These non-coding RNAs may 'guide' specific protein complexes, including PcG proteins, to specific transcripts or genomic loci (similar to X-chromosome inactivation). These non-coding RNAs have been implicated in transcriptional interference phenomena occurring in different loci of the genome but also along HOX clusters (Petruk *et al.*, 2006). In contrast to the hypothesis that intergenic transcription counteracts Polycomb binding to target loci, it is conceivable that processive transcription of non-coding RNAs may prevent binding of transcriptional activators and TrxG to regulatory DNA regions of PcG target genes, thus favouring PcG protein binding. In this scenario, RNAi components may be required to 'keep on' transcriptional interference and maintain silencing of PcG target genes. Ago1 travelling along with RNA Pol II may mark the non-coding nascent transcript as a signal for a transcriptional interference phenomenon. Alternatively, the association of Ago1 with RNA pol II, may 'freeze' RNA pol II in an inactive unphosphorylated state that prevents its interaction with transcription activators and elongation factors.

Another possible role of RNAi components in regulated RNA pol II transcription of Polycomb target genes may be linked to a surveillance mechanism. A dsRNA-binding protein, TRBP2, interacting usually with Ago2, has been found in complex with RNA pol II and Ago1 at the Polycomb target promoter CCR5. Antisense siRNAs are sufficient to drive TGS in human cells. This suggests that, if PcG-mediated transcriptional silencing would transiently fail in a subset of cells, this might lead to activation of sense and antisense transcription in the promoter region of the gene. The TRBP2 and Ago1 RNAi components present at the promoter would strengthen binding of the Polycomb machinery and trimethylation of H3K27, to bring back the gene in a silent state. The production of siRNAs and activation of the RNAi pathway may be a very rare event in the cell and this may explain why non-coding endogenous siRNAs mapping to HOX cluster have not been identified so far.

In a similar way, the miRNAs located across the HOX genes could have evolved as a surveillance mechanism. So far, only few miRNAs (miR-196 in human and miR-iab-4 and miR-10 in *Drosophila*) located in these clusters have been shown to target HOX genes located anteriorly. However, no current data is available to explain how miRNA expression across the HOX cluster is regulated and whether Polycomb plays a role in it. In this case, both TGS and PTGS mechanisms could act to control not only HOX gene expression but also miRNAs that have HOX genes as targets, in a feedback loop mechanism that is still to be unveiled.

In flies, RNAi components are not necessary for Polycomb recruitment at their endogenous targets, but they have a role in Polycomb nuclear organization and Dicer2, Piwi and Ago1 colocal-ize with some Polycomb bodies in the nucleus (Grimaud *et al.*, 2006). Thus, RNAi components could directly interact with components of PcG proteins and this interaction may be necessary to maintain the nuclear organization of Polycomb bodies. Nevertheless, whether RNAi mutants lose Polycomb-mediated silencing and the chromatin modifications associated at PREs of endogenous loci in RNAi mutants has not been investigated so far. One way in which RNAi components could act in Polycomb-mediated gene silencing could be to favour the spreading of the histone H3K27me3 mark at silent chromatin domains, and this may in turn affect their nuclear organization throughout development. In accordance with this model, data from Grimaud *et al.*, (2006) suggest that the RNAi components do not impair the establishment of nuclear organization of PcG target elements during the initial phases in embryogenesis. However, these initial gene contacts normally increase during development, and it is this process that requires RNAi components. Understanding the molecular underpinnings of this process, as well as of the general function of the RNAi machinery in the cell nucleus, is a fascinating challenge for future research.

Acknowledgements

We would like to thank Mythily Ganapathi for comments on the text. M.P. is supported by the European Union FP6 (Network of Excellence The Epigenome) and an EMBO long-term fellowship. G.C. was supported by grants of the CNRS, the Human Frontier Science Program Organization, the European Union FP6 (Network of Excellence The Epigenome and STREP 3D Genome), by the Indo-French Centre for Promotion of Advanced Research, by the Agence Nationale de la Recherche, and by the Ministère de l'Enseignement Supérieur, ACI BCMS.

References

Akhtar, A., Zink, D., and Becker, P.B. (2000). Chromodomains are protein-RNA interaction modules. Nature *407*, 405–409.

Ambros, V., Lee, R.C., Lavanway, A., Williams, P.T., and Jewell, D. (2003). MicroRNAs and other tiny endogenous RNAs in C. elegans. Curr. Biol. *13*, 807–818.

Aravin, A., Gaidatzis, D., Pfeffer, S., Lagos-Quintana, M., Landgraf, P., Iovino, N., Morris, P., Brownstein, M.J., Kuramochi-Miyagawa, S., Nakano, T., *et al.* (2006).

A novel class of small RNAs bind to MILI protein in mouse testes. Nature *442*, 203–207.

Aravin, A.A., Klenov, M.S., Vagin, V.V., Bantignies, F., Cavalli, G., and Gvozdev, V.A. (2004). Dissection of a natural RNA silencing process in the *Drosophila melanogaster* germ line. Mol. Cell. Biol. *24*, 6742–6750.

Aravin, A.A., Lagos-Quintana, M., Yalcin, A., Zavolan, M., Marks, D., Snyder, B., Gaasterland, T., Meyer, J., and Tuschl, T. (2003). The small RNA profile during *Drosophila melanogaster* development. Dev. Cell *5*, 337–350.

Aravin, A.A., Naumova, N.M., Tulin, A.V., Vagin, V.V., Rozovsky, Y.M., and Gvozdev, V.A. (2001). Double-stranded RNA-mediated silencing of genomic tandem repeats and transposable elements in the D. melanogaster germline. Curr. Biol. *11*, 1017–1027.

Bae, E., Calhoun, V.C., Levine, M., Lewis, E.B., and Drewell, R.A. (2002). Characterization of the intergenic RNA profile at abdominal-A and Abdominal-B in the *Drosophila* bithorax complex. Proc. Natl. Acad. Sci. USA *99*, 16847–16852.

Bantignies, F., and Cavalli, G. (2006). Cellular memory and dynamic regulation of polycomb group proteins. Curr. Opin. Cell. Biol. *18*, 275–283.

Bantignies, F., Grimaud, C., Lavrov, S., Gabut, M., and Cavalli, G. (2003). Inheritance of Polycomb-dependent chromosomal interactions in *Drosophila*. Genes Dev. *17*, 2406–2420.

Bartel, D.P. (2004). MicroRNAs: genomics, biogenesis, mechanism, and function. Cell *116*, 281–297.

Bender, L.B., Cao, R., Zhang, Y., and Strome, S. (2004). The MES-2/MES-3/MES-6 complex and regulation of histone H3 methylation in *C. elegans*. Curr. Biol. *14*, 1639–1643.

Bernstein, E., and Allis, C.D. (2005). RNA meets chromatin. Genes Dev. *19*, 1635–1655.

Bernstein, E., Duncan, E.M., Masui, O., Gil, J., Heard, E., and Allis, C.D. (2006). Mouse polycomb proteins bind differentially to methylated histone H3 and RNA and are enriched in facultative heterochromatin. Mol. Cell. Biol. *26*, 2560–2569.

Borsani, O., Zhu, J., Verslues, P.E., Sunkar, R., and Zhu, J.K. (2005). Endogenous siRNAs derived from a pair of natural *cis*-antisense transcripts regulate salt tolerance in *Arabidopsis*. Cell *123*, 1279–1291.

Brennecke, J., Aravin, A.A., Stark, A., Dus, M., Kellis, M., Sachidanandam, R., and Hannon, G.J. (2007). Discrete small RNA-generating loci as master regulators of transposon activity in *Drosophila*. Cell *128*, 1089–1103.

Brennecke, J., Stark, A., Russell, R.B., and Cohen, S.M. (2005). Principles of microRNA-target recognition. PLoS Biol. *3*, e85.

Brodersen, P., and Voinnet, O. (2006). The diversity of RNA silencing pathways in plants. Trends Genet. *22*, 268–280.

Buhler, M., Verdel, A., and Moazed, D. (2006). Tethering RITS to a nascent transcript initiates RNAi- and heterochromatin-dependent gene silencing. Cell *125*, 873–886.

Caplen, N.J., Parrish, S., Imani, F., Fire, A., and Morgan, R.A. (2001). Specific inhibition of gene expression by small double-stranded RNAs in invertebrate and vertebrate systems. Proc. Natl. Acad. Sci. USA *98*, 9742–9747.

Castanotto, D., Tommasi, S., Li, M., Li, H., Yanow, S., Pfeifer, G.P., and Rossi, J.J. (2005). Short hairpin RNA-directed cytosine (CpG) methylation of the RASSF1A gene promoter in HeLa cells. Mol. Ther. *12*, 179–183.

Chendrimada, T.P., Gregory, R.I., Kumaraswamy, E., Norman, J., Cooch, N., Nishikura, K., and Shiekhattar, R. (2005). TRBP recruits the Dicer complex to Ago2 for microRNA processing and gene silencing. Nature *436*, 740–744.

Cheng, J., Kapranov, P., Drenkow, J., Dike, S., Brubaker, S., Patel, S., Long, J., Stern, D., Tammana, H., Helt, G., et al. (2005). Transcriptional maps of 10 human chromosomes at 5-nucleotide resolution. Science *308*, 1149–1154.

Chopra, V.S., and Mishra, R.K. (2006). 'Mir'acles in hox gene regulation. Bioessays *28*, 445–448.

Djupedal, I., Portoso, M., Spahr, H., Bonilla, C., Gustafsson, C.M., Allshire, R.C., and Ekwall, K. (2005). RNA Pol II subunit Rpb7 promotes centromeric transcription and RNAi-directed chromatin silencing. Genes Dev. *19*, 2301–2306.

Dudley, N.R., Labbe, J.C., and Goldstein, B. (2002). Using RNA interference to identify genes required for RNA interference. Proc. Natl. Acad. Sci. USA *99*, 4191–4196.

Duina, A.A., and Winston, F. (2004). Analysis of a mutant histone H3 that perturbs the association of Swi/Snf with chromatin. Mol. Cell. Biol. *24*, 561–572.

Elbashir, S.M., Lendeckel, W., and Tuschl, T. (2001). RNA interference is mediated by 21- and 22-nucleotide RNAs. Genes Dev. *15*, 188–200.

Fire, A., Xu, S., Montgomery, M.K., Kostas, S.A., Driver, S.E., and Mello, C.C. (1998). Potent and specific genetic interference by double-stranded RNA in *Caenorhabditis elegans*. Nature *391*, 806–811.

Fong, Y., Bender, L., Wang, W., and Strome, S. (2002). Regulation of the different chromatin states of autosomes and X chromosomes in the germ line of C. elegans. Science *296*, 2235–2238.

Fukagawa, T., Nogami, M., Yoshikawa, M., Ikeno, M., Okazaki, T., Takami, Y., Nakayama, T., and Oshimura, M. (2004). Dicer is essential for formation of the heterochromatin structure in vertebrate cells. Nat. Cell. Biol. *6*, 784–791.

Gilfillan, G.D., Dahlsveen, I.K., and Becker, P.B. (2004). Lifting a chromosome: dosage compensation in *Drosophila melanogaster*. FEBS Lett. *567*, 8–14.

Girard, A., Sachidanandam, R., Hannon, G.J., and Carmell, M.A. (2006). A germline-specific class of small RNAs binds mammalian Piwi proteins. Nature *442*, 199–202.

Grimaud, C., Bantignies, F., Pal-Bhadra, M., Ghana, P., Bhadra, U., and Cavalli, G. (2006). RNAi components are required for nuclear clustering of Polycomb group response elements. Cell *124*, 957–971.

Grivna, S.T., Beyret, E., Wang, Z., and Lin, H. (2006). A novel class of small RNAs in mouse spermatogenic cells. Genes Dev. *20*, 1709–1714.

Grun, D., Wang, Y.L., Langenberger, D., Gunsalus, K.C., and Rajewsky, N. (2005). microRNA target predictions across seven *Drosophila* species and comparison to mammalian targets. PLoS Comp. Biol. *1*, e13.

Gunawardane, L.S., Saito, K., Nishida, K.M., Miyoshi, K., Kawamura, Y., Nagami, T., Siomi, H., and Siomi, M.C. (2007). A slicer-mediated mechanism for repeat-associated siRNA 5′ end formation in *Drosophila*. Science *315*, 1587–1590.

Hagstrom, K., Muller, M., and Schedl, P. (1996). Fab-7 functions as a chromatin domain boundary to ensure proper segment specification by the *Drosophila* bithorax complex. Genes Dev. *10*, 3202–3215.

Hamilton, A., Voinnet, O., Chappell, L., and Baulcombe, D. (2002). Two classes of short interfering RNA in RNA silencing. EMBO J. *21*, 4671–4679.

Han, J., Kim, D., and Morris, K.V. (2007). Promoter-associated RNA is required for RNA-directed transcriptional gene silencing in human cells. Proc. Natl. Acad. Sci. USA *104*, 12422–12427.

Haussecker, D., and Proudfoot, N.J. (2005). Dicer-dependent turnover of intergenic transcripts from the human beta-globin gene cluster. Mol. Cell. Biol. *25*, 9724–9733.

Henderson, I.R., Zhang, X., Lu, C., Johnson, L., Meyers, B.C., Green, P.J., and Jacobsen, S.E. (2006). Dissecting *Arabidopsis thaliana* DICER function in small RNA processing, gene silencing and DNA methylation patterning. Nat. Genet. *38*, 721–725.

Hornstein, E., Mansfield, J.H., Yekta, S., Hu, J.K., Harfe, B.D., McManus, M.T., Baskerville, S., Bartel, D.P., and Tabin, C.J. (2005). The microRNA miR-196 acts upstream of Hoxb8 and Shh in limb development. Nature *438*, 671–674.

Janowski, B.A., Huffman, K.E., Schwartz, J.C., Ram, R., Nordsell, R., Shames, D.S., Minna, J.D., and Corey, D.R. (2006). Involvement of AGO1 and AGO2 in mammalian transcriptional silencing. Nat. Struct. Mol. Biol. *13*, 787–792.

Jorgensen, R. (1990). Altered gene expression in plants due to *trans* interactions between homologous genes. Trends Biotechnol. *8*, 340–344.

Kagey, M.H., Melhuish, T.A., and Wotton, D. (2003). The polycomb protein Pc2 is a SUMO E3. Cell *113*, 127–137.

Kanellopoulou, C., Muljo, S.A., Kung, A.L., Ganesan, S., Drapkin, R., Jenuwein, T., Livingston, D.M., and Rajewsky, K. (2005). Dicer-deficient mouse embryonic stem cells are defective in differentiation and centromeric silencing. Genes Dev. *19*, 489–501.

Kato, H., Goto, D.B., Martienssen, R.A., Urano, T., Furukawa, K., and Murakami, Y. (2005). RNA polymerase II is required for RNAi-dependent heterochromatin assembly. Science *309*, 467–469.

Kennerdell, J.R., and Carthew, R.W. (2000). Heritable gene silencing in *Drosophila* using double-stranded RNA. Nat. Biotechnol. *18*, 896–898.

Ketting, R.F., Haverkamp, T.H., van Luenen, H.G., and Plasterk, R.H. (1999). Mut-7 of *C. elegans*, required for transposon silencing and RNA interference, is a homolog of Werner syndrome helicase and RNaseD. Cell *99*, 133–141.

Kim, D.H., Villeneuve, L.M., Morris, K.V., and Rossi, J.J. (2006). Argonaute-1 directs siRNA-mediated transcriptional gene silencing in human cells. Nat. Struct. Mol. Biol. *13*, 793–797.

Kim, V.N. (2005). Small RNAs: classification, biogenesis, and function. Mol. Cell *19*, 1–15.

Kohlmaier, A., Savarese, F., Lachner, M., Martens, J., Jenuwein, T., and Wutz, A. (2004). A chromosomal memory triggered by Xist regulates histone methylation in X inactivation. PLoS Biol. *2*, E171.

Lagos-Quintana, M., Rauhut, R., Lendeckel, W., and Tuschl, T. (2001). Identification of novel genes coding for small expressed RNAs. Science *294*, 853–858.

Lai, E.C., Tomancak, P., Williams, R.W., and Rubin, G.M. (2003). Computational identification of *Drosophila* microRNA genes. Genome Biol. *4*, R42.

Lanctot C, Cheutin T, Cremer M, Cavalli G, Cremer T. (2007). Dynamic genome architecture in the nuclear space: regulation of gene expression in three dimensions. Nat. Rev. Genet. *8*, 104–115.

Lau, N.C., Seto, A.G., Kim, J., Kuramochi-Miyagawa, S., Nakano, T., Bartel, D.P., and Kingston, R.E. (2006). Characterization of the piRNA complex from rat testes. Science *313*, 363–367.

Lee, R.C., Hammell, C.M., and Ambros, V. (2006). Interacting endogenous and exogenous RNAi pathways in *Caenorhabditis elegans*. RNA *12*, 589–597.

Lee, Y.S., Nakahara, K., Pham, J.W., Kim, K., He, Z., Sontheimer, E.J., and Carthew, R.W. (2004). Distinct roles for *Drosophila* Dicer-1 and Dicer-2 in the siRNA/miRNA silencing pathways. Cell *117*, 69–81.

Lee, S.R., and Collins, K. (2006). Two classes of endogenous small RNAs in *Tetrahymena thermophila*. Genes Dev. *20*, 28–33.

Lei, E.P., and Corces, V.G. (2006). RNA interference machinery influences the nuclear organization of a chromatin insulator. Nat. Genet. *38*, 936–941.

Lippman, Z., Gendrel, A.V., Black, M., Vaughn, M.W., Dedhia, N., McCombie, W.R., Lavine, K., Mittal, V., May, B., Kasschau, K.D., *et al.* (2004). Role of transposable elements in heterochromatin and epigenetic control. Nature *430*, 471–476.

Liu, Y., Taverna, S.D., Muratore, T.L., Shabanowitz, J., Hunt, D.F., and Allis, C.D. (2007). RNAi-dependent H3K27 methylation is required for heterochromatin formation and DNA elimination in *Tetrahymena*. Genes Dev. *21*, 1530–1545.

Llave, C., Kasschau, K.D., Rector, M.A., and Carrington, J.C. (2002). Endogenous and silencing-associated small RNAs in plants. Plant Cell *14*, 1605–1619.

Maison, C., Bailly, D., Peters, A.H., Quivy, J.P., Roche, D., Taddei, A., Lachner, M., Jenuwein, T., and Almouzni, G. (2002). Higher-order structure in pericentric heterochromatin involves a distinct pattern of histone modification and an RNA component. Nat. Genet. *30*, 329–334.

Martens, J.A., Laprade, L., and Winston, F. (2004). Intergenic transcription is required to repress the *Saccharomyces cerevisiae* SER3 gene. Nature *429*, 571–574.

Mette, M.F., Aufsatz, W., van der Winden, J., Matzke, M.A., and Matzke, A.J. (2000). Transcriptional silencing and promoter methylation triggered by double-stranded RNA. EMBO J. *19*, 5194–5201.

Mochizuki, K., Fine, N.A., Fujisawa, T., and Gorovsky, M.A. (2002). Analysis of a piwi-related gene implicates small RNAs in genome rearrangement in *Tetrahymena*. Cell *110*, 689–699.

Mochizuki, K., and Gorovsky, M.A. (2004). Conjugation-specific small RNAs in Tetrahymena have predicted properties of scan (scn) RNAs involved in genome rearrangement. Genes Dev. *18*, 2068–2073.

Morris, K.V., Chan, S.W., Jacobsen, S.E., and Looney, D.J. (2004). Small interfering RNA-induced transcriptional gene silencing in human cells. Science *305*, 1289–1292.

Napoli, C., Lemieux, C., and Jorgensen, R. (1990). Introduction of a chimeric chalcone synthase gene into petunia results in reversible co-suppression of homologous genes in *trans*. Plant Cell 2, 279–289.

Okamura, K., Ishizuka, A., Siomi, H., and Siomi, M.C. (2004). Distinct roles for Argonaute proteins in small RNA-directed RNA cleavage pathways. Genes Dev. *18*, 1655–1666.

Pal-Bhadra, M., Bhadra, U., and Birchler, J.A. (1997). Cosuppression in *Drosophila*: gene silencing of Alcohol dehydrogenase by white-Adh transgenes is Polycomb dependent. Cell *90*, 479–490.

Pal-Bhadra, M., Bhadra, U., and Birchler, J.A. (2002). RNAi related mechanisms affect both transcriptional and posttranscriptional transgene silencing in *Drosophila*. Mol. Cell 9, 315–327.

Pal-Bhadra, M., Leibovitch, B.A., Gandhi, S.G., Rao, M., Bhadra, U., Birchler, J.A., and Elgin, S.C. (2004). Heterochromatic silencing and HP1 localization in *Drosophila* are dependent on the RNAi machinery. Science *303*, 669–672.

Park, C.W., Chen, Z., Kren, B.T., and Steer, C.J. (2004). Double-stranded siRNA targeted to the huntingtin gene does not induce DNA methylation. Biochem. Biophys. Res. Commun. *323*, 275–280.

Peng, J.C., and Karpen, G.H. (2007). H3K9 methylation and RNA interference regulate nucleolar organization and repeated DNA stability. Nat. Cell. Biol. *9*, 25–35.

Peragine, A., Yoshikawa, M., Wu, G., Albrecht, H.L., and Poethig, R.S. (2004). SGS3 and SGS2/SDE1/RDR6 are required for juvenile development and the production of *trans*-acting siRNAs in *Arabidopsis*. Genes Dev. *18*, 2368–2379.

Petruk, S., Sedkov, Y., Riley, K.M., Hodgson, J., Schweisguth, F., Hirose, S., Jaynes, J.B., Brock, H.W., and Mazo, A. (2006). Transcription of bxd noncoding RNAs promoted by trithorax represses Ubx in *cis* by transcriptional interference. Cell *127*, 1209–1221.

Piccin, A., Salameh, A., Benna, C., Sandrelli, F., Mazzotta, G., Zordan, M., Rosato, E., Kyriacou, C.P., and Costa, R. (2001). Efficient and heritable functional knock-out of an adult phenotype in *Drosophila* using a GAL4-driven hairpin RNA incorporating a heterologous spacer. Nucleic Acids Res. *29*, E55–55.

Plath, K., Fang, J., Mlynarczyk-Evans, S.K., Cao, R., Worringer, K.A., Wang, H., de la Cruz, C.C., Otte, A.P., Panning, B., and Zhang, Y. (2003). Role of histone H3 lysine 27 methylation in X inactivation. Science *300*, 131–135.

Plath, K., Talbot, D., Hamer, K.M., Otte, A.P., Yang, T.P., Jaenisch, R., and Panning, B. (2004). Developmentally regulated alterations in Polycomb repressive complex 1 proteins on the inactive X chromosome. J. Cell Biol. *167*, 1025–1035.

Rank, G., Prestel, M., and Paro, R. (2002). Transcription through intergenic chromosomal memory elements of the *Drosophila* bithorax complex correlates with an epigenetic switch. Mol. Cell. Biol. *22*, 8026–8034.

Reinhart, B.J., and Bartel, D.P. (2002). Small RNAs correspond to centromere heterochromatic repeats. Science *297*, 1831.

Rinn, J.L., Kertesz, M., Wang, J.K., Squazzo, S.L., Xu, X., Brugmann, S.A., Goodnough, L.H., Helms, J.A., Farnham, P.J., Segal, E., and Chang, H.Y. (2007). Functional demarcation of active and silent chromatin domains in human HOX loci by noncoding RNAs. Cell *129*, 1311–1323.

Ronshaugen, M., Biemar, F., Piel, J., Levine, M., and Lai, E.C. (2005). The *Drosophila* microRNA iab-4 causes a dominant homeotic transformation of halteres to wings. Genes Dev. *19*, 2947–2952.

Saffery, R., Sumer, H., Hassan, S., Wong, L.H., Craig, J.M., Todokoro, K., Anderson, M., Stafford, A., and Choo, K.H. (2003). Transcription within a functional human centromere. Mol. Cell *12*, 509–516.

Sanchez-Elsner, T., Gou, D., Kremmer, E., and Sauer, F. (2006). Noncoding RNAs of trithorax response elements recruit *Drosophila* Ash1 to Ultrabithorax. Science *311*, 1118–1123.

Sasaki, T., Shiohama, A., Minoshima, S., and Shimizu, N. (2003). Identification of eight members of the Argonaute family in the human genome small star, filled. Genomics *82*, 323–330.

Schmitt, S., Prestel, M., and Paro, R. (2005). Intergenic transcription through a polycomb group response element counteracts silencing. Genes Dev. *19*, 697–708.

Schuettengruber, B., Chourrout, D., Vervoort, M., Leblanc, B., and Cavalli, G. (2007) Genome Regulation by Polycomb and Trithorax proteins, Cell *128*, 735–745.

Schwartz, Y.B., and Pirrotta, V. (2007). Polycomb silencing mechanisms and the management of genomic programmes. Nat. Rev. Genet. *8*, 9–22.

Sessa, L., Breiling, A., Lavorgna, G., Silvestri, L., Casari, G., and Orlando, V. (2007). Noncoding RNA synthesis and loss of Polycomb group repression accompanies the colinear activation of the human HOXA cluster. RNA *13*, 223–239.

Sigova, A., Rhind, N., and Zamore, P.D. (2004). A single Argonaute protein mediates both transcriptional and posttranscriptional silencing in *Schizosaccharomyces pombe*. Genes Dev. *18*, 2359–2367.

Silva, J., Mak, W., Zvetkova, I., Appanah, R., Nesterova, T.B., Webster, Z., Peters, A.H., Jenuwein, T., Otte, A.P., and Brockdorff, N. (2003). Establishment of histone h3 methylation on the inactive X chromosome requires transient recruitment of Eed-Enx1 polycomb group complexes. Dev. Cell 4, 481–495.

Svoboda, P. (2004). Long dsRNA and silent genes strike back: RNAi in mouse oocytes and early embryos. CytoGenet. Genome Res. *105*, 422–434.

Tabara, H., Sarkissian, M., Kelly, W.G., Fleenor, J., Grishok, A., Timmons, L., Fire, A., and Mello, C.C. (1999). The rde-1 gene, RNA interference, and transposon silencing in C. elegans. Cell 99, 123–132.

Taverna, S.D., Coyne, R.S., and Allis, C.D. (2002). Methylation of histone h3 at lysine 9 targets programmed DNA elimination in tetrahymena. Cell 110, 701–711.

Ting, A.H., Schuebel, K.E., Herman, J.G., and Baylin, S.B. (2005). Short double-stranded RNA induces transcriptional gene silencing in human cancer cells in the absence of DNA methylation. Nat. Genet. 37, 906–910.

Tran, R.K., Zilberman, D., de Bustos, C., Ditt, R.F., Henikoff, J.G., Lindroth, A.M., Delrow, J., Boyle, T., Kwong, S., Bryson, T.D., et al. (2005). Chromatin and siRNA pathways cooperate to maintain DNA methylation of small transposable elements in Arabidopsis. Genome Biol. 6, R90.

Vagin, V.V., Sigova, A., Li, C., Seitz, H., Gvozdev, V., and Zamore, P.D. (2006). A distinct small RNA pathway silences selfish genetic elements in the germline. Science 313, 320–324.

van der Krol, A.R., Mur, L.A., Beld, M., Mol, J.N., and Stuitje, A.R. (1990a). Flavonoid genes in petunia: addition of a limited number of gene copies may lead to a suppression of gene expression. Plant Cell 2, 291–299.

van der Krol, A.R., Mur, L.A., de Lange, P., Mol, J.N., and Stuitje, A.R. (1990b). Inhibition of flower pigmentation by antisense CHS genes: promoter and minimal sequence requirements for the antisense effect. Plant Mol. Biol. 14, 457–466.

Vastenhouw, N.L., and Plasterk, R.H. (2004). RNAi protects the Caenorhabditis elegans germline against transposition. Trends Genet. 20, 314–319.

Vazquez, F. (2006). Arabidopsis endogenous small RNAs: highways and byways. Trends Plant Sci 11, 460–468.

Vazquez, F., Vaucheret, H., Rajagopalan, R., Lepers, C., Gasciolli, V., Mallory, A.C., Hilbert, J.L., Bartel, D.P., and Crete, P. (2004). Endogenous trans-acting siRNAs regulate the accumulation of Arabidopsis mRNAs. Mol. Cell 16, 69–79.

Vazquez, J., Muller, M., Pirrotta, V., and Sedat, J.W. (2006). The Mcp element mediates stable long-range chromosome-chromosome interactions in Drosophila. Mol. Biol. Cell 17, 2158–2165.

Verdel, A., Jia, S., Gerber, S., Sugiyama, T., Gygi, S., Grewal, S.I., and Moazed, D. (2004). RNAi-mediated targeting of heterochromatin by the RITS complex. Science 303, 672–676.

Volpe, T.A., Kidner, C., Hall, I.M., Teng, G., Grewal, S.I., and Martienssen, R.A. (2002). Regulation of heterochromatic silencing and histone H3 lysine-9 methylation by RNAi. Science 297, 1833–1837.

Wassenegger, M., and Krczal, G. (2006). Nomenclature and functions of RNA-directed RNA polymerases. Trends Plant Sci. 11, 142–151.

Watanabe, T., Takeda, A., Tsukiyama, T., Mise, K., Okuno, T., Sasaki, H., Minami, N., and Imai, H. (2006). Identification and characterization of two novel classes of small RNAs in the mouse germline: retrotransposon-derived siRNAs in oocytes and germline small RNAs in testes. Genes Dev. 20, 1732–1743.

Weinberg, M.S., Villeneuve, L.M., Ehsani, A., Amarzguioui, M., Aagaard, L., Chen, Z.X., Riggs, A.D., Rossi, J.J., and Morris, K.V. (2006). The antisense strand of small interfering RNAs directs histone methylation and transcriptional gene silencing in human cells. RNA 12, 256–262.

Xu, L., Fong, Y., and Strome, S. (2001). The Caenorhabditis elegans maternal-effect sterile proteins, MES-2, MES-3, and MES-6, are associated in a complex in embryos. Proc. Natl. Acad. Sci. USA 98, 5061–5066.

Yekta, S., Shih, I.H., and Bartel, D.P. (2004). MicroRNA-directed cleavage of HOXB8 mRNA. Science 304, 594–596.

Yigit, E., Batista, P.J., Bei, Y., Pang, K.M., Chen, C.C., Tolia, N.H., Joshua-Tor, L., Mitani, S., Simard, M.J., and Mello, C.C. (2006). Analysis of the C. elegans Argonaute family reveals that distinct Argonautes act sequentially during RNAi. Cell 127, 747–757.

Zhang, H., Azevedo, R.B., Lints, R., Doyle, C., Teng, Y., Haber, D., and Emmons, S.W. (2003). Global regulation of Hox gene expression in C. elegans by a SAM domain protein. Dev. Cell 4, 903–915.

Zhang, H., Smolen, G.A., Palmer, R., Christoforou, A., van den Heuvel, S., and Haber, D.A. (2004). Sumo modification is required for in vivo Hox gene regulation by the Caenorhabditis elegans Polycomb group protein SOP-2. Nat. Genet. 36, 507–511.

Zhang, T., Sun, Y., Tian, E., Deng, H., Zhang, Y., Luo, X., Cai, Q., Wang, H., Chai, J., and Zhang, H. (2006). RNA-binding proteins SOP-2 and SOR-1 form a novel PcG-like complex in C. elegans. Development 133, 1023–1033.

Zhou, J., Barolo, S., Szymanski, P., and Levine, M. (1996). The Fab-7 element of the bithorax complex attenuates enhancer-promoter interactions in the Drosophila embryo. Genes Dev. 10, 3195–3201.

Heterochromatin Assembly and Transcriptional Gene Silencing under the Control of Nuclear RNAi: Lessons from Fission Yeast

4

Aurélia Vavasseur, Leila Touat-Todeschini and André Verdel

Abstract

Heterochromatin is a prevalent chromatin state among eukaryotes that has critical functions in chromosome segregation, control of genomic stability and epigenetic regulation of gene expression. Here, we review studies conducted in the fission yeast *Schizosaccharomyces pombe*, which reveal that two RNAi complexes, the RNA induced transcriptional gene silencing (RITS) complex and the RNA-directed RNA polymerase complex (RDRC), are part of a RNAi machinery involved in the initiation, propagation and maintenance of heterochromatin assembly. It appears that these two complexes localize in a siRNA-dependent manner on chromosomes, at the site of heterochromatin assembly. Moreover, these studies reveal an unprecedented and central role for RNA polymerase II (RNApII) in RNAi-dependent heterochromatin assembly. RNApII synthesizes a nascent transcript that is believed to serve as a RNA platform to recruit, RITS, RDRC and possibly other complexes required for heterochromatin assembly. Finally, recent findings indicate that RNAi as well as an exosome-dependent RNA degradation process contribute to heterochromatic gene silencing. These findings challenge the widely accepted view that heterochromatic gene silencing is caused only by chromatin compaction. As RNAi-dependent chromatin modifications have been observed throughout the eukaryotic kingdom the mechanisms reviewed here are susceptible to occur in a large range of eukaryotes.

Introduction

In eukaryotic cells, genomic DNA associates with histones, a family of small and highly conserved proteins, to form chromatin. The nucleosome represents the basic and repetitive unit of chromatin. Each nucleosome is formed of a histone protein octamer, consisting of two dimers of histones H2A and H2B associated to a tetramer of histones H3 and H4, around which 146 base pairs of DNA wrap in nearly two turns (Luger *et al.*, 1997). Each of these histones possesses a N-terminal extremity that protrudes out of the nucleosome and that can be modified in a post-translational and reversible manner on multiple residues. In a given genomic region, these modifications can modulate the binding of non-histone proteins to chromatin as well as the accessibility to the DNA wrapped around the histone octamer (Jenuwein and Allis, 2001). By doing so, histone modifications have the potential to change the transcriptional competency of their associated DNA. Methylation of DNA is another chromatin modification that can similarly change the transcription status (Klose *et al.*, 2006). In multiple eukaryotic organisms, RNAi or RNAi-related mechanisms mediate such chromatin modifications with a high degree of sequence specificity.

In the fission yeast *Schizosaccharomyces pombe*, *Drosophila melanogaster* and *Arabidopsis thaliana*, RNA interference (RNAi) is directly involved in formation of heterochromatin or heterochromatin-like structures, and accumulating evidences suggest that this role is conserved in other eukaryotes, such as vertebrates (Deshpande *et al.*, 2005; Fukagawa *et al.*, 2004; Kanellopoulou *et al.*, 2005; May *et al.*, 2005; Pal-Bhadra *et al.*, 2004; Verdel and Moazed, 2005; Volpe *et al.*, 2002). In parallel, transfection of mammalian cell lines with double-stranded siRNA specific for promoter regions can trigger a RNAi-dependent transcriptional gene silencing, which correlates with the appearance of heterochromatic marks such as DNA and specific histone methylations (Kim *et al.*, 2006; Morris *et al.*, 2004; Ting *et al.*, 2005; Weinberg *et al.*, 2006). Heterochromatin and euchromatin represent the two major states of chromatin in a eukaryotic cell during interphase. These states differ in their DNA content, structure and function.

Heterochromatin, which represents the main chromatin form found in higher eukaryotic genomes, is poor in coding genes, is highly enriched in non-coding and repetitive DNA, and it adopts a more compact structure that is believed to restrict access to its DNA (Craig, 2005; Grewal and Jia, 2007). Oppositely, euchromatin contains most of the active genes and is more prone to associate with DNA-binding proteins and complexes such as transcription factors and RNA polymerase II (RNApII). Differences in the nature and composition of post-translational histone modifications also exist between these two states of chromatin. Heterochromatin is hypoacetylated and highly enriched in one specific post-translational histone modification, histone H3 methylated on lysine 9 (H3K9me), whereas euchromatin is relatively hyperacetylated and enriched in other histone marks, such as histone H3 methylated on lysine 4 (H3K4me). On the functional level, heterochromatin plays critical roles in chromosome segregation and in genomic stability. Indeed, assembly of heterochromatin at pericentromeric DNA surrounding the central core of centromeres (to which the kinetochore binds) is critical to recruit high levels of cohesins, as they bind to heterochromatin structural components, and thereafter ensure proper chromosome separation during cell division (Bernard *et al.*, 2001; Pidoux and Allshire, 2004). Additionally, heterochromatin formation greatly reduces the potential of genomic sites containing repetitive DNA or mobile elements such as retrotransposons, to provoke undesired

homologous recombination or deleterious genomic integration and invasion, respectively (Grewal and Moazed, 2003). Heterochromatin or heterochromatin-like structures exert also another critical function by imposing an epigenetically inherited transcriptional silencing that is essential for cellular differentiation and development (Grewal and Moazed, 2003).

Over the last decade, studies conducted in *S. pombe* have greatly contributed to our current understanding of the molecular events important for heterochromatin formation (Horn and Peterson, 2006; Martienssen *et al.*, 2005; Verdel and Moazed, 2005). Key events include methylation of lysine 9 of histone H3 by the histone methyltransferase Clr4 (the homologue of Suv39h in vertebrates), as well as recruitment to chromatin of Swi6 (a HP1-like protein) and Chp1 proteins, which directly bind methylated H3K9 (H3K9me) thanks to a motif termed chromodomain (Nakayama *et al.*, 2001; Partridge *et al.*, 2002). The general consensus is that these events occur in an ordered fashion. Histone deacetylation, by Sir2, Clr3 and Clr6 histone deacetylases (HDACs), is followed by methylation of lysine 9 of histone H3 by Clr4. H3K9me then serves as a docking site to recruit Swi6 and Chp1. Swi6 bound to H3K9me is in turn believed to recruit the same HDACs and Clr4 to start a new cycle of histone modifications and histone binding on adjacent nucleosomes (Nakayama *et al.*, 2001; Yamada *et al.*, 2005). Hence, this Swi6-dependent mechanism can be responsible for the self-propagation of heterochromatin structural elements along the chromatin fibre and can also play a critical role in the epigenetic maintenance of heterochromatin.

The discovery nearly 5 years ago that RNAi is involved in the formation of heterochromatin has been followed by a series of biochemical and molecular genetics studies revealing important mechanistic features of RNAi-mediated heterochromatin formation. A RNAi machinery, including RITS and RDRC complexes, promotes several steps essential for formation, propagation and maintenance of heterochromatin, as well as for the transcriptional silencing imposed by heterochromatin assembly. In this chapter, we present an overview of our current understanding of this deep implication of RNAi in heterochromatin assembly and heterochromatic transcriptional silencing.

A role for RNAi in initiation and maintenance of heterochromatin assembly

The essential role of RNAi in the assembly of heterochromatin was revealed in S. pombe by Martienssen, Grewal and co-workers (Volpe et al., 2002). In fission yeast, and conversely to most eukaryotic models, key RNAi proteins Dicer (Dcr1), RNA-directed RNA polymerase (Rdp1) and Argonaute (Ago1) are encoded by single copy genes. Quite surprisingly at the time, it was found that deletion of either *dcr1*, *rdp1* or *ago1* gene leads to loss of heterochromatin normally present on DNA repeats surrounding

the central region of the centromere (Fig. 4.1). In addition, non-coding RNA emanating from forward and reverse strands of pericentromeric repeats accumulate in those mutants, reflecting a concomitant loss of transcriptional silencing (Volpe et al., 2003; Volpe et al., 2002). A direct implication of RNAi in heterochromatin assembly was strongly favoured by the fact that Rdp1 associates specifically with these pericentromeric repeats (Volpe et al., 2003; Volpe et al., 2002). In parallel, the cloning of small RNAs in the size range of a siRNA revealed the existence of siRNAs emanating from the same heterochromatic repeats (Reinhart and Bartel, 2002). It was proposed from these first results that centromeric double-stranded RNA (dsRNA) would activate RNAi and be processed into siRNAs through

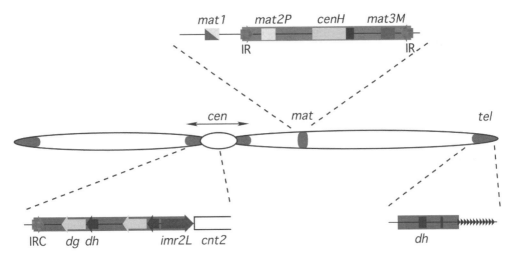

■ Heterochromatic region

Figure 4.1 The main heterochromatic regions in fission yeast. S. pombe forms large domains of heterochromatin at three genomic locations that are the pericentromeric, subtelomeric and mating-type regions. For simplicity, only chromosome II is represented, which contains the three types of heterochromatic domains. Centromeres of S. pombe (*cen*) share a common DNA organization, a central core region (*cnt*), where the kinetochore forms, surrounded by two regions composed of different repeats, termed *innermost* (*imr*) and *outermost* (*otr*). The *otr* region contains a number of *dg* and *dh* repeats that varies from one pericentromeric region to another. Inverted repeats (IR and IRC) and/or tRNA genes are present at both extremities of pericentromeric heterochromatin. These DNA domains seem to be important to stop spreading of heterochromatin. The mating-type locus (*mat*) spans over 20 kb among which more than 15 kb are embedded in heterochromatin. Two of the three genes present in this locus (*mat2P* and *mat3P*) are transcriptionally silenced by heterochromatin, whereas *mat1* located in euchromatin is expressed and determines the mating-type (P or M) of the cell. The mating-type locus shares with centromeric *otr* regions a DNA domain known as *cenH* that is formed of portions of *dg* and *dh* pericentromeric repeats. *cenH* constitutes the mating-type nucleation centre for RNAi-dependent heterochromatin assembly. Subtelomeric regions from the left arm of chromosome I as well as the right arm of chromosome II also contain fragments of *dh* repeats. Similarly, these repeats have been proposed to be nucleation centres for heterochromatin assembly at telomeres (*tel*). For the other subtelomeric regions, as the genome sequencing is not yet fully completed, the presence of *dh* or *dg* repeats still needs to be determined. The small black triangles represent the 6 nucleotide repeats present at each telomere.

the action of the ribonuclease III Dcr1 (Volpe *et al.*, 2002). It was also suggested that siRNAs participate by an unknown mechanism in the formation of heterochromatin.

Further investigations showed that RNAi-mediated heterochromatin assembly is not restricted to pericentromeric DNA but in fact acts also on the other major heterochromatic regions present in the *S. pombe* genome, which are the subtelomeric regions and the mating-type domain (Fig. 4.1). siRNAs associated to RITS were found to match these regions (Cam *et al.*, 2005). These heterochromatic regions turn to have in common DNA domains of several hundred nucleotides that share around 97% sequence identity (Wood *et al.*, 2002). Interestingly, these DNA domains can initiate heterochromatin formation in a RNAi-dependent manner when they are integrated in an ectopic region of the genome, indicating that they act as nucleation centres for heterochromatin formation (Hall *et al.*, 2002; Partridge *et al.*, 2002; Sadaie *et al.*, 2004). Consistent with this, at the mating-type locus, such domain (termed *cenH*; see Fig. 4.1) has been found to be the nucleation centre for the RNAi-mediated heterochromatin formation occurring in this region (Hall *et al.*, 2002). In a similar manner, the homologous domains present in the subtelomeric part of chromosomes might also act as nucleation centre for RNAi-driven heterochromatin assembly.

Initial studies on the role of RNAi in heterochromatin assembly revealed a major distinction between pericentromeric repeats and the other heterochromatic sites. Single deletion of either *dcr1*, *rdp1* or *ago1* gene does not significantly impact on the maintenance of heterochromatin marks or on the transcriptional silencing at the mating-type and subtelomeric regions (Hall *et al.*, 2002). Subsequent investigations revealed that this discrepancy comes from the fact that RNAi-independent mechanisms also govern heterochromatin assembly in these genomic regions (Jia *et al.*, 2004; Kanoh *et al.*, 2005; Petrie *et al.*, 2005). These mechanisms involve DNA-binding proteins that recruit the components required to nucleate heterochromatin formation without any need for RNAi. Within the mating type region, the proteins are Atf1 and Pcr1, which belong to the ATF/CREB (Cyclic AMP Responsive

Element Binding protein) transcription factor family (Jia *et al.*, 2004). Atf1 and Pcr1 guide heterochromatin formation through interaction with their consensus DNA-binding sequence. Similarly, the protein Taz1, which binds to the 6 nucleotide telomeric repeat (TTAGGG), nucleates heterochromatin assembly at telomeres (Kanoh *et al.*, 2005).

Heterochromatin is a chromatin state that has the property to be epigenetically maintained at a specific place in the genome once its formation has been initiated. One important aspect in investigating the role of RNAi has been to determine at which step(s) of heterochromatin initiation and/or maintenance RNAi is involved. In the absence of either Atf1 or Pcr1 DNA-binding proteins and of any marks of heterochromatin at the mating type interval (Fig. 4.1), RNAi is sufficient to trigger heterochromatin assembly, indicating that RNAi is capable of initiating heterochromatin formation very efficiently (Jia *et al.*, 2004). Furthermore, once initiated and in the absence of RNAi-independent heterochromatin formation mechanism, RNAi becomes absolutely required for maintenance of heterochromatic marks (H3K9me and Swi6) and transcriptional silencing at the mating-type region. This is reminiscent to what occurs at centromeres, where RNAi is necessary for efficient preservation of these marks and for heterochromatic transcriptional silencing (Verdel *et al.*, 2004; Volpe *et al.*, 2003; Volpe *et al.*, 2002). From these results, it appears that RNAi plays an active role in initiation of heterochromatin but also in its maintenance. Implication of RNAi in maintenance of heterochromatin questions the molecular mechanisms responsible for the epigenetic inheritance of heterochromatin. The protein Swi6 is believed to play a central role in maintaining heterochromatin structure, particularly after genome duplication, by directly recruiting HDACs and the H3K9 methyltransferase Clr4 to heterochromatic regions (Nakayama *et al.*, 2001; Yamada *et al.*, 2005). However, under certain circumstances this Swi6-dependent mechanism is not sufficient to maintain the heterochromatic structure and it needs to be assisted by RNAi (Jia, 2004). One intriguing possibility is that RNAi contributes to heterochromatin maintenance by constantly reinitiating heterochromatin assembly. Such highly dynamic mechanism contrasts with the general belief that, once established, heterochromatin is a relatively static structure that can be maintained without the need to reinitiate its formation.

RITS, a RNAi complex that mediates heterochromatin assembly through its siRNA-driven recruitment to chromatin

In parallel to molecular genetics investigations that provided important insights into the role of RNAi in heterochromatin formation, purification of RNAi complexes has strongly contributed to our current knowledge on the molecular mechanisms by which RNAi mediates heterochromatin assembly. Purification of Chp1 protein, which was known to act early in the assembly of heterochromatin and to be essential for this process, led to the isolation of the

RNA-induced transcriptional silencing complex (RITS, Fig. 4.2) (Verdel *et al.*, 2004; Verdel and Moazed, 2005). RITS contains, besides Chp1, Ago1, an RNAi protein that belongs to the Argonaute family, and Tas3, a protein with no specific function. Additionally, RITS also contains small RNAs. Investigations on the nature of these small RNAs revealed that, RITS is devoid of small RNAs in a *dcr1*-deficient strain. Moreover, these small RNAs possess the 5′-phosphate and 3′ hydroxy terminal signatures of Dicer-cleavage products, which strongly suggested that RITS-associated small RNAs are *bona fide* siRNAs (Cam *et al.*, 2005; Reinhart and Bartel, 2002). Hence, just from the analysis of RITS composition it appeared, for the first time, that RNAi is physically linked to heterochromatin formation as this complex is formed by the association of two essential RNAi elements, Ago1 and siRNA, with a key structural component of heterochromatin, Chp1.

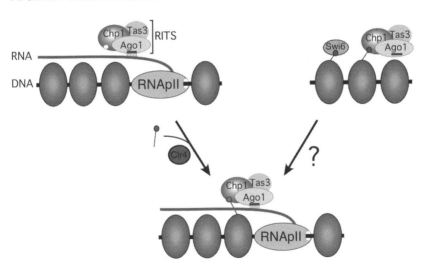

Figure 4.2 RITS small interfering RNA (siRNA)- and H3K9me-driven recruitment to chromatin. RITS is a ribonucleoprotein complex formed by the association of a siRNA with three proteins, Chp1, Tas3 and Ago1. It is thought that RITS possesses two chromatin recognition modules, which are the siRNA (dark small line) bound to Ago1 and Chp1 chromodomain (white circle present on Chp1). RITS can use the siRNA to bind chromatin in a sequence specific-manner. siRNA-driven recruitment of RITS involves a base-pairing interaction between the siRNA and a nascent transcript synthesized by RNA polymerase II (RNApII). In parallel, it has been proposed that methylation of histone H3 lysine 9 (H3K9) by the histone methyltransferase Clr4 provides a docking site that can be sufficient to recruit RITS. In a situation like establishment of heterochromatin assembly, an intimate interplay might take place between RITS two recognition modules. The siRNA would be critical to initially target RITS to the proper genomic region, whereas Chp1 chromodomain would stabilize RITS binding by anchoring the complex to a methylated H3K9 (H3K9me, small grey lollipop shape). The question mark points out that it remains to be determined if RITS can first bind to chromatin via H3K9me and subsequently to a nascent transcript. See text for further details.

Functional characterization of Tas3 revealed that, as previously shown for Chp1 and Ago1, it is essential for pericentromeric heterochromatin formation and silencing, indicating that all subunits of RITS play a critical role in RNAi-mediated heterochromatin formation (Verdel *et al.*, 2004). Furthermore, analyses of Tas3 chromatin localization by chromatin immuno-precipitation (ChIP) experiments showed that Tas3, like Chp1 and Ago1, cross-links to the major heterochromatic sites present in *S. pombe* (Cam *et al.*, 2005; Noma *et al.*, 2004; Verdel *et al.*, 2004).

Investigations have been conducted to determine how RITS gets recruited to chromatin. A biochemical analysis of RITS composition after its purification from a *dcr1*-deficient strain showed that the complex still forms in absence of siRNA and therefore that siRNAs do not play a major role in the architecture of RITS. However, in the absence of siRNA, RITS no longer binds to heterochromatic DNA (Noma *et al.*, 2004; Verdel *et al.*, 2004). It was further found that RITS-associated siRNAs match DNA sequences present only in heterochromatic regions, which pointed out that the nucleotide sequence of the siRNA is probably important for RITS binding to chromatin (Cam *et al.*, 2005). From these results, it was proposed that a siRNA could guide RITS to chromatin thanks to a base-pairing interaction between the siRNA and its genomic target (Verdel *et al.*, 2004). Experiments based on tethering RITS to chromatin have strongly supported the likelihood of such mechanism (Buhler *et al.*, 2006). Once RITS is tethered to a given euchromatic region, it initiates formation of heterochromatin as well as production of new siRNAs matching this region. Interestingly, and presumably thanks to this production of *neosiRNAs*, RITS can then act in *trans* to target another euchromatic region that shares a high sequence identity with the genomic site where RITS was initially tethered. This *trans* chromatin targeting mechanism strongly favours the siRNA-driven target-recognition mechanism for RITS recruitment to chromatin. It should be pointed out that, however, this *trans* targeting has so far only been seen in a *eri1* deficient strain. The *eri1* gene encodes for a ribonuclease believed to be a negative regulator of RNAi that degrades

siRNAs (Iida *et al.*, 2006). As a matter of fact, in the absence of Eri1, siRNAs accumulate. Therefore, although RITS is capable of targeting new genomic regions thanks to its associated siRNA, in a wild type situation it seems that RITS-mediated heterochromatin formation in *trans* is strongly limited by the potential RNAi negative regulator Eri1.

A critical role for the histone methyltransferase Clr4 in recruiting and stabilizing RITS to chromatin

ChIP experiments have indicated that Clr4, the H3K9 specific methyltransferase, is necessary to cross-link RITS to chromatin (Noma *et al.*, 2004). This double requirement on both Dcr1 and Clr4 for RITS recruitment to chromatin has lead to the proposal that RITS possesses two chromatin recognition modules that are the Chp1 chromodomain and the Ago1-bound siRNA (Fig. 4.2). However, contribution of each module to RITS recruitment to chromatin remained to be determined. Tas3 contains a motif formed by GW (glycin and tryptophan) repeats that is also found in the protein GW182, which in *Drosophila melanogaster* and human interacts with members of the Argonaute family (that includes Ago1) (Behm-Ansmant *et al.*, 2006). Recently, point mutations in Tas3 GW motif were shown to abolish Tas3 interaction with Ago1 but not with Chp1 (Partridge *et al.*, 2007). Interestingly, in this mutant background, all three subunits of RITS still localize to heterochromatic DNA, supporting the idea that RITS probably possesses two chromatin recognition modules, one present in the Chp1-Tas3 dimer (the chromodomain) and one in Ago1 (the siRNA). As expected, Chp1 and Tas3 binding to chromatin requires H3K9me at the site of recruitment. Importantly, in a situation where no H3K9me is present Chp1 and Tas3 do not initiate heterochromatin assembly unless they are capable of interacting with Ago1. Hence, two different situations seem to emerge depending on whether RITS acts to establish or maintain heterochromatin formation (Fig. 4.2). In the situation where RITS establishes heterochromatin assembly, the siRNA would act as the primary recognition module for RITS initial recruitment, whereas when the heterochromatic mark H3K9me is already present RITS chromodomain (located on Chp1) seems sufficient for its recruitment.

In fact, RITS recruitment to chromatin during establishment of heterochromatin seems to involve a close interplay between RITS two recognition modules (Fig. 4.2). Ago1-bound siRNA is believed to be the first module involved, because it can target RITS to the proper genomic region through a nucleotide-based recognition mechanism that does not require any preceding chromatin modification. However, ChIP experiments have indicated that even in the establishment situation RITS requires Clr4 to be cross-linked to chromatin (Partridge *et al.*, 2007). The reason for this Clr4-dependency might be that once bound to chromatin through the siRNA-based mechanism RITS absolutely needs Chp1 chromodomain to anchor on H3K9me and thereby to stably interact with chromatin. Interestingly, such mechanism provides a molecular explanation to early genetics experiments that had found a mutual dependency between Chp1 recruitment to chromatin and H3K9 methylation by Clr4 (Partridge *et al.*, 2002). Further studies are however needed to validate such mechanism and to determine how then RITS siRNA-driven recruitment to chromatin would favour Clr4-dependent H3K9 methylation on the surrounding nucleosomes so that it can bind to H3K9me.

Clr4 was found to interact with Rik1, a WD repeat protein known to be essential for both RNAi-dependent and -independent heterochromatin assembly (Sadaie *et al.*, 2004). The C-terminal part of Rik1 also shares homology to CPSF-A, a protein belonging to the cleavage RNA polyadenylation specificity factor that is believed to interact with the polyadenylation signal. Rik1 potential RNA-binding property could be important for connecting RITS and Clr4, as RNA transcripts are essential for RITS recruitment to chromatin (see next section for details on the rÔole of RNA in recruiting RITS). In parallel, Clr4 and Rik1 are both subunits of a Cul4 E3 ubiquitin ligase complex (Hong *et al.*, 2005; Horn *et al.*, 2005; Jia *et al.*, 2005; Li *et al.*, 2005; Thon *et al.*, 2005). The exact function of this Cul4 complex remains to be determined, but

it is anticipated that its characterization might provide precious indications on its link to RITS and RNAi.

Essential roles for RNA polymerase II, nascent transcripts and dsRNA synthesis in heterochromatin formation

A major aspect of RNAi-mediated heterochromatin formation concerns the mechanism by which RITS-associated siRNAs recognize the complementary genomic regions. Two principal mechanisms were initially proposed (Verdel *et al.*, 2004; Verdel and Moazed, 2005). In the first mechanism, RITS binds directly a genomic DNA region complementary to the siRNA nucleotidic sequence. This mechanism implies that the DNA double-helix is unwound to allow a DNA-siRNA base-pairing interaction. The second mechanism involves a chromatin-bound RNA that serves as a platform on which RITS-associated siRNA base-pairs and forms a RNA-RNA hybrid. The first piece of evidence in favour of one of these mechanisms came from the identification of a second RNAi complex, named the RNA-Directed RNA-polymerase Complex (RDRC) (Motamedi *et al.*, 2004).

Purification of RDRC revealed that it is composed of three core subunits, Rdp1, a RNAi protein with a putative RNA-Directed RNA-polymerase (RDR) activity, Hrr1, a probable RNA helicase, and Cid12, a member of the poly(A) polymerase family. *In vitro* experiments showed that RDRC possesses a RDR activity, thereby implying that RDRC interacts with RNA. Point mutations in Rdp1 catalytic site or truncation in the extreme C-terminal part of Rdp1 destroy this polymerase activity and have confirmed that Rdp1 is the subunit responsible for the RDR activity of RDRC (Motamedi *et al.*, 2004; Sugiyama *et al.*, 2005). *In vivo*, strains that carry these mutations exhibit a complete loss of pericentomeric heterochromatin and transcriptional silencing. This further indicated that the RDR activity of RDRC is essential for RNAi-mediated heterochromatin formation and that RDRC also interacts with RNA *in vivo*. Importantly, RDRC purification also revealed a physical connection between RDRC and

RITS. As RDRC binds also RNA, experiments were conducted to test RITS binding to RNA (Motamedi et al., 2004). These experiments showed that indeed RITS and RDRC bind to RNA in vivo. Furthermore, these two complexes selectively associate with RNA emanating from a heterochromatic region but not from a euchromatic region. Immunofluorescence and co-immunoprecipitation experiments pointed out that RITS and RDRC seem to colocalize only at heterochromatic regions, suggesting that both RITS-RDRC interaction as well as their respective interaction with RNA occur only on or at the close periphery of heterochromatin. Consistent with these observations, RITS bindings to RNA and to RDRC are also dependent on the histone methyltransferase Clr4 (Motamedi et al., 2004). Thus, it appeared that RITS and RDRC most probably interact with a chromatin-bound RNA. For sequence specificity reasons it was suggested that the target RNA is a nascent transcript (Motamedi et al., 2004).

Subsequent investigations have indicated that RNA polymerase II (RNApII) is present in heterochromatic regions and that point mutations in RNApII subunits (identified by genetic screens) disrupt heterochromatin formation (Djupedal et al., 2005; Kato et al., 2005). In particular, a mutation in RNApII subunit Rpb7 strongly alleviates RNApII elongation step but not its recruitment to a heterochromatic region (Djupedal et al., 2005). These results have thus reinforced the possibility that RITS and RDRC get recruited to chromatin by interacting with a nascent transcript. Importantly, an additional study based on the analysis of another RNApII point mutation located in the subunit Rpb2 showed that severe heterochromatin formation defects can occur without altering the transcription of heterochromatic regions by RNApII (Kato et al., 2005). This further indicated that RNApII itself (and not just the nascent transcript) might be important for RNAi-mediated heterochromatin formation. The requirement for RNApII and its transcription activity was not anticipated, as it implies that, in what sounds as a paradox, active transcription is required for silencing transcription of a genomic region by RNAi-mediated heterochromatin formation.

The possibility that RITS recruitment to chromatin and RITS-mediated heterochroma-tin formation involves a nascent transcript was directly tested in vivo (Buhler et al., 2006). An artificial tethering system was built by fusing to Tas3 (a subunit of RITS) a protein motif (λN), which binds to a RNA specific structure (Box B). In parallel, the DNA sequence encoding the Box B RNA motif was integrated into ura4, a euchromatic gene. A detailed analysis of this tethering system revealed that it is capable of recruiting RITS to ura4-BoxB gene and that it triggers formation of heterochromatin and transcriptional silencing at this locus. Importantly, transcription of ura4-BoxB gene is absolutely required for both RITS recruitment and the appearance of heterochromatin marks. Altogether, these experiments demonstrated that RITS binding to a nascent transcript synthesized by RNApII is a mechanism that can contribute to its stable recruitment on chromatin. The tethering system also demonstrated that RNAi is required at a step(s) downstream of RITS binding to the nascent transcript. Indeed, in this system it appears that although RITS can bind the ura4-BoxB RNA in a siRNA-independent manner, it still requires a functional RNAi process to trigger heterochromatin assembly and gene silencing at ura4-BoxB gene (Buhler et al., 2006).

It should be noted that, although it appears clearly from all the experiments previously described that RITS can be recruited to chromatin thanks to its interaction with a nascent transcript, none of these experiments rule out the possible requirement for a siRNA-DNA interaction in RITS binding to chromatin. Interestingly, in plants RNA-directed DNA methylation involves a RNAi-driven mechanism in which a siRNA can guide DNA methylation with a precision approaching the nucleotide level (Huettel et al., 2007; Pelissier and Wassenegger, 2000). This high precision of targeting favours a target recognition mechanism based on a direct interaction between the siRNA and the DNA sequence to be methylated, indicating that such a mechanism might yet exist.

In parallel of its role in recruiting RITS to chromatin, the nascent transcript could also serve as a platform for RDRC-dependent dsRNA synthesis (Fig. 4.3). This possibility is supported by the fact that Rdp1, the catalytic subunit of RDRC was found to bind heterochromatic regions (Volpe et al., 2002). In addition, Rdp1

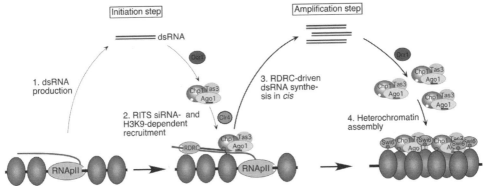

Figure 4.3 A two step model for RNAi-dependent heterochromatin assembly. In the first step, production of dsRNA activates RNAi and are cleaved into small interfering RNA by Dcr1, a ribonuclease of type III. Transcription of both forward and reverse DNA strands by RNApII could produce the first dsRNAs. Once loaded onto RITS, the complex is recruited to chromatin thanks to a siRNA-nascent transcript base-pairing mechanism (see Fig. 4.2 and text for more details on RITS recruitment to chromatin). The second step begins by the recruitment of RDRC thanks to its interaction with RITS. RDRC then starts synthesis of dsRNA in *cis* by using the nascent transcript as a template. This synthesis can represent a mean to locally recruit more RITS complexes, which might be crucial for triggering RNAi-dependent heterochromatin assembly.

binds to non-coding RNA emanating from pericentromeric repeats but not from a euchromatic region like the actin gene and, importantly, this binding requires the H3K9 methyltransferase Clr4 (Motamedi *et al.*, 2004). Therefore, RDRC-dependent dsRNA synthesis probably occurs in the context of a heterochromatin structure. Due to the fact that RITS also binds to these centromeric RNAs and that it interacts with RDRC in a Clr4- and Dcr1-dependent manner, it has been proposed that RITS serves as a priming complex to recruit RDRC to nascent transcripts and to initiate dsRNA synthesis (Motamedi *et al.*, 2004). Once bound to the nascent transcript, RDRC can start dsRNA production in *cis*. In absence of RDRC, no or very little RITS-bound siRNAs can be detected, indicating that RDRC plays a major role in the production of RITS-associated siRNAs (Motamedi *et al.*, 2004). RDRC-dependent production of siRNA in *cis* might be the molecular mechanism that overcomes the threshold concentration of RITS bound to a nucleation centre and required to trigger heterochromatin assembly (Fig. 4.3).

Revisiting heterochromatin dynamics and the mechanisms involved in heterochromatic gene silencing

Heterochromatin is classically viewed as a condensed and static structure, which imposes a transcriptional silencing by reducing DNA accessibility. Recently, the characterization of a complex, containing an ATP-dependent remodelling factor homologous to Snf2 as well as a the Clr3 HDAC, termed SHREC (Snf2/HDAC-containing REpressor Complex), has supported the view that in *S. pombe* also heterochromatin assembly involves a chromatin remodelling step that makes it less accessible to RNApII (Fig. 4.4) (Sugiyama *et al.*, 2007). However, other studies on RNAi-mediated heterochromatin assembly mechanisms are challenging this view as they indicate that heterochromatin does not necessarily need to be less accessible to silence transcription and that heterochromatin might be actually a structure far more dynamic than previously anticipated. First, RNApII occupancy was found to not always diminish in a region associated with heterochromatin. Indeed, ChIP experiments show that levels of RNApII present on transgenes integrated in pericentromeric repeats do not necessarily vary whether or not heterochromatin assembly takes place (Buhler *et al.*, 2006). Interestingly, transcription from these transgenes is efficiently silenced although RNApII is still there, suggesting that in this case silencing of the transgenes is imposed by another mechanism than restriction of RNApII access. Second, the fact that active transcription plays a central role in RNAi-mediated heterochromatin assembly, as mentioned previously, favours the view that heterochromatin can be a dynamic chromatin state. This is in agreement with studies done on HP1 proteins (homologues of Swi6) that have shown a high mobility of these proteins

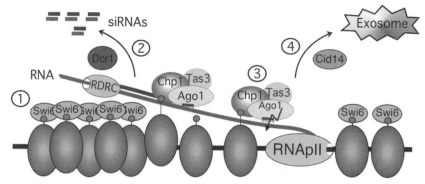

Figure 4.4 Mechanisms of heterochromatic gene silencing. Classically, heterochromatin assembly is thought to silence transcription by restricting access to DNA. Such mechanism seems to take place in *S. pombe* (1). In parallel, two RNAi-linked activities might actively silence transcription by degrading nascent transcripts emanating from a heterochromatic region. Indeed, nascent transcripts are used as templates by RDRC to produce dsRNA, which are in turn degraded by Dcr1 (2). In addition, RITS might use its Ago1 slicer activity to cleave the nascent transcript (3) (the dark arrow shaped as a lightning represents Ago1 slicing the transcript). Finally, another RNA processing machinery that involves the polyA polymerase Cid14 and the exosome also contributes to heterochromatic gene silencing (4).

in stable heterochromatic domains (Cheutin *et al.*, 2003; Festenstein *et al.*, 2003).

If heterochromatic DNA can be transcribed, how then is its transcription silenced? Several RNAi-driven degradation processes can potentially be important for heterochromatic silencing (Fig. 4.4). Theoretically, RDRC-dependent dsRNA synthesis could be coupled to Dcr1-dependent processing, which would hence contribute to silence RNApII transcription by degrading the nascent transcript (corresponding to one strand of the dsRNA) and by producing more siRNAs that will reinforce RITS recruitment to chromatin (Fig. 4.3). Another RNAi-mediated processing activity was also suggested to contribute to heterochromatic gene silencing. Ago1 possesses a siRNA-dependent endonuclease activity known as slicer activity (Buker *et al.*, 2007; Irvine *et al.*, 2006; Fig. 4.4). Ago1 slicer activity is important for Ago1 association with chromatin and for spreading of H3K9me and RITS into a transgene sequence preintegrated in a pericentromeric repeat. Moreover, evidence for Ago1 slicing of the nascent transcript has been provided (Irvine *et al.*, 2006). According to these results, it was suggested that Ago1-associated nuclease activity could contribute to heterochromatic silencing by slicing nascent transcripts. However, the importance of this step is currently questioned by recent investigations, which revealed that Ago1 slicer activity is required at a step upstream of RITS binding to nascent tran-

scripts (Buker *et al.*, 2007). These investigations show that Ago1 is present in a second complex, named Argonaute siRNA Chaperone (ARC), which contains Arb1 and Arb2, two other proteins with unknown function. Similarly to RITS, ARC also contains siRNAs. Yet, in ARC these siRNAs are mainly double-stranded whereas siRNAs present in RITS are mostly single stranded. Moreover, ARC does not associate with heterochromatic sites conversely to RITS. *In vivo* analyses based on Ago1 point mutations that abolish Ago1 slicer activity indicate that this activity is required for elimination of one strand (the passenger strand) of the duplex siRNA in RITS. This elimination does not occur in ARC complex because the protein Arb1 inhibits Ago1 slicer activity. It has been proposed that ARC complex might be an intermediary component in the process that transfers the double-stranded siRNA from Dcr1 to RITS (Buker *et al.*, 2007). Once loaded onto RITS, the passenger strand is sliced by Ago1 and eliminated, which frees the other strand that can then target RITS to the nascent transcript. Therefore, as Ago1 slicer activity happens to be critical for elimination of the passenger strand, which corresponds to a step upstream of RITS binding to the nascent transcript, it still remains to be determined whether Ago1 slicing of nascent transcripts is also critical for RNAi-mediated heterochromatic silencing.

A last study has recently revealed that processing of heterochromatic RNAs and hence heterochromatic gene silencing can involve another process than RNAi and chromatin remodelling/compaction (Buhler *et al.*, 2007). Silencing of reporter genes integrated in all major heterochromatic regions requires a newly characterized poly(A) polymerase, Cid14 (Fig. 4.4). The exact mechanism by which Cid14 contributes to transcriptional silencing is currently not known. However, it appears that Cid14 is in a complex with Mtr4 and Air1 proteins. This complex turned to be the *S. pombe* counterpart of the TRAMP complex present in *S. cerevisiae*, and which is known to degrade in an exosome-dependent manner transcripts emanating from cryptic promoters (LaCava *et al.*, 2005; Vanacova *et al.*, 2005; Wyers *et al.*, 2005). Therefore, in parallel to RNAi-mediated RNA degradation, heterochromatic gene silencing might involve an alternative process that actively degrades heterochromatic RNA through the exosome. This exosome-dependent mechanism probably plays a major role in silencing transcription as in a *cid14*-deletion strain heterochromatic RNAs from transgenes integrated in pericentromeric repeats, in the mating-type locus and in subtelomeric region all accumulate. Similarly, mutations that disrupt the exosome RNA-processing activity also alleviate heterochromatic silencing (Buhler *et al.*, 2007; Murakami *et al.*, 2007). Remarkably, the relief of silencing in exosome mutants does not come along with disappearance of H3K9me, Swi6 and Chp1 heterochromatin marks, indicating that heterochromatic silencing can be strongly alleviated without the necessity to displace the core structural components of heterochromatin. As it has been the case, since the discovery of a role for RNAi in heterochromatin assembly, this last result might just be another illustration that many more surprises shall continue to come out of the ongoing dissection of the role of RNAi or other RNA-based processes in heterochromatin assembly in the fission yeast *Schizosaccharomyces pombe*.

Conclusion

Studies in *S. pombe* have shown that RNAi is deeply involved in heterochromatin assembly and heterochromatic gene silencing. It appears that RNAi complexes such as RITS and RDRC, which can act in parallel of DNA-binding proteins to nucleate heterochromatin assembly, are likely to be recruited to chromatin by a base-pairing interaction between a siRNA and a nascent transcript that acts as a recruiting platform and which is synthesized by RNApII. RITS uses both the Ago1-bound siRNA and Chp1 chromodomain to stably interact with specific chromosome regions. Furthermore, an intimate interplay between these two chromatin recognition modules of RITS seems critical for RNAi-mediated establishment of heterochromatin assembly. In parallel, RDRC can use nascent transcripts as templates to produce dsRNA directly at the site of heterochromatin formation. Collectively, and quite unexpectedly, these studies reveal a more dynamic view of heterochromatin assembly, in which transcription of the genomic region to be heterochromatinized plays a central role. Among the important aspects of RNAi-mediated heterochromatin assembly that remain to be elucidated, a critical issue will be to determine how histone-modifying activities such as the histone methyltransferase Clr4 and HDACs are recruited by the RNAi machinery to assemble heterochromatin. Finally, RNAi and an exosome-dependent process can actively degrade RNAs emanating from heterochromatic DNA to silence transcription arising from genomic regions embedded in heterochromatin. Their essential roles in heterochromatic gene silencing strongly support the view that chromatin compaction is not the only mean by which heterochromatin silences transcription. An important part of this silencing may actually be under the control of RNA degradation machineries.

Acknowledgements
We thank ElCherif Ibrahim, Fabienne Hans and Thierry Gautier for their helpful comments. Aurelia Vavasseur is supported by a doctoral fellowship from the French ministry of research. Our work is supported by Human Frontier Science Program Organization and the French ARC (Association pour la Recherche contre le Cancer).

References
Behm-Ansmant, I., Rehwinkel, J., Doerks, T., Stark, A., Bork, P., and Izaurralde, E. (2006). mRNA degradation by miRNAs and GW182 requires both CCR4:NOT deadenylase and DCP1:DCP2 decapping complexes. Genes Dev. *20*, 1885–1898.

Bernard, P., Maure, J.F., Partridge, J.F., Genier, S., Javerzat, J.P., and Allshire, R.C. (2001). Requirement of heterochromatin for cohesion at centromeres. Science *294*, 2539–2542.

Buhler, M., Haas, W., Gygi, S.P., and Moazed, D. (2007). RNAi-dependent and -independent RNA turnover mechanisms contribute to heterochromatic gene silencing. Cell *129*, 707–721.

Buhler, M., Verdel, A., and Moazed, D. (2006). Tethering RITS to a nascent transcript initiates RNAi- and heterochromatin-dependent gene silencing. Cell *125*, 873–886.

Buker, S.M., Iida, T., Buhler, M., Villen, J., Gygi, S.P., Nakayama, J., and Moazed, D. (2007). Two different Argonaute complexes are required for siRNA generation and heterochromatin assembly in fission yeast. Nat. Struct. Mol. Biol. *14*, 200–207.

Cam, H.P., Sugiyama, T., Chen, E.S., Chen, X., FitzGerald, P.C., and Grewal, S.I. (2005). Comprehensive analysis of heterochromatin- and RNAi-mediated epigenetic control of the fission yeast genome. Nat. Genet. *37*, 809–819.

Cheutin, T., McNairn, A.J., Jenuwein, T., Gilbert, D.M., Singh, P.B., and Misteli, T. (2003). Maintenance of stable heterochromatin domains by dynamic HP1 binding. Science *299*, 721–725.

Craig, J.M. (2005). Heterochromatin – many flavours, common themes. Bioessays *27*, 17–28.

Deshpande, G., Calhoun, G., and Schedl, P. (2005). *Drosophila* argonaute-2 is required early in embryogenesis for the assembly of centric/centromeric heterochromatin, nuclear division, nuclear migration, and germ-cell formation. Genes Dev. *19*, 1680–1685.

Djupedal, I., Portoso, M., Spahr, H., Bonilla, C., Gustafsson, C.M., Allshire, R.C., and Ekwall, K. (2005). RNA Pol II subunit Rpb7 promotes centromeric transcription and RNAi-directed chromatin silencing. Genes Dev. *19*, 2301–2306.

Festenstein, R., Pagakis, S.N., Hiragami, K., Lyon, D., Verreault, A., Sekkali, B., and Kioussis, D. (2003). Modulation of heterochromatin protein 1 dynamics in primary Mammalian cells. Science *299*, 719–721.

Fukagawa, T., Nogami, M., Yoshikawa, M., Ikeno, M., Okazaki, T., Takami, Y., Nakayama, T., and Oshimura, M. (2004). Dicer is essential for formation of the heterochromatin structure in vertebrate cells. Nat. Cell. Biol. *6*, 784–791.

Grewal, S.I., and Jia, S. (2007). Heterochromatin revisited. Nat. Rev. Genet. *8*, 35–46.

Grewal, S.I., and Moazed, D. (2003). Heterochromatin and epigenetic control of gene expression. Science *301*, 798–802.

Hall, I.M., Shankaranarayana, G.D., Noma, K., Ayoub, N., Cohen, A., and Grewal, S.I. (2002). Establishment and maintenance of a heterochromatin domain. Science *297*, 2232–2237.

Hong, E.J., Villen, J., Gerace, E.L., Gygi, S.P., and Moazed, D. (2005). A cullin E3 ubiquitin ligase complex associates with Rik1 and the Clr4 histone H3-K9 methyltransferase and is required for RNAi-mediated heterochromatin formation. RNA Biol. *2*, 106–111.

Horn, P.J., Bastie, J.N., and Peterson, C.L. (2005). A Rik1-associated, cullin-dependent E3 ubiquitin ligase is essential for heterochromatin formation. Genes Dev. *19*, 1705–1714.

Horn, P.J., and Peterson, C.L. (2006). Heterochromatin assembly: a new twist on an old model. Chromosome Res. *14*, 83–94.

Huettel, B., Kanno, T., Daxinger, L., Bucher, E., van der Winden, J., Matzke, A.J., and Matzke, M. (2007). RNA-directed DNA methylation mediated by DRD1 and Pol IVb: A versatile pathway for transcriptional gene silencing in plants. Biochim. Biophys. Acta *1769*, 358–374.

Iida, T., Kawaguchi, R., and Nakayama, J. (2006). Conserved ribonuclease, Eri1, negatively regulates heterochromatin assembly in fission yeast. Curr. Biol. *16*, 1459–1464.

Irvine, D.V., Zaratiegui, M., Tolia, N.H., Goto, D.B., Chitwood, D.H., Vaughn, M.W., Joshua-Tor, L., and Martienssen, R.A. (2006). Argonaute slicing is required for heterochromatic silencing and spreading. Science *313*, 1134–1137.

Jenuwein, T., and Allis, C.D. (2001). Translating the histone code. Science *293*, 1074–1080.

Jia, S., Kobayashi, R., and Grewal, S.I. (2005). Ubiquitin ligase component Cul4 associates with Clr4 histone methyltransferase to assemble heterochromatin. Nat. Cell. Biol. *7*, 1007–1013.

Jia, S., Noma, K., and Grewal, S.I. (2004). RNAi-independent heterochromatin nucleation by the stress-activated ATF/CREB family proteins. Science *304*, 1971–1976.

Kanellopoulou, C., Muljo, S.A., Kung, A.L., Ganesan, S., Drapkin, R., Jenuwein, T., Livingston, D.M., and Rajewsky, K. (2005). Dicer-deficient mouse embryonic stem cells are defective in differentiation and centromeric silencing. Genes Dev. *19*, 489–501.

Kanoh, J., Sadaie, M., Urano, T., and Ishikawa, F. (2005). Telomere binding protein Taz1 establishes Swi6 heterochromatin independently of RNAi at telomeres. Curr. Biol. *15*, 1808–1819.

Kato, H., Goto, D.B., Martienssen, R.A., Urano, T., Furukawa, K., and Murakami, Y. (2005). RNA polymerase II is required for RNAi-dependent heterochromatin assembly. Science *309*, 467–469.

Kim, D.H., Villeneuve, L.M., Morris, K.V., and Rossi, J.J. (2006). Argonaute-1 directs siRNA-mediated transcriptional gene silencing in human cells. Nat. Struct. Mol. Biol. *13*, 793–797.

Klose, R.J., and Bird, A.P. (2006). Genomic DNA methylation: the mark and its mediators. Trends Biochem. Sci. *31*, 89–97.

LaCava, J., Houseley, J., Saveanu, C., Petfalski, E., Thompson, E., Jacquier, A., and Tollervey, D. (2005). RNA degradation by the exosome is promoted by a nuclear polyadenylation complex. Cell *121*, 713–724.

Li, F., Goto, D.B., Zaratiegui, M., Tang, X., Martienssen, R., and Cande, W.Z. (2005). Two novel proteins, dos1 and dos2, interact with rik1 to regulate heterochromatic RNA interference and histone modification. Curr Biol *15*, 1448–1457.

Luger, K., Mader, A.W., Richmond, R.K., Sargent, D.F., and Richmond, T.J. (1997). Crystal structure of the nucleosome core particle at 2.8 A resolution. Nature *389*, 251–260.

Martienssen, R.A., Zaratiegui, M., and Goto, D.B. (2005). RNA interference and heterochromatin in the fission yeast *Schizosaccharomyces pombe*. Trends Genet. *21*, 450–456.

May, B.P., Lippman, Z.B., Fang, Y., Spector, D.L., and Martienssen, R.A. (2005). Differential regulation of strand-specific transcripts from *Arabidopsis* centromeric satellite repeats. PLoS Genet. *1*, e79.

Morris, K.V., Chan, S.W., Jacobsen, S.E., and Looney, D.J. (2004). Small interfering RNA-induced transcriptional gene silencing in human cells. Science *305*, 1289–1292.

Motamedi, M.R., Verdel, A., Colmenares, S.U., Gerber, S.A., Gygi, S.P., and Moazed, D. (2004). Two RNAi complexes, RITS and RDRC, physically interact and localize to noncoding centromeric RNAs. Cell *119*, 789–802.

Murakami, H., Goto, D.B., Toda, T., Chen, E.S., Grewal, S.I., Martienssen, R.A., and Yanagida, M. (2007). Ribonuclease activity of Dis3 is required for mitotic progression and provides a possible link between heterochromatin and kinetochore function. PLoS ONE 2, e317.

Nakayama, J., Rice, J.C., Strahl, B.D., Allis, C.D., and Grewal, S.I. (2001). Role of histone H3 lysine 9 methylation in epigenetic control of heterochromatin assembly. Science *292*, 110–113.

Noma, K., Sugiyama, T., Cam, H., Verdel, A., Zofall, M., Jia, S., Moazed, D., and Grewal, S.I. (2004). RITS acts in *cis* to promote RNA interference-mediated transcriptional and post-transcriptional silencing. Nat. Genet. *36*, 1174–1180.

Pal-Bhadra, M., Leibovitch, B.A., Gandhi, S.G., Rao, M., Bhadra, U., Birchler, J.A., and Elgin, S.C. (2004). Heterochromatic silencing and HP1 localization in *Drosophila* are dependent on the RNAi machinery. Science *303*, 669–672.

Partridge, J.F., DeBeauchamp, J.L., Kosinski, A.M., Ulrich, D.L., Hadler, M.J., and Noffsinger, V.J. (2007). Functional separation of the requirements for establishment and maintenance of centromeric heterochromatin. Mol. Cell *26*, 593–602.

Partridge, J.F., Scott, K.S., Bannister, A.J., Kouzarides, T., and Allshire, R.C. (2002). *cis*-acting DNA from fission yeast centromeres mediates histone H3 methylation and recruitment of silencing factors and cohesin to an ectopic site. Curr. Biol. *12*, 1652–1660.

Pelissier, T., and Wasseneger, M. (2000). A DNA target of 30 bp is sufficient for RNA-directed DNA methylation. RNA *6*, 55–65.

Petrie, V.J., Wuitschick, J.D., Givens, C.D., Kosinski, A.M., and Partridge, J.F. (2005). RNA interference (RNAi)-dependent and RNAi-independent association of the Chp1 chromodomain protein with distinct heterochromatic loci in fission yeast. Mol. Cell. Biol. *25*, 2331–2346.

Pidoux, A.L., and Allshire, R.C. (2004). Kinetochore and heterochromatin domains of the fission yeast centromere. Chromosome Res. *12*, 521–534.

Reinhart, B.J., and Bartel, D.P. (2002). Small RNAs correspond to centromere heterochromatic repeats. Science *297*, 1831.

Sadaie, M., Iida, T., Urano, T., and Nakayama, J. (2004). A chromodomain protein, Chp1, is required for the establishment of heterochromatin in fission yeast. EMBO J. *23*, 3825–3835.

Sugiyama, T., Cam, H., Verdel, A., Moazed, D., and Grewal, S.I. (2005). RNA-dependent RNA polymerase is an essential component of a self-enforcing loop coupling heterochromatin assembly to siRNA production. Proc. Natl. Acad. Sci. USA *102*, 152–157.

Sugiyama, T., Cam, H.P., Sugiyama, R., Noma, K., Zofall, M., Kobayashi, R., and Grewal, S.I. (2007). SHREC, an effector complex for heterochromatic transcriptional silencing. Cell *128*, 491–504.

Thon, G., Hansen, K.R., Altes, S.P., Sidhu, D., Singh, G., Verhein-Hansen, J., Bonaduce, M.J., and Klar, A.J. (2005). The Clr7 and Clr8 directionality factors and the Pcu4 cullin mediate heterochromatin formation in the fission yeast *Schizosaccharomyces pombe*. Genetics *171*, 1583–1595.

Ting, A.H., Schuebel, K.E., Herman, J.G., and Baylin, S.B. (2005). Short double-stranded RNA induces transcriptional gene silencing in human cancer cells in the absence of DNA methylation. Nat. Genet. *37*, 906–910.

Vanacova, S., Wolf, J., Martin, G., Blank, D., Dettwiler, S., Friedlein, A., Langen, H., Keith, G., and Keller, W. (2005). A new yeast poly(A) polymerase complex involved in RNA quality control. PLoS Biol. 3, e189.

Verdel, A., Jia, S., Gerber, S., Sugiyama, T., Gygi, S., Grewal, S.I., and Moazed, D. (2004). RNAi-mediated targeting of heterochromatin by the RITS complex. Science *303*, 672–676.

Verdel, A., and Moazed, D. (2005). RNAi-directed assembly of heterochromatin in fission yeast. FEBS Lett. *579*, 5872–5878.

Volpe, T., Schramke, V., Hamilton, G.L., White, S.A., Teng, G., Martienssen, R.A., and Allshire, R.C. (2003). RNA interference is required for normal centromere function in fission yeast. Chromosome Res. *11*, 137–146.

Volpe, T.A., Kidner, C., Hall, I.M., Teng, G., Grewal, S.I., and Martienssen, R.A. (2002). Regulation of heterochromatic silencing and histone H3 lysine-9 methylation by RNAi. Science *297*, 1833–1837.

Weinberg, M.S., Villeneuve, L.M., Ehsani, A., Amarzguioui, M., Aagaard, L., Chen, Z.X., Riggs, A.D., Rossi, J.J., and Morris, K.V. (2006). The antisense strand of small interfering RNAs directs histone methylation and transcriptional gene silencing in human cells. RNA *12*, 256–262.

Wood, V., Gwilliam, R., Rajandream, M.A., Lyne, M., Lyne, R., Stewart, A., Sgouros, J., Peat, N., Hayles, J., Baker, S., *et al.* (2002). The genome sequence of *Schizosaccharomyces pombe*. Nature *415*, 871–880.

Wyers, F., Rougemaille, M., Badis, G., Rousselle, J.C., Dufour, M.E., Boulay, J., Regnault, B., Devaux, F., Namane, A., Seraphin, B., *et al.* (2005). Cryptic pol II transcripts are degraded by a nuclear quality control pathway involving a new poly(A) polymerase. Cell *121*, 725–737.

Yamada, T., Fischle, W., Sugiyama, T., Allis, C.D., and Grewal, S.I. (2005). The nucleation and maintenance of heterochromatin by a histone deacetylase in fission yeast. Mol. Cell 20, 173–185.

RNA-mediated Gene Regulation in *Drosophila*

5

Harsh H. Kavi, Harvey R. Fernandez, Weiwu Xie
and James A. Birchler

Abstract

Short RNAs are increasingly recognized to play multiple roles in affecting gene expression at many levels as illustrated by work in *Drosophila*. Here we review the biochemical parameters of RNA interference, the technique that uses double-stranded RNA, which is cleaved by Dicer to produce small interfering RNAs (siRNAs) as guides to cleave homologous mRNAs. This process is believed to occur in the cytoplasm and is used in the endogenous process of viral resistance. In addition, many of the same gene products are also involved in transcriptional gene silencing processes. This was first documented for cosuppression of white-Alcohol dehydrogenase transgenes, which is associated with the Polycomb repressive complex of chromatin proteins. Genetic studies of RNA silencing genes also implicate a role in heterochromatin silencing. Some gene products involved in RNAi are also involved in the formation of repeat associated small RNAs (rasiRNAs), whose formation appears to be Dicer independent and critical for repressing transposon expression particularly in the germline. Roles for small RNAs are also implicated in chromatin insulator activity, the integrity of the nucleolus and long-range associations of homeotic genes.

Introduction

A series of biochemical and genetic experiments in *Drosophila* has contributed to an understanding the mechanism and the biological role of RNA interference (RNAi). The small interfering RNAs (siRNA) that perform a pivotal role in RNA interference (Zamore *et al.*, 2000) have been implicated in transgene silencing (Pal Bhadra *et al.*, 2002), heterochromatin structure formation (Pal-Bhadra *et al.*, 2004), viral defence (Wang *et al.*, 2006), transposon activity in the germline (Aravin *et al.*, 2001; Kalmykova *et al.*, 2005) and other important biological phenomena in *Drosophila* (Lei and Corces, 2006; Peng and Karpen, 2006). In this chapter, we highlight the mechanism and biological roles of small RNAs in gene regulation.

RNAi mechanism – biochemical and structural studies

RNAi (RNA interference) involves the introduction of dsRNA into the cell that results in degradation of a specific mRNA molecule. The specificity is attributed to the base pair complementation between the antisense strand of the dsRNA introduced and the specific mRNA target molecule. The degradation is by virtue of endonucleases in the multi-protein complex with which the siRNA guide strand associates. The trigger for RNAi (i.e. dsRNA) is either introduced artificially by injection or it is synthesized *in vivo*, for example, from the dense heterochromatic repeat regions near the centromeres (Volpe *et al.*, 2002). Transgenic flies that harbour an inverted exonic region will also trigger RNAi *in vivo* (Lee and Carthew, 2003). Recent experiments in *Drosophila* S2 cells have revealed that the uptake of dsRNA in the culture medium is by an active receptor-mediated endocytosis pathway (Saleh *et al.*, 2006). These studies also reveal that the uptake of long dsRNA is more efficient than the uptake of small dsRNA and siRNA. The scavenger receptors that play a role in innate immunity are involved with the uptake of dsRNA by the cells, thus reflecting a possible connection between RNAi and innate immune responses.

The dsRNA trigger is cleaved into 21–23 nt small interfering RNAs (siRNA) by the enzyme Dicer (Bernstein *et al.*, 2001). Unlike in mammals, the *Drosophila* Dicer cleaves dsRNA in an ATP-dependant manner. The enzyme has a C-terminal dsRNA binding domain, a PAZ domain, a DUF283 (Domain of Unknown Function), a helicase domain and two ribonuclease domains- the larger IIIa domain and the smaller IIIb domain (Tahbaz *et al.*, 2004; Zhang *et al.*, 2004). Based upon the structural studies of human Dicer, it has been proposed that the two ribonuclease domains form a dimer resulting in a single catalytic centre.

The PAZ and helicase domains have been implicated in the binding of the 3′ end of the substrate and the spatial orientation of the substrate for efficient catalysis, respectively. The Ribonuclease domain (RIII a) coordinates with the PIWI domain of Argonaute (Ago) proteins in the process of RISC assembly. *Drosophila* has two Dicer paralogues, *Dcr-1* and *Dcr-2*, which are involved in miRNA and siRNA biogenesis, respectively. *Dcr-1* lacks the PAZ domain while *Dcr-2* lacks the helicase domain (Lee *et al.*, 2004). Also, flies have a third distinct type of Ribonuclease III known as Drosha that is involved in the generation of miRNA precursor molecules.

The siRNAs produced during the cleavage of dsRNA by Dicer are the hallmark of the RNAi process. The structural integrity of the siRNA is essential for effective degradation of the target mRNA (Nykanen *et al.*, 2001; Zamore *et al.*, 2000). The siRNA duplex has a 5′-PO4 and a two nucleotide overhang at the 3′ end bearing a hydroxyl group. These structural features reflect the mechanism of Dicer catalysis. The siRNA duplex has different thermodynamic stabilities at either end, a property exploited by the RISC complex to select the bona fide antisense (guide) strand. The other strand of the duplex that does not participate in target mRNA degradation is destroyed and is known as the passenger strand (Liu *et al.*, 2003; Tomari *et al.*, 2004b).

The sense strand of the siRNA duplex is relatively more resistant to chemical modifications than the antisense strand. Similarly mutations in the 5′ end of the sense strand of the duplex are well tolerated relative to those in the central and the 3′ end. This highlights the importance of the antisense strand in target elimination. The presence of a phosphate group near the 5′ end stabilizes the RISC complex and results in the cleavage of the correct scissile phosphodiester bond on the target molecule. The 5′-PO$_4$ group is critical for various protein-protein interactions during RISC assembly. Its absence or replacement with a bulky moiety such as a methyl group results in impairment of the RNAi process. The catalytic efficiency of the RISC complex and hence RNAi is compromised when the length of the 3′ end is varied by addition or deletion of nucleotides. The nucleotides near the 5′ end of the siRNA strand determine the K_m (i.e. affinity for the target mRNA) while those near the 3′ end determine the K_{cat} (catalytic efficiency) of the RISC complex (Haley and Zamore, 2004). The sugar moiety can be modified by attachment of 2′-O methyl, fluoro etc. to provide more stability to the siRNA without adversely affecting the catalytic efficiency. Certain modifications like replacement of the 5′ terminal residues with deoxyribose results in the selective degradation of its cognate strand (Pham and Sontheimer, 2005). Such modifications result in reduced incidences of off-target effects.

The catalytic property of the RISC complex that degrades the mRNA molecules resides in a class of chemically basic proteins known as Argonaute. A wealth of data has been accumulated about the mechanism of RNAi based upon structural studies of Ago protein crystals from prokaryotes such as *Pyrococcus furiosus* and *Aquifes aeolicus* (Ma *et al.*, 2005; Parker *et al.*, 2004, 2005; Yuan *et al.*, 2005). These studies revealed that Ago proteins have a PAZ domain of about 130 amino acids and a PIWI domain of about 300 amino acids. The 3′ end of the antisense siRNA strand is embedded in the PAZ domain while the 5′-PO$_4$ group interacts with a highly basic mid-domain in the centre of the PIWI domain. Dicer makes contact with the

PIWI domain of the AGO protein by virtue of its ribonuclease domains.

The interpretation of crystal structure of hAgo1 from human T293 cells has shown the presence of an oligonucleotide binding fold in the PAZ domain to which the two nucleotide overhangs at the 3′ end bind (Song *et al.*, 2004). The PIWI domain has a structure that closely resembles the catalytic domain of RNase H. Based upon these studies hAgo2 was identified as the 'slicer' component of the RISC complex. The crystal structure studies also revealed the presence of a catalytic triad of three amino acids DDH (Asp-Asp-His) and an additional QH (Gln-His) in the PIWI domain. The mutations in these conserved catalytic domains completely abolished cleavage activity mediated by hAgo2.

In the case of *Drosophila*, AGO2 performs the degradation of mRNA while translational repression mediated by miRNA has been attributed to AGO1. The defects in AGO2 mutants overlapped with those of Dicer mutants, thus reflecting functional coordination between Dicer and AGO2 in the RNAi pathway. Recent biochemical and genetic studies have, however, revealed that *Drosophila* AGO1 sequence closely resembles hAgo2, thus implicating 'slicer' activity for both AGO2 and AGO1. The study also highlighted the importance of the PIWI domain in the cleavage of the target molecule while the PAZ domain is dispensable for the same function.

AGO2, besides acting as the 'slicer', also performs the unwinding of the siRNA duplex and subsequently cleaves the passenger strand. This results in a new species of RNA intermediate, which bears a nicked passenger strand bound to the guide strand. The failure to cleave the passenger strand, as exhibited by AGO mutants or substitution of chemically modified passenger strand resistant to nucleases, results in defective RISC assembly and abolishment of RNAi. The *Drosophila* genome has three additional Argonaute members: *AGO3*, which is uncharacterized, *piwi* and *aubergine*. These latter two proteins play an important role in germline stem cell renewal and in heterochromatin formation. Recently *piwi* was shown to possess 'slicer' activity *in vitro* (Saito *et al.*, 2006).

RISC assembly

RISC is a multi-protein complex, which possesses the endonuclease activity resulting in target mRNA degradation. The *Drosophila* RISC complex contains AGO2, Dmp68, DFXR (fragile X-related protein), TUDOR-SN (nuclease with Tudor domains), ARMITAGE (assembly protein) and VIG (Vasa intronic gene). The active RISC complex contains the guide siRNA strand assembled with other protein components in a stepwise order (Sontheimer, 2005).

The first step involves the formation of the RDI Complex (R2D2-Dcr2). R2D2 is a sensor protein that binds to the 5′ end of the siRNA duplex nearer to the thermodynamically stable end. *r2d2* mutant embryos are abnormal and fail to exhibit RNAi due to defects in RISC assembly.

DCR-2 binds to the 3′ end of the other strand nearer to the thermodynamically unstable end. The RDI complex is then converted to a transient R2/RLC (RISC Loading Complex). The RLC is associated with AGO2 that unwinds the siRNA duplex and cleaves the passenger strand resulting in guide strand assembly into the RISC complex (Matranga *et al.*, 2005). The RLC then further associates with additional protein components to form the holo-RISC complex. The holo-RISC complex (80S) is a catalytically active endonuclease dependent upon the presence of $Mg+2$ (Pham *et al.*, 2004; Schwarz *et al.*, 2003; Schwarz *et al.*, 2004; Tomari *et al.*, 2004a). The cleavage of the target mRNA scissile phosphodiester bonds takes place across the 10th and 11th nucleotide on the guide strand, beginning from the 5′ end of the guide strand.

RNAi components repress transposable element expression

To protect the *Drosophila* genome, there a need to suppress transposable element (TE) activity. Many of these elements and some other repeat sequences have been shown to be silenced in the germline via RNAi. Like in plants, slightly longer rasiRNA (repeat-associated siRNA, 24–29 nt) are produced from the repeats and a distinct RNA based machinery not requiring DCR-2, AGO2 and R2D2, is used to repress TEs (Vagin *et al.*, 2006).

Aravin and colleagues (Aravin *et al.*, 2003) first isolated and cloned rasiRNAs (repeat associated siRNAs) from *Drosophila* testes and early embryos. These rasiRNAs were from 38 different elements and covered 40% of all known TEs in the genome. Among those, the most frequently present rasiRNAs were for *roo*, a LTR retrotransposon with the highest abundance in the genome. Immuno-precipitation by anti-PIWI recovered rasiRNAs from ovary lysate, which were also cloned (Saito *et al.*, 2006). About 40% of the rasiRNAs matched 34 transposable elements and the others were homologous to the heterochromatic regions. The data show that TEs are generally inhibited by RNA-mediated processes.

Besides the fact that rasiRNAs differ in size from siRNAs, they are predominately antisense (Pelisson *et al.*, 2006; Vagin *et al.*, 2006). In contrast, siRNAs generated from long dsRNA by RNase III enzymes (e.g. DCR-2, DCR-1 and DROSHA) include sense and antisense molecules without reference to the target mRNA orientation (Zamore *et al.*, 2000; Schwarz *et al.*, 2003). This feature can be detected by microarrays containing melting-temperature normalized 22 nt probes for all theoretically possible sense and antisense siRNAs to a specific dsRNA. When the technique was applied to analyse *roo* rasiRNAs from the female germline, large numbers of antisense rasiRNAs were shown to be non-homogeneously distributed along the *roo* sequence but with only limited number of peaks likely to be sense rasiRNAs (Vagin *et al.*, 2006). These data are consistent with the observation that the cloned *roo* rasiRNAs are mostly antisense with a few sense ones (Aravin *et al.*, 2003; Saito *et al.*, 2006). The results raise the possibility that the rasiRNAs are not produced by RNase III. This point is further supported by three lines of evidence:

1 Neither *Dcr-2* or *Dcr-1* mutants increased transposons expression or rasiRNA level, although double mutants of the dicers could not be tested (Vagin *et al.*, 2006).
2 None of the known Drosophila RNase III enzymes can cleave long dsRNA into small RNAs 24–29 nt in length (Bernstein et al., 2001; Liu et al., 2003; Saito et al., 2005).

3 The 3′-ends of rasiRNAs lack either a 2′ or 3′ hydroxyl group demonstrated by failed reaction with NaIO$_4$, which differs from miRNAs (Vagin *et al.*, 2006).

In mammalian testes, a new group of siRNAs was recently identified, which are 29–31 nt long and which interact with PIWI. Thus, they are referred to as PIWI interacting RNA (piRNA) (reviewed by Carthew, 2006). In *Drosophila*, PIWI is mostly located in the ovary, embryo and probably only to the hub of the testis (Saito *et al.*, 2006). In the ovary, PIWI is associated with rasiRNAs. Similar to AGO1 and AGO2, PIWI was demonstrated *in vitro* to be able to cleave RNA when mixed with homologous single-stranded small RNA with length 21 nt or 30 nt (Saito *et al.*, 2006). The other PIWI-like proteins, AGO3 and Aubergine (AUB), are also only expressed in germline cells (Williams and Rubin, 2002). AUB was shown to bind rasiRNA as well (Vagin *et al.*, 2006). PIWI and AUB may be partially redundant for *roo* silencing in that *roo* rasiRNAs clearly bound to both proteins but the *aub* mutant does not affect *roo* expression and the *piwi* mutant moderately increases *roo* expression three fold (Vagin *et al.*, 2006).

A group of RNA helicases, which are part of the RNAi machinery were found to be necessary for repression of TEs and other repetitive sequences. The helicases are encoded by the genes, *homeless* (*spn-E*), *armitage* (*armi*) and perhaps *vasa*. The expression of retrotransposons *mdg1*, *1731* and the *F* element in the testes and ovaries was shown to increase in the *homeless* mutants (Aravin *et al.*, 2001). The results of *in situ* hybridization show that *homeless* and *piwi* caused accumulation of the LTR elements *mdg1*, *1731* and *copia* transcripts in nurse cells of the ovaries but not in the oocyte (Kalmykova *et al.*, 2005). On the other hand, in the developing oocyte, mutants of *aub*, *homeless* (*hls*), *armi* and *vasa* increase the expression of the non-LTR *HeT-A* and *I* elements (Vagin *et al.*, 2004). *HeT-A* and *TART* are the non-LTR retrotransposons that comprise the *Drosophila* telomeres and are used for maintaining telomere length. Different from *HeT-A*, *TART* transcription was shown to be substantially increased in the nurse cells at late stages in the *aub* and *hls* mutants (Savitsky *et al.*,

2006). In addition, *TART* transcripts and siRNA were shown to be at intermediate levels with one allele of the *hls* mutant, but only the homozygote mutants affect *HeT-A* (Savitsky *et al.*, 2006).

How the rasiRNA pathway works on the molecular level to silence the expression of the repetitive elements is largely unknown. In plants, rasiRNA has been suggested to repress expression at the transcriptional level (TGS) (Lippman *et al.*, 2004). In *Drosophila*, similar mechanisms may be occurring. PIWI protein is located in the nucleus (Saito *et al.*, 2006) whereas post-transcriptional silencing (PTGS) is believed to be executed in cytoplasm. Additional clues come from the investigations of two aggressive transposons: P, a DNA transposon, and I, a non-LTR retrotransposon (for a more detailed review, see Kavi *et al.*, 2005). Both of these elements cause hybrid dysgenesis – a phenomenon in which offspring show high rates of chromosomal abnormality and sterility when a maternal strain is crossed to a paternal strain containing the potential active P or I element. This syndrome results from hyperactivation of the related element transposition. Evidence indicates that an RNAi-based cosuppression mechanism, which involves both PTGS and TGS, inactivates the I element (Jensen *et al.*, 1999). P insertions in heterochromatic regions show strong repression of hybrid dysgenesis. The repression depends on *aubergine*, and probably heterochromatin elements (Reiss *et al.*, 2004), suggesting a transcriptional level control.

RNAi and antiviral immune defence mechanism in *Drosophila*

Two core components of the RNAi pathway, Dcr-2 and AGO2, mediate host defence mechanisms against RNA viruses in *Drosophila* (Galiana-Arnoux *et al.*, 2006; Wang *et al.*, 2006). Experiments conducted with *Drosophila* C virus, flock house virus and cricket paralysis virus showed that viral transcripts accumulated to a very high level compared to wild type in *AGO2* null mutants. The experiments also showed that *AGO2*, *Dcr-2* and *r2d2* homozygous mutants are more susceptible to viral infection, and hence show much higher rate of mortality than wild type flies. In the absence of an effective RNAi

machinery, viral replication proceeds unchecked. The dsRNA synthesized as an intermediate during viral infection acts as a trigger. An interesting observation from these studies was the presence of a suppressor of the RNAi pathway encoded by the *Drosophila* C virus. The DCV encodes a dsRNA-binding protein that inhibits processing of long dsRNA by Dicer-2.

RNAi is widely assumed to have an antiviral role in organisms because of its feature of sequence-specific target recognition (Qu and Morris, 2005). Many viruses encode a protein to minimize the RNAi repression from their hosts. These RNAi suppression genes were shown to function in different kingdoms reflecting the wide conservation of RNAi mechanisms (Li *et al.*, 2002; Li *et al.*, 2004). For example, the *Drosophila* pathogen, the flock house virus (FHV), expresses protein B2 sharing similar anti-viral function with the 2b protein of the cucumber mosaic virus (CMV) but without significant sequence similarity (Ding *et al.*, 1995). B2 was demonstrated to be an RNAi silencing suppressor at first in plants, where it is functionally exchangeable with 2b, then in *Drosophila* S2 cells and embryos. When viral B2 is absent or mutated, the viral RNA was significantly decreased, and on the other hand the homologous siRNA was increased. Therefore, B2 is hypothesized to help the viral genome proliferate by inhibiting the host siRNA-mediated silencing (Li *et al.*, 2002). By analysis of the protein structure, B2 was shown to bind dsRNA tightly, thus blocking Dicer cleavage, preventing siRNA assembly to RISC (Chao *et al.*, 2005). Accordingly, repression of *Argonaute2* (*AGO2*) expression in S2 cells increased the viral RNA accumulation, as did *Dcr-2* and *AGO2* mutant embryos (Li *et al.*, 2002; Wang *et al.*, 2006).

Viral RNA was highly accumulated in adult flies when either FHV particles were injected or a transgene expressing RNA1 (one of the two molecules in the viral genome, which encodes the RNA-dependent RNA polymerase and B2) was present (Galiana-Arnoux *et al.*, 2006; Wang *et al.*, 2006). The accumulation depends on a functionally intact B2 gene. However, B2 becomes dispensable when either one of the siRNA pathway components in the somatic cells (namely *Dcr-2*, *AGO2* and *r2d2*) is mutated. The data indicate that the siRNA mechanism plays a central role for innate immunity to viruses and the viruses counter by encoding proteins that inhibit RNAi.

As expected, B2 transgenic flies increased susceptibility to *Drosophila* C virus (DCV, a major pathogen of fruit flies) infection (van Rij *et al.*, 2006). Similarly, the DCV infected flies with *Dcr-2* or *AGO2* mutations were hypersensitive with dramatically increased mortality and viral RNA level. In S2 cells using a marker gene, which is regularly silenced by long dsRNA-initiated RNAi, DCV infection of the cells was demonstrated to cause strong suppression of the silencing, suggesting a RNAi suppressor is also encoded by this virus. Further analysis shows that the suppressor specifically inhibits the siRNA generation likely through binding of the long dsRNA, thus limiting substrate access of the DCR-2 enzyme, but does not affect the subsequent steps mediated by AGO-2 (van Rij *et al.*, 2006). Supporting this idea is the fact that a canonical dsRNA binding domain is found at the N terminus of the DCV protein. Separate expression of the 99aa N-terminal truncation containing this domain was able to efficiently repress RNAi (van Rij *et al.*, 2006).

Both FHV and DCV are bipartite plus-stranded RNA viruses. A more divergent virus, the cricket paralysis virus (CrPV), was tested for its interaction with RNAi (Wang *et al.*, 2006). CrPV contains a non-segmented plus strand RNA genome and is picorna-like. The flies are susceptible to CrPV infection. The *Dcr-2* or *r2d2* mutations strongly enhanced the susceptibility and the viral RNA was detected more rapidly and to higher level of accumulation. CrPV is also found to encode an RNAi suppression function at its N-terminus, which contains a dsRNA binding domain (Wang *et al.*, 2006). Along with confirmed antiviral activity of RNAi to another virus SINV (solenopsis invicta virus, picorna-like) (Galiana-Arnoux *et al.*, 2006), a scenario emerges that the siRNA pathway components *Dcr-2*, *AGO2* and *r2d2* form a general viral defence system for *Drosophila*.

RNA silencing and Polycomb

An involvement of RNA in at least some Polycomb-based silencing grew out of studies of transgene silencing. A *white* eye colour promoter was fused to the *Alcohol dehydrogenase* structural gene as a promoter reporter system (*w-Adh*). Paired copies of this transgene exhibited less expression than unpaired copies, indicating that it exhibits pairing sensitive silencing. With increasing dosage of the transgene, total RNA levels declined (Pal Bhadra *et al.*, 1997). Single unpaired transgenes did not show detectable Polycomb complex association, but at higher silencing doses, the complex was present. The level of silencing was also modulated by mutations in *Polycomb* and *Polycomb-like*.

A reciprocal transgene involving the *Adh* promoter and the *white* structural gene (*Adh-w*) was silenced in the presence of *w-Adh* (Pal Bhadra *et al.*, 1999) despite the fact that there was no homology between them. Silencing was greater with increasing numbers of *w-Adh* transgenes. This silencing requires the presence of the endogenous *Adh* gene to be present in the nucleus for the silencing to occur, suggesting a relay of the silencing signal. Active *Adh-w* copies did not accumulate Polycomb complex, but did when silenced by *w-Adh*.

These cases of silencing were shown to occur at the transcriptional level as might be anticipated from the association with the Polycomb complex (Pal Bhadra *et al.*, 2002). Mutation in the RNAi gene, *piwi*, alleviated the transcriptional silencing as well as post-transcriptional silencing. Thus, a connection was drawn between RNA silencing processes and Polycomb-mediated transcriptional silencing.

Given the involvement of Polycomb in pairing sensitive silencing, tests of whether the RNA silencing machinery was involved in this type of silencing proved positive (Pal Bhadra *et al.*, 2004a). In this case, an *engrailed-white* fusion transgene was tested. The *piwi* and *homeless* mutations caused the pairing sensitive silencing to be stronger, but had no effect on single copies of the *en-w* transgenes.

Transgenes exhibiting pairing sensitive silencing can be involved in long-range contacts in the cell (Grimaud *et al.*, 2006), in particular the *Fab-7* Polycomb response element from the *bithorax* segment identity gene. Moreover, endogenous loci associated with Polycomb have been found in association, at least temporally during development. The pairing sensitive silencing of this transgene was affected by mutation in selected 'RNAi' genes. The association of the endogenous loci was disrupted in some 'RNAi' mutants, but the Polycomb complex was not disrupted nor did the endogenous loci exhibit mutant effects. It was suggested that RNAs stabilize these intranuclear contacts (Grimaud *et al.*, 2006).

Overall, the evidence suggests a role of RNA in some aspects of Polycomb complex action. Transgene silencing that is Polycomb mediated is typically suppressed by RNAi mutations, but the latter do not exhibit phenotypic effects typical of the Polycomb Group. Thus, the role of RNA in Polycomb complex action is yet to be fully defined.

RNAi and chromatin insulator structure

Recent evidence suggests an involvement of the RNAi genes in chromatin insulator function (Lei and Corces, 2006). Chromatin insulators compartmentalize the genome in such a way that the transcriptionally inhibitory environment of heterochromatin does not spread into the surrounding euchromatin region. The insulators also inhibit the interaction between enhancer and promoter regions of a gene, thus influencing gene expression. In *Drosophila*, the gypsy insulator sequence has binding sites for the protein *suppressor of Hairy wing* [*su(Hw)*]. It has been proposed that gypsy insulators bring two regions of DNA in proximity, thereby creating DNA loops and distinct higher order chromatin structures. In a recent study (Lei and Corces, 2006), the *piwi* and *aubergine* mutations were found to disrupt the higher order insulator nuclear organization. It was postulated that the small RNAs generated by the RNAi machinery might be involved in tethering the different proteins (su(Hw), CP190) to the gypsy element or by itself promoting nuclear architecture of the gypsy insulator. Interestingly, *Lip* (Dmp68), an RNA helicase, acts negatively on the insulator and mutations in *Lip* improve insulator function. It has been proposed that *Lip* is recruited to the site of the chromatin insulator loop where, by virtue of its helicase activity, it unwinds the protein/RNA complex, thus disrupting/remodelling the nuclear organization of chromatin insulators. In such a manner, the RNAi genes influence the structure of the nuclear matrix.

RNAi and heterochromatin

The involvement of the RNAi apparatus in heterochromatin formation and function was initially discovered in fission yeast (Volpe *et al.*, 2002). Mutants of *argonaute*, *dicer*, and *RNA dependent RNA polymerase* disrupted the silencing of transgenes in heterochromatic locations, as well as having reduced levels of the heterochromatic mark histone H3 lysine-9 methylation (H3-mK9) (Volpe *et al.*, 2002). In addition, transcripts from centromeric repeats were present in the mutants, suggesting that the RNAi pathway processed them in wild type cells. This was complemented by the discovery of the presence of siRNAs corresponding to heterochromatic repeats (Reinhart and Bartel, 2002) – including one that was shown to be transcribed by Volpe *et al.* (2002), further implicating the RNAi pathway in the function and/or formation of heterochromatin. Subsequent work led to the isolation of a complex called RITS (RNA induced initiation of transcriptional gene silencing), which contained a protein involved in RNAi (Ago1), a heterochromatin protein (Chp1) and a third protein Tas3, in addition to siRNAs homologous to centromeric repeats, which were required for correct targeting of the complex to heterochromatic regions (Verdel *et al.*, 2004). Ago1 was later found to interact with the carboxy terminal domain (CTD) of RNA polymerase II (RNA polII), mutation of which led to the loss of centromeric siRNAs and heterochromatic histone modifications, illustrating the importance of transcription of these repeats for heterochromatin formation (Kato *et al.*, 2005; Schramke *et al.*, 2005). Ago1 endonucleolytic cleavage ('slicer') activity was later shown to be necessary for heterochromatic silencing of reporter genes mediated through siRNAs (Irvine *et al.*, 2006).

The association between RNAi and heterochromatin formation in *Drosophila* was demonstrated through the examination of PEV and transgene silencing in various RNAi mutants (Pal-Bhadra *et al.*, 2004b). A *white* eye colour reporter gene present in tandem repeats which exhibited variegated expression in wild type flies showed a suppression of silencing in mutants for *piwi* and *homeless* (*hls*). Similarly, there was a derepression of the variegated expression of a heterochromatic reporter gene in mutants for *piwi*, *hls* and *aub*. The staining of polytene chromosomes for two markers of heterochromatin, heterochromatin protein 1 (HP1) and histone H3-lysine 9 methylation (H3-mK9), was examined in these mutants to determine the impact of their loss on heterochromatin structure. Mutations in *piwi* and *aub* caused partial loss of H3-mK9 staining, while mutation of *hls* resulted in a much more pronounced reduction. The *hls* mutants also had a redistribution of HP1, associating with all the chromosome arms.

In an attempt to identify DNA sequence elements that are involved in the determination of heterochromatin or euchromatin domains, the transposable element *1360*, or *hoppel*, was found to be involved in the formation of heterochromatin on the *Drosophila* fourth chromosome (Sun et al., 2004). This element was demonstrated to serve as an initiator of heterochromatin formation, which spread for up to 10 kb. The variegation of reporter gene expression induced by proximity to *1360* was suppressed by mutation of HP1. This variegation was suggested by the authors to rely on the RNAi apparatus. Subsequent work by this group confirmed this hypothesis by using a *white* reporter gene fused to a copy of *1360* (Haynes et al., 2006). One aim was to determine whether a single copy of *1360* would be able to act as a *cis*-acting silencer. The isolation of a variegating strain, which exhibited a suppression of the PEV once the *1360* element was removed, affirmed this. A number of transgenic lines containing this reporter were obtained; however, only one showed this variegated expression pattern. Analysis of the genomic locations of the insertions showed that the reporters without any silencing were present in TE poor regions, whereas the variegating reporter was present in a TE rich region. This variegation was dependent on HP1 and the presence of siRNAs specific for *1360* led them to investigate whether it was dependent on RNAi. PEV was indeed found to be sensitive to mutations in RNAi genes including the Argonaute family genes *aubergine* and *piwi*. Mutation of *hls*, which is required for TE

silencing, also suppressed PEV, as did mutations in *Dicer-1*. The authors proposed that the *1360* siRNAs could function to target transcriptional silencing using a RISC-like complex.

As noted above, another function of the PIWI protein in gene silencing was recently reported by Saito and colleagues (2006). The protein was isolated as part of a complex and was found to contain rasiRNAs. It was demonstrated that PIWI had slicer activity *in vitro* and this activity, combined with the previously reported function of the protein in LTR transposon silencing, implies that PIWI might have a role in heterochromatic gene silencing in *Drosophila* analogous to that of Argonaute in fission yeast as reported (Irvine et al., 2006).

The Argonaute2 protein has also been demonstrated to be required for proper heterochromatin assembly in *Drosophila* embryos (Deshpande et al., 2005). Mutants for AGO2 were shown to have changes in HP1 association with centric heterochromatin, as well as defects in H3-mK9 methylation, in precellular blastoderm embryos. Furthermore, there was a suppression of PEV of a *white* reporter gene in a proportion of AGO2 mutants, with the results suggesting that AGO2 was involved in establishing silenced heterochromatin early in development.

The *Rm62* gene (also called *Lighten-up*) encodes a dsRNA helicase that is involved in RNAi, as well as heterochromatin function. Mutation of *Rm62* results in the inhibition of RNAi targeted to a reporter gene in *Drosophila* S2 cells (Ishizuka et al., 2002). A previous study had demonstrated that *Rm62* (or *Lip*) encoded a suppressor of PEV and that mutation of *Lip* resulted in an increase in transcription of a number of retrotransposons (Csink et al., 1994). These results collectively illustrate another link between the RNAi apparatus and heterochromatin function.

RNAi role in nucleolar structural integrity and rDNA repeat stability

The nucleolus is the site of ribosome assembly and contains the rDNA repeats, which encode the major ribosomal RNAs. The rDNA repeats are localized in a dense heterochromatin landscape. Experiments performed in *Drosophila* show that the nucleolus organization and the integrity of rDNA stability is strongly influenced by RNAi genes such as *Dcr-2* (Peng and Karpen, 2006). The RNAi defective cells showed a large number of ectopic nucleoli and the formation of extra chromosomal circular DNA. This phenotype has been attributed to the decrease in H3-di-mK9 at the repeated regions of DNA. The result is chromatin decondensation and a high rate of recombination between the rDNA repeat regions that is usually suppressed due to chromatin condensation in wild type cells. Thus, the RNAi pathway strongly influences gene expression by virtue of its role in stabilization of rDNA repeats and nucleolus structure determination (Peng and Karpen, 2006).

RNAi, RNA editing and DNA repair

The *Drosophila* protein DDP1 (*Drosophila* dodeca-satellite-binding protein) contains multiple KH domains, which are single stranded nucleic acid binding motifs and associates with pericentric heterochromatin (Cortes *et al.*, 1999). While it was initially isolated through its binding to a centromeric satellite repeat, it was found to display a more general binding to heterochromatic regions in polytene chromosomes, as well as colocalizing closely with HP1. Further characterization of the protein showed that it was a suppressor of PEV and that *ddp1* mutants had a reduction in H3-mK9 as well as HP1 deposition at the chromocentre, suggesting a role in heterochromatin function (Huertas *et al.*, 2004). In a study conducted to isolate proteins that bound to promiscuously edited RNA molecules, a protein complex was isolated of which one member was the mammalian homologue of DDP1, vigilin (Wang *et al.*, 2005). Promiscuous editing refers to the non-specific conversion of up to 50% of the adenosine residues on a long dsRNA substrate to inosine (Zhang and Carmichael, 2001) – as opposed to the site-specific editing of an mRNA that occurs to modify the function of the encoded protein. Wang and colleagues (2005) used chromatin immunoprecipitation (ChIP) to demonstrate that vigilin associates with heterochromatin in mammalian cells, as well as confirming that DDP1 in *Drosophila* S2 cells colocalized with HP1. Another component of the complex was identified as a subunit of the DNA-dependent protein kinase, which phosphorylates a number of cellular targets. It was shown that HP1 was phosphorylated by the complex; this modification is important for heterochromatin assembly and the silencing activity of HP1 (Eissenberg *et al.*, 1994; Zhao *et al.*, 2001).

The promiscuous editing of dsRNA molecules is carried out by adenosine deaminases (ADARs) and ADAR1 was another of the proteins present in the complex bound to the edited RNAs (Wang *et al.*, 2005). The fact that the substrate for ADARs are potentially the same substrates that the RNAi apparatus acts upon, namely dsRNA, suggests the possibility that the two pathways might interact with each other. Since RNAi generally requires precise base pairing interactions, the editing of a dsRNA molecule would be predicted to interfere with RNAi processing. Conversely, the processing of a dsRNA into siRNAs would mean that the RNA molecule is too short to be able to be used as a substrate for ADARs. Indeed it had previously been shown using *Drosophila* cell extracts that the promiscuous editing of an RNA substrate reduced the RNAi-mediated processing to siRNAs (Scadden and Smith, 2001). Further support for this concept was demonstrated in *C. elegans* when it was found that transgene silencing was induced in *adar; adar2* double mutants, implying that dsRNA from these transgenes were edited in wildtype animals, thereby preventing this silencing (Knight and Bass, 2002). This potential prevention of silencing in wild type animals as a result of editing was approached from another angle. It was shown that the aberrant phenotype of animals that were mutant for the *adar* genes was rescued by the simultaneous mutation of genes involved in RNAi (Tonkin and Bass, 2003). The authors suggested the possibility that ADARs could play a role in regulating whether a dsRNA could enter the RNAi pathway. While it is intuitive that the two pathways of RNA editing and RNAi might work separately, this picture was clouded when it was reported that Tudor staphylococcal nuclease, which is a subunit of RISC, was one protein that bound and cleaved hyper edited dsRNAs in *Xenopus laevis* extracts (Scadden, 2005). While further research will be required to determine the precise relationship between the two pathways, it is at least clear that they are interrelated in some way. The possibility that RNA editing could have an impact on heterochromatin formation as suggested by Wang and colleagues (2005) is supported by the observation that the majority of RNA editing in humans occurs in repetitive elements and introns and not in coding regions (Nishikura, 2006).

As previously mentioned, the study by Wang and colleagues (2005) uncovered a DNA dependent protein kinase as part of the complex binding the edited RNAs. This protein is also an important DNA repair protein. The authors suggest that this result illustrates a link between DNA repair and RNA-mediated gene silencing. A direct link between these two pathways had been previously reported in other organisms. The *bru1* gene in *Arabidopsis* was isolated in a screen for genes involved in DNA repair, in which mutants had greatly increased genotoxic stress responses, but in addition, they also showed a derepression of transcriptional gene silencing of a reporter gene (Takeda *et al.*, 2004). A genetic screen for genes involved in transcriptional gene silencing in *Chlamydomonas reinhardtii* uncovered two genes, *mut-9* and *mut-11*, which, when mutated, alleviated transcriptional gene silencing of a reporter, but also led to increased sensitivity to DNA damaging agents (Jeong Br *et al.*, 2002). Other genes in *Drosophila* have been identified as having dual roles in DNA repair and heterochromatin function with the latter pathway, therefore implying an overlap with RNAi related functions. The *Drosophila* ribosomal protein P0, which is an apurinic/apyrimidinic endonuclease with possible roles in DNA repair, is potentially involved in heterochromatin function due to its ability to suppress PEV (Frolov and Birchler, 1998). The *Drosophila parp* gene encodes a DNA repair protein in addition to also being necessary for proper heterochromatin formation (Tulin *et al.*, 2002). Similarly, the *Ataxia Telangiectasia Mutated* (*ATM*) gene, which encodes a protein kinase central to the response to many types of DNA damage, has been found to interact with HP1 (Oikemus *et al.*, 2004). Therefore the overlap between the pathways of RNAi, heterochromatin formation and DNA repair are becoming increasingly evident.

Concluding remarks

As the above narrative illustrates, small RNAs and the gene products of the so-called RNAi machinery are involved in many processes in the cell that impact gene expression. At one level, a post-transcriptional degradation of double-stranded RNAs exists in the cytoplasm. This process might exist for the major purpose of counteracting virus expression. At another level, transcriptional silencing occurs in the nucleus. This process might involve double-stranded RNAs but the rasiRNAs from transposable elements do not have characteristics typical of dicer products, suggesting other processes for generation of small RNAs. If this is indeed the case, then the scenario is raised that small RNAs are likely to be more global in gene regulation, as might be suggested by their involvement with chromatin insulators and Polycomb Response Elements, than just as a defence against viruses and transposable elements. This idea might suggest a deeper evolutionary origin for RNA based silencing than as a defence mechanism against viruses and transposons.

References

Aravin, A.A., Lagos-Quintana, M., Yalcin, A., Zavolan, M., Marks, D., Snyder, B., Gaasterland, T., Meyer, J., and Tuschl, T. (2003). The small RNA profile during *Drosophila melanogaster* development. Dev. Cell 5, 337–50.

Aravin, A.A., Naumova, N.M., Tulin, A.V., Vagin, V.V., Rozovsky, Y.M., and Gvozdev, V.A. (2001). Double-stranded RNA-mediated silencing of genomic tandem repeats and transposable elements in the D. melanogaster germline. Curr. Biol. *11*, 1017–1027.

Bernstein, E., Caudy, A.A., Hammond, S.M., and Hannon, G.J. (2001). Role for a bidentate ribonuclease in the initiation step of RNA interference. Nature *409*, 363–366.

Carthew, R.W. (2006). Molecular biology. A new RNA dimension to genome control. Science *313*, 305–6.

Chao, J.A., Lee, J.H., Chapados, B.R., Debler, E.W., Schneemann, A., and Williamson, J.R. (2005). Dual modes of RNA-silencing suppression by Flock House virus protein B2. Nat. Struct. Mol. Biol. *12*, 952–7.

Cortes, A., Huertas, D., Fanti, L., Pimpinelli, S., Marsellach, F.X., Pina, B., and Azorin, F. (1999). DDP1, a single-stranded nucleic acid-binding protein of *Drosophila*, associates with pericentric heterochromatin and is functionally homologous to the yeast Scp160p, which is involved in the control of cell ploidy. EMBO J. *18*, 3820–3833.

Csink, A.K., Linsk, R., and Birchler, J.A. (1994). The *Lighten up* (*Lip*) gene of *Drosophila melanogaster*, a modifier of retroelement expression, position effect variegation and white locus insertion alleles. Genetics *138*, 153–163.

Deshpande, G., Calhoun, G., and Schedl, P. (2005). *Drosophila* argonaute-2 is required early in embryogenesis for the assembly of centric/centromeric heterochromatin, nuclear division, nuclear migration, and germ-cell formation. Genes Dev. *19*, 1680–1685.

Ding, S.W., Li, W.X., and Symons, R.H. (1995). A novel naturally occurring hybrid gene encoded by a plant RNA virus facilitates long distance virus movement. EMBO J. *14*, 5762–72.

Eissenberg, J.C., Ge, Y.W., and Hartnett, T. (1994). Increased phosphorylation of HP1, a heterochromatin-associated protein of *Drosophila*, is correlated with heterochromatin assembly. J. Biol. Chem. *269*, 21315–21321.

Frolov, M.V., and Birchler, J.A. (1998). Mutation in P0, a dual function ribosomal protein/apurinic/apyrimidinic endonuclease, modifies gene expression and position effect variegation in *Drosophila*. Genetics *150*, 1487–1495.

Galiana-Arnoux, D., Dostert, C., Schneemann, A., Hoffmann, J.A., and Imler, J.L. (2006). Essential function *in vivo* for Dicer-2 in host defense against RNA viruses in *Drosophila*. Nature Immunol. *7*, 590–597.

Grimaud, C., Bantignies, F., Pal Bhadra, M., Bhadra, U., and Cavalli, G. (2006). RNAi components are required for nuclear clustering of Polycomb Group Response Elements. Cell *124*, 957–971.

Haley, B., and Zamore, P.D. (2004). Kinetic analysis of the RNAi enzyme complex. Nat. Struct. Mol. Biol. *11*, 599–606.

Haussecker, D., and Proudfoot, N.J. (2005). Dicer-dependent turnover of intergenic transcripts from the human beta-globin gene cluster. Mol. Cell. Biol. *25*, 9724–9733.

Haynes, K.A., Caudy, A.A., Collins, L., and Elgin, S.C. (2006). Element 1360 and RNAi components contribute to HP1-dependent silencing of a pericentric reporter. Curr. Biol. *16*, 2222–2227.

Huertas, D., Cortes, A., Casanova, J., and Azorin, F. (2004). *Drosophila* DDP1, a multi-KH-domain protein, contributes to centromeric silencing and chromosome segregation. Curr. Biol. *14*, 1611–1620.

Irvine, D.V., Zaratiegui, M., Tolia, N.H., Goto, D.B., Chitwood, D.H., Vaughn, M.W., Joshua-Tor, L., and Martienssen, R.A. (2006). Argonaute slicing is required for heterochromatic silencing and spreading. Science *313*, 1134–1137.

Ishizuka, A., Siomi, M.C., and Siomi, H. (2002). A *Drosophila* fragile X protein interacts with components of RNAi and ribosomal proteins. Genes Dev. *16*, 2497–2508.

Jensen, S., Gassama, M.P., and Heidmann, T. (1999). Taming of transposable elements by homology-dependent gene silencing. Nat. Genet. *21*, 209–12.

Jeong Br, B.R., Wu-Scharf, D., Zhang, C., and Cerutti, H. (2002). Suppressors of transcriptional transgenic silencing in Chlamydomonas are sensitive to DNA-damaging agents and reactivate transposable elements. Proc. Natl. Acad. Sci. USA *99*, 1076–1081.

Kalmykova, A.I., Klenov, M.S., and Gvozdev, V.A. (2005). Argonaute protein PIWI controls mobilization of retrotransposons in the *Drosophila* male germline. Nucleic Acids Res. *33*, 2052–2059.

Kato, H., Goto, D.B., Martienssen, R.A., Urano, T., Furukawa, K., and Murakami, Y. (2005). RNA polymerase II is required for RNAi-dependent heterochromatin assembly. Science *309*, 467–469.

Kavi, H.H., Fernandez, H.R., Xie, W., and Birchler, J.A. (2005). RNA silencing in *Drosophila*. FEBS Lett. *579*, 5940–5949.

Knight, S.W., and Bass, B.L. (2002). The role of RNA editing by ADARs in RNAi. Mol. Cell *10*, 809–817.

Lee, Y.S., and Carthew, R.W. (2003). Making a better RNAi vector for *Drosophila*: use of intron spacers. Methods *30*, 322–329.

Lee, Y.S., Nakahara, K., Pham, J.W., Kim, K., He, Z., Sontheimer, E.J., and Carthew, R.W. (2004). Distinct roles for *Drosophila* Dicer-1 and Dicer-2 in the siRNA/miRNA silencing pathways. Cell *117*, 69–81.

Lei, E.P., and Corces, V.G. (2006). RNA interference machinery influences the nuclear organization of a chromatin insulator. Nat. Genet. *38*, 936–941.

Li, H., Li, W.X., and Ding, S.W. (2002). Induction and suppression of RNA silencing by an animal virus. Science *296*, 1319–21.

Li, W.X., Li, H., Lu, R., Li, F., Dus, M., Atkinson, P., Brydon, E.W., Johnson, K.L., Garcia-Sastre, A., Ball, L.A., Palese, P., and Ding, S.W. (2004). Interferon antagonist proteins of influenza and vaccinia viruses are suppressors of RNA silencing. Proc. Natl. Acad. Sci. USA *101*, 1350–5.

Lippman, Z., Gendrel, A.V., Black, M., Vaughn, M.W., Dedhia, N., McCombie, W.R., Lavine, K., Mittal, V., May, B., Kasschau, K.D., Carrington, J.C., Doerge, R.W., Colot, V., and Martienssen, R. (2004). Role of transposable elements in heterochromatin and epigenetic control. Nature *430*, 471–6.

Liu, Q., Rand, T.A., Kalidas, S., Du, F., Kim, H.E., Smith, D.P., and Wang, X. (2003). R2D2, a bridge between the initiation and effector steps of the *Drosophila* RNAi pathway. Science *301*, 1921–1925.

Ma, J.B., Yuan, Y.R., Meister, G., Pei, Y., Tuschl, T., and Patel, D.J. (2005). Structural basis for 5′-end-specific recognition of guide RNA by the A. fulgidus Piwi protein. Nature *434*, 666–670.

Matranga, C., Tomari, Y., Shin, C., Bartel, D.P., and Zamore, P.D. (2005). Passenger-strand cleavage facilitates assembly of siRNA into Ago2-containing RNAi enzyme complexes. Cell *123*, 607–620.

Nishikura, K. (2006). Editor meets silencer: crosstalk between RNA editing and RNA interference. Nat. Rev. Mol. Cell. Biol. *7*, 919–931.

Nykanen, A., Haley, B., and Zamore, P.D. (2001). ATP requirements and small interfering RNA structure in the RNA interference pathway. Cell *107*, 309–321.

Oikemus, S.R., McGinnis, N., Queiroz-Machado, J., Tukachinsky, H., Takada, S., Sunkel, C.E., and Brodsky, M.H. (2004). *Drosophila* atm/telomere fusion is required for telomeric localization of HP1 and telomere position effect. Genes Dev. *18*, 1850–1861.

Pal-Bhadra, M., Bhadra, U., and Birchler, J.A. (1997). Cosuppression in *Drosophila*: gene silencing of Alcohol dehydrogenase by *white-Adh* transgenes is Polycomb dependent. Cell *90*, 479–490.

Pal-Bhadra, M., Bhadra, U., and Birchler, J.A. (1999). Cosuppression of nonhomologous transgenes in *Drosophila* involves mutually related endogenous sequences. Cell *99*, 35–46.

Pal-Bhadra, M., Bhadra, U., and Birchler, J.A. (2002). RNAi related mechanisms affect both transcriptional and posttranscriptional transgene silencing in *Drosophila*. Mol. Cell 9, 315–327.

Pal-Bhadra, M., Bhadra, U., and Birchler, J.A. (2004a). Interrelationship of RNAi and transcriptional gene silencing in *Drosophila*. Cold Spring Harbor Symposium on Quantitative Biology 69, 433–438.

Pal-Bhadra, M., Leibovitch, B.A., Gandhi, S.G., Rao, M., Bhadra, U., Birchler, J.A., and Elgin, S.C. (2004b). Heterochromatic silencing and HP1 localization in *Drosophila* are dependent on the RNAi machinery. Science *303*, 669–672.

Parker, J.S., Roe, S.M., and Barford, D. (2004). Crystal structure of a PIWI protein suggests mechanisms for siRNA recognition and slicer activity. EMBO J. *23*, 4727–4737.

Parker, J.S., Roe, S.M., and Barford, D. (2005). Structural insights into mRNA recognition from a PIWI domain-siRNA guide complex. Nature *434*, 663–666.

Pelisson, A., Sarot, E., Payen-Groschene, G., and Bucheton, A. (2006). A novel rasiRNA-mediated silencing pathway downregulates sense gypsy transcripts in the somatic cells of the *drosophila* ovary. J. Virol. Published online on Nov. 29, 2006.

Peng, J.C., and Karpen, G.H. (2006). H3K9 methylation and RNA interference regulate nucleolar organization and repeated DNA stability. Nat. Cell Biol. 9, 25–35.

Pham, J.W., Pellino, J.L., Lee, Y.S., Carthew, R.W., and Sontheimer, E.J. (2004). A Dicer-2-dependent 80s complex cleaves targeted mRNAs during RNAi in *Drosophila*. Cell *117*, 83–94.

Pham, J.W., and Sontheimer, E.J. (2005). Molecular requirements for RNA-induced silencing complex assembly in the *Drosophila* RNA interference pathway. J. Biol. Chem. *280*, 39278–39283.

Qu, F., and Morris, T.J. (2005). Suppressors of RNA silencing encoded by plant viruses and their role in viral infections. FEBS Lett. *579*, 5958–64.

Reinhart, B.J., and Bartel, D.P. (2002). Small RNAs correspond to centromere heterochromatic repeats. Science *297*, 1831.

Reiss, D., Josse, T., Anxolabehere, D., and Ronsseray, S. (2004). *aubergine* mutations in *Drosophila melanogaster* impair P cytotype determination by telomeric P elements inserted in heterochromatin. Mol Genet. Genomics *272*, 336–43.

Saito, K., Nishida, K.M., Mori, T., Kawamura, Y., Miyoshi, K., Nagami, T., Siomi, H., and Siomi, M.C. (2006). Specific association of Piwi with rasiRNAs derived from retrotransposon and heterochromatic regions in the *Drosophila* genome. Genes Dev. *20*, 2214–2222.

Saito, K., Ishizuka, A., Siomi, H., and Siomi, M.C. (2005). Processing of pre-microRNAs by the Dicer-1-Loquacious complex in *Drosophila* cells. PLoS Biol. *3*, e235.

Saleh, M.C., van Rij, R.P., Hekele, A., Gillis, A., Foley, E., O'Farrell, P.H., and Andino, R. (2006). The endocytic pathway mediates cell entry of dsRNA to induce RNAi silencing. Nature cell biology *8*, 793–802.

Savitsky, M., Kwon, D., Georgiev, P., Kalmykova, A., and Gvozdev, V. (2006). Telomere elongation is under the control of the RNAi-based mechanism in the *Drosophila* germline. Genes Dev. *20*, 345–54.

Scadden, A.D. (2005). The RISC subunit Tudor-SN binds to hyper-edited double-stranded RNA and promotes its cleavage. Nat. Struct. Mol. Biol. *12*, 489–496.

Scadden, A.D., and Smith, C.W. (2001). RNAi is antagonized by A→I hyper-editing. EMBO Rep. *2*, 1107–1111.

Schramke, V., Sheedy, D.M., Denli, A.M., Bonila, C., Ekwall, K., Hannon, G.J., and Allshire, R.C. (2005). RNA-interference-directed chromatin modification coupled to RNA polymerase II transcription. Nature *435*, 1275–1279.

Schwarz, D.S., Hutvagner, G., Du, T., Xu, Z., Aronin, N., and Zamore, P.D. (2003). Asymmetry in the assembly of the RNAi enzyme complex. Cell *115*, 199–208.

Schwarz, D.S., Tomari, Y., and Zamore, P.D. (2004). The RNA-induced silencing complex is a Mg2+-dependent endonuclease. Curr. Biol. *14*, 787–791.

Song, J.J., Smith, S.K., Hannon, G.J., and Joshua-Tor, L. (2004). Crystal structure of Argonaute and its implications for RISC slicer activity. Science *305*, 1434–1437.

Sontheimer, E.J. (2005). Assembly and function of RNA silencing complexes. Nat. Rev. *6*, 127–138.

Sun, F.L., Haynes, K., Simpson, C.L., Lee, S.D., Collins, L., Wuller, J., Eissenberg, J.C., and Elgin, S.C. (2004). *cis*-Acting determinants of heterochromatin formation on *Drosophila melanogaster* chromosome four. Mol. Cell. Biol. *24*, 8210–8220.

Tahbaz, N., Kolb, F.A., Zhang, H., Jaronczyk, K., Filipowicz, W., and Hobman, T.C. (2004). Characterization of the interactions between mammalian PAZ PIWI domain proteins and Dicer. EMBO Rep. *5*, 189–194.

Takeda, S., Tadele, Z., Hofmann, I., Probst, A.V., Angelis, K.J., Kaya, H., Araki, T., Mengiste, T., Mittelsten Scheid, O., Shibahara, K., Scheel, D., and Paszkowski, J. (2004). BRU1, a novel link between responses to DNA damage and epigenetic gene silencing in *Arabidopsis*. Genes Dev. *18*, 782–793.

Tomari, Y., Du, T., Haley, B., Schwarz, D.S., Bennett, R., Cook, H.A., Koppetsch, B.S., Theurkauf, W.E., and Zamore, P.D. (2004a). RISC assembly defects in the *Drosophila* RNAi mutant armitage. Cell *116*, 831–841.

Tomari, Y., Matranga, C., Haley, B., Martinez, N., and Zamore, P.D. (2004b). A protein sensor for siRNA asymmetry. Science *306*, 1377–1380.

Tonkin, L.A., and Bass, B.L. (2003). Mutations in RNAi rescue aberrant chemotaxis of ADAR mutants. Science *302*, 1725.

Tulin, A., Stewart, D., and Spradling, A.C. (2002). The *Drosophila* heterochromatic gene encoding poly(ADP-ribose) polymerase (PARP) is required to modulate chromatin structure during development. Genes Dev. *16*, 2108–2119.

Vagin, V.V., Klenov, M.S., Kalmykova, A.I., Stolyarenko, A.D., Kotelnikov, R.N., and Gvozdev, V.A. (2004). The RNA Interference Proteins and Vasa Locus are Involved in the Silencing of Retrotransposons in the Female Germline of *Drosophila melanogaster*. RNA Biology *1*, 54–58.

Vagin, V.V., Sigova, A., Li, C., Seitz, H., Gvozdev, V., and Zamore, P.D. (2006). A distinct small RNA pathway silences selfish genetic elements in the germline. Science *313*, 320–4.

van Rij, R.P., Saleh, M.C., Berry, B., Foo, C., Houk, A., Antoniewski, C., and Andino, R. (2006). The RNA silencing endonuclease Argonaute 2 mediates specific antiviral immunity in *Drosophila melanogaster*. Genes Dev. *20*, 2985–95.

Verdel, A., Jia, S., Gerber, S., Sugiyama, T., Gygi, S., Grewal, S.I., and Moazed, D. (2004). RNAi-mediated targeting of heterochromatin by the RITS complex. Science *303*, 672–676.

Volpe, T.A., Kidner, C., Hall, I.M., Teng, G., Grewal, S.I., and Martienssen, R.A. (2002). Regulation of heterochromatic silencing and histone H3 lysine-9 methylation by RNAi. Science *297*, 1833–1837.

Wang, Q., Zhang, Z., Blackwell, K., and Carmichael, G.G. (2005). Vigilins bind to promiscuously A-to-I-edited RNAs and are involved in the formation of heterochromatin. Curr. Biol. *15*, 384–391.

Wang, X.H., Aliyari, R., Li, W.X., Li, H.W., Kim, K., Carthew, R., Atkinson, P., and Ding, S.W. (2006). RNA interference directs innate immunity against viruses in adult *Drosophila*. Science *312*, 452–454.

Williams, R.W., and Rubin, G.M. (2002). ARGONAUTE1 is required for efficient RNA interference in *Drosophila* embryos. Proc. Natl. Acad. Sci. USA *99*, 6889–94.

Yuan, Y.R., Pei, Y., Ma, J.B., Kuryavyi, V., Zhadina, M., Meister, G., Chen, H.Y., Dauter, Z., Tuschl, T., and Patel, D.J. (2005). Crystal structure of A. aeolicus argonaute, a site-specific DNA-guided endoribonuclease, provides insights into RISC-mediated mRNA cleavage. Molecular Cell *19*, 405–419.

Zamore, P.D., Tuschl, T., Sharp, P.A., and Bartel, D.P. (2000). RNAi: double-stranded RNA directs the ATP-dependent cleavage of mRNA at 21 to 23 nucleotide intervals. Cell *101*, 25–33.

Zhang, H., Kolb, F.A., Jaskiewicz, L., Westhof, E., and Filipowicz, W. (2004). Single processing center models for human Dicer and bacterial RNase III. Cell *118*, 57–68.

Zhang, Z., and Carmichael, G.G. (2001). The fate of dsRNA in the nucleus: a p54(nrb)-containing complex mediates the nuclear retention of promiscuously A-to-I edited RNAs. Cell *106*, 465–475.

Zhao, T., Heyduk, T., and Eissenberg, J.C. (2001). Phosphorylation site mutations in heterochromatin protein 1 (HP1) reduce or eliminate silencing activity. J. Biol. Chem. *276*, 9512–9518.

MicroRNA-mediated Regulation of Gene Expression

Lena J. Chin and Frank J. Slack

Abstract

MiRNAs are short, ~22 nucleotide regulatory RNAs, first discovered in *Caenhorhabditis elegans*. Since then, hundreds of miRNAs have been identified in plants and animals. Based on the current number of predicted miRNAs, one to three percent of genomic DNA is believed to encode these small, regulatory RNAs. MiRNAs inhibit protein synthesis by binding to their target mRNAs and regulating gene expression in a post-transcriptional manner. The exact mechanism by which target gene expression is down-regulated is unclear; however, experimental evidence has led to several different theories to explain miRNA-mediated mRNA repression. These possible mechanisms include target degradation, localization to P-bodies, inhibition of translation initiation or elongation, mRNA deadenylation, and mRNA destabilization.

Introduction

In the 1990s, a new class of regulatory RNAs was discovered in *Caenorhabditis elegans*. These ~22 nucleotide-long RNAs became known as microRNAs (miRNAs). The first miRNAs, *lin-4* and *let-7*, were identified through studies of genes involved in developmental timing (Lee *et al.*, 1993; Reinhart *et al.*, 2000). *lin-4* and *let-7* were found to control the transitions between different larval and adult developmental stages by directly interacting with the 3′ untranslated regions (UTRs) of their target genes. Initially, miRNAs were considered a *C. elegans*-specific phenomenon, but it was soon discovered that *let-7* is highly conserved (Pasquinelli *et al.*, 2000).

To date, hundreds of miRNAs have been identified in plants and animals. Some of these miRNAs were found through forward genetics and cloning; however, many miRNAs have been identified through bioinformatic approaches. These programs take into account a variety of criteria, including sequence conservation, secondary structure, and base pairing (Grad *et al.*, 2003; Lagos-Quintana *et al.*, 2001; Lagos-Quintana *et al.*, 2002; Lau *et al.*, 2001; Lim *et al.*, 2003). MiRBase, the miRNA registry, currently lists 93 miRNAs in *Drosophila melanogaster*, 533 miRNAs in humans, 337 miRNAs in *Danio rerio*, and 135 miRNAs in *Caenorhabditis elegans* (Release 10.0). Similarly, bioinformatics programs have been constructed to predict miRNA targets (Enright *et al.*, 2003; John *et al.*, 2004; Kiriakidou *et al.*, 2004; Lall *et al.*, 2006; Lewis *et al.*, 2003; Sethupathy *et al.*, 2006). It is thought that miRNA genes account for about 1–3% of all genes and that each miRNA can potentially regulate hundreds of targets. The important roles miRNAs have are just being understood; recent research has shown that miRNAs are involved in aspects of metabolism, aging, and development, including haematopoiesis, apoptosis, and cell proliferation (Boehm and Slack, 2005; Brennecke *et al.*, 2003; Xu *et al.*, 2003; Yekta *et al.*, 2004).

MiRNAs regulate gene expression by binding to their target mRNAs in one of two places: the coding sequence or the 3′ UTR. In plants, miRNAs bind within the coding sequence of their mRNA targets, generally doing so with perfect complementarity and resulting in cleavage of the target mRNA. In animals, however, post-transcriptional down-regulation of target genes is a result of miRNAs binding with imperfect complementarity to their 3′ UTRs. The nature of these interactions presents an added layer of complexity, and a major challenge, to the accurate prediction of the targets of each miRNA. It is generally thought that perfect, or near perfect, complementarity between the mRNA and the 5′ end of the miRNA is the most important criterion of target recognition. These important base pairs of the miRNA are known as the 'seed' sequence and generally include the first eight nucleotides (Lewis et al., 2003). Seed sequences are used in classifying miRNAs into families; miRNAs with high homology at their 5′ ends are grouped together.

Discovery of the first miRNAs

lin-4, the first miRNA identified, was initially studied for its role in developmental timing in C. elegans. In 1993, Lee et al mapped lin-4 and determined that it did not code for a protein. Northern blot and RNase protection assays identified two lin-4 transcripts, a long and a short transcript. A 22-nucleotide stretch of sequence identity existed at the 5′ ends of both transcripts. These 22 nucleotides made up the mature lin-4 miRNA. lin-4 was known to be a negative regulator of lin-14, a novel gene involved in controlling developmental timing. Close investigation of the lin-4 and lin-14 sequence revealed seven points of hybridization, known as lin-4 complementary sites, in the lin-14 3′ UTR (Lee et al., 1993).

Like lin-4, let-7 was first identified through its role in developmental timing in C. elegans. Reinhart et al found that mutant let-7 worms undergo an extra larval division (L5 stage) in their seam cells (a class of hypodermal tissue) before reaching the adult stage. lin-41 loss-of-function mutants suppressed this phenotype; lin-41 codes for a protein containing a RING finger, B box, coiled coil (RBCC) domain. They found that let-7 is a non-protein-coding gene that makes a 21-nucleotide miRNA. let-7 is expressed at low levels in the L3 stage and at high levels in the L4 and adult stages, indicative of its role in determining adult fates (Reinhart et al., 2000). Furthermore, Slack et al saw that expression of a GFP::LIN-41 reporter decreased at the L4 stage,

when let-7 expression was high, and determined that there were sequences complementary to let-7 within the lin-41 3′ UTR. A hypodermal-specific lac-Z reporter containing the lin-41 3′ UTR was down-regulated at the adult stage in a let-7-dependent manner (Slack et al., 2000). let-7 was later shown to be highly conserved in animals, revealing that the existence of miRNAs was not a C. elegans-specific phenomenon (Pasquinelli et al., 2000). The discovery of these two genes introduced the scientific community to a new class of regulatory RNAs.

MicroRNA biogenesis

MiRNA genes, which can be found within introns of genes in animals, are transcribed by RNA polymerase II (Lee et al., 2004a) (Fig. 6.1). Often, multiple miRNA genes exist in clusters and are transcribed together; however, this is more common in animals than in plants (Cai et al., 2004). The transcript generated is referred to as the primary-miRNA (pri-miRNA), which can be several kilobases long, has a 5′ cap and a poly-A tail, and can consist of one or more hairpin structures (Cai et al., 2004). In animals, the Microprocessor protein complex, consisting of Drosha and DGCR8/Pasha, cleaves these hairpins, producing segments ~70 nucleotides long (Denli et al., 2004; Gregory et al., 2004; Landthaler et al., 2004; Zeng et al., 2005). Drosha is an RNase III-like endonuclease with two RNase III domains (RIIIDs) and a double-stranded RNA-binding domain (dsRBD) (Han et al., 2004). DiGeorge syndrome critical region gene 8 (DGCR8), like its Drosophila and C. elegans homolog, Pasha, is thought to be involved in recognizing the pri-miRNA (Gregory et al., 2004; Han et al., 2004). The processing complex recognizes the stem–loop structure of the pri-miRNA and excises it by cutting at the base of the hairpin (Lee et al., 2003). Drosha cuts about two helical turns after the terminal loop of the hairpin, leaving a 5′ phosphate and about a two-nucleotide 3′ overhang, as is typical for RNase III enzymes. Each processed hairpin is now referred to as a precursor miRNA (pre-miRNA). This processing is slightly different in plants since they lack homologues to Drosha and DGCR8/Pasha. It is thought that in Arabidopsis, the Dicer protein DCL1 is the only RNase III enzyme

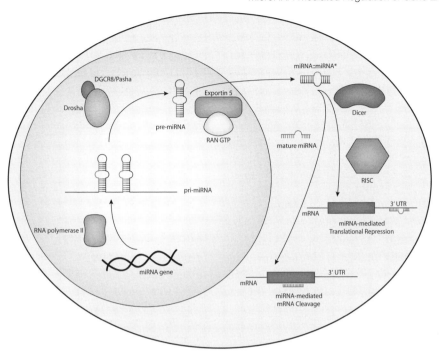

Figure 6.1 MicroRNA biogenesis. RNA polymerase II transcribes the miRNA gene(s), which results in the formation of the pri-miRNA. Processing of the pri-miRNA by the Microprocessor complex consisting of Drosha and DGCR8/Pasha gives rise to the pre-miRNA. This is then exported out of the nucleus by Exportin5 and RanGTP. Dicer processes the pre-miRNA into the miRNA::miRNA* duplex. The duplex enters into the miRISC and the mature miRNA-bound miRISC goes on to bind to its target mRNAs. In animals, binding occurs in the 3' UTR, leading to miRNA-mediated translational repression. In plants, binding occurs in the coding sequence, usually leading to miRNA-mediated mRNA cleavage.

required for miRNA processing (Kurihara and Watanabe, 2004; Papp *et al.*, 2003). Plant pri-miRNAs are generally longer as well.

Pre-miRNAs are exported out of the nucleus by Exportin 5 in animals, or Hasty in plants, in a RanGTP-dependant manner (Bohnsack *et al.*, 2004; Bollman *et al.*, 2003; Lund *et al.*, 2004; Yi *et al.*, 2003). Once in the cytoplasm, the pre-miRNAs are processed into mature miRNAs by Dicer (Grishok *et al.*, 2001; Hutvágner *et al.*, 2001). This RNase III-like enzyme has two RIIIDs, a dsRBD, a DEAD-box helicase domain, and a Piwi, Argonaute, and Zwille (PAZ) domain. The PAZ domain binds 3' ends of small RNAs and is found in members of the Dicer and Argonaute protein families (Lingel *et al.*, 2004; Ma *et al.*, 2004; Yan *et al.*, 2003). In *Drosophila*, Dicer-1 is the RNase III-like enzyme involved in miRNA biogenesis, while in plants, DCL1 plays this role (Lee *et al.*, 2004b; Park *et al.*, 2002). Ultimately, the result of Dicer activity is ~22 nucleotide, double-stranded complex known as the miRNA::miRNA* complex. The free energy

of the ends of the complex will determine which strand will become the mature miRNA that is used to target mRNA transcripts. The strand whose 5' end is bound with the lower free energy becomes the guide strand, or mature miRNA (Khvorova *et al.*, 2003). The miRNA* strand, or passenger strand, does not target mRNAs and gets degraded, although it is unclear exactly how the miRNA* strand is removed (Khvorova *et al.*, 2003; Lau *et al.*, 2001; Schwarz *et al.*, 2003).

Numerous Dicer-associated proteins have been identified. In *Drosophila*, R2D2, a Dicer-associated protein, has been shown to bind to the more thermodynamically stable end of the miRNA::miRNA* duplex (Liu *et al.*, 2003; Tomari *et al.*, 2004). In the case of *Drosophila* siRNA, the duplex is loaded onto Argonaute2 (Ago2) and the passenger strand gets cleaved and removed, activating the associated protein complex (Rand *et al.*, 2005). Ago2 is a member of the Dicer-associated Argonaute family of proteins. Research has shown that the HIV-1 transactivating response (TAR) RNA-binding protein (TRBP) is needed in recruiting the Dicer complex to Ago2 (Chendrimada *et al.*, 2005). In humans, miRNAs can associate with Ago1, -2, -3, and -4, but Ago2, which is often referred to as Slicer, is the only one with cleavage capabilities (Liu *et al.*, 2004; Meister *et al.*, 2004; Song *et al.*, 2004). Remarkably, Ago2 does not cleave the miRNA* strand before entering into the associated protein complex, whereas the siRNA passenger strand would get cleaved (Matranga *et al.*, 2005). The mature miRNA is now found in a protein complex known as the miRNA-containing ribonucleoprotein complex (miRNP) or miRNA-containing RNA-induced silencing complex (miRISC) (Mourelatos *et al.*, 2002). Finally, the RISC-complexed miRNA binds to its target mRNA and inhibits protein production (Hutvágner and Zamore, 2002).

As previously mentioned, depending on the type of organism, miRNAs can bind with perfect or imperfect complementarity. It is important to note that G:U base pairings are tolerated (Lee *et al.*, 1993; Reinhart *et al.*, 2000). It is thought that more than one copy of a miRNA or combinations of miRNAs can bind to an mRNA

to regulate expression (Doench *et al.*, 2003). The manner in which miRNAs down-regulate expression of their target genes is unclear, since there appears to be multiple possibilities. It was initially thought that miRNAs inhibited translation in animals in a manner that did not involve mRNA loss, while mRNA was cleaved in plants (Llave *et al.*, 2002; Wightman *et al.*, 1993). However, recent advances in this field have made it apparent that miRNAs may be repressing genes through a variety of mechanisms (see below, Fig. 6.2).

MiRNA-directed mRNA cleavage in plants

Plant miRNAs bind to their target coding regions with almost perfect, if not perfect, complementarity. Unlike the original findings in worms and the observations discussed above, miRNAs in plants generally cleave their target mRNAs. 5′-rapid amplification of cDNA ends (RACE) of 3′ cleavage products show that this cleavage occurs in the middle of the miRNA::mRNA target binding site (Llave *et al.*, 2002; Mallory *et al.*, 2005; Souret *et al.*, 2004).

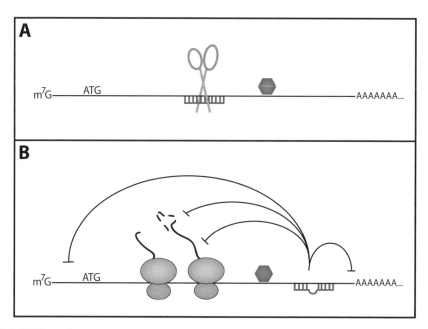

Figure 6.2 MiRNA-mediated target gene regulation. (A) MiRNA-mediated target gene regulation in plants. Plant miRNAs bind with perfect complementarity to the coding region of their target genes. In general, plant miRNAs bind to their target mRNAs and bring about mRNA cleavage. (B) MiRNA-mediated target gene regulation in animals. Animal miRNAs bind with imperfect complementarity to the 3′ UTR of their target genes. When animal miRNAs bind to their targets, a variety of different mechanisms of target gene repression that may follow. Some of these mechanisms include inhibiting translation initiation or elongation, promoting polypeptide degradation, and promoting deadenylation or mRNA instability.

For example, in *Arabidopsis*, miR39 is found at high levels in inflorescence tissues. It targets three *Scarecrow-like* genes, a family of transcription factors also found in inflorescence tissues, through perfectly complementary internal sites. 5′ RACE assays determined that there is sequence-specific target cleavage in the middle of the miR39 binding site (Llave *et al.*, 2002).

Cleavage products have also been found for a miR159 target, MYB33. Achard *et al* found that three *Arabidopsis* GAMYB-like genes, AtMYB33, AtMYB65, and AtMYB101, have internal sequences with almost-perfect complementarity to miR159. Transcription of genes activated by gibberellin (GA), a phytohormone involved in plant development, is regulated by GAMYB binding to response elements in the 5′ end of these GA-activated genes. When miR159 was overexpressed with MYC-MYB33 in *Nicotiana benthamiana*, full-length MYC-MYB33 mRNA levels were significantly reduced, and there was a detectable smaller transcript that hybridized to a MYB33 probe. This smaller species was only present when the miR159 binding site remained unaltered. As expected, MYC-MYB33 protein could not be detected in the presence of miR159 (Achard *et al.*, 2004). These results suggest that miR159 is directing MYB33 mRNA cleavage by directly binding to its internal complementary sequence.

Another example of miRNA-mediated mRNA cleavage is found in miR160, which has complementary sites in *AUXIN RESPONSE FACTOR10* (*ARF10*), *ARF16*, and *ARF17* and directs their mRNA cleavage. Mallory *et al* identified fragments of *ARF10*, *ARF16*, and *ARF17* mRNAs and used 5′ RACE to map the 5′ ends. Each fragment ended at the same point, revealing that miR160 regulated cleavage of its targets between the nucleotides corresponding to bases 10 and 11 of the miRNA. These cleavage products were seen in all tissue types where full-length mRNAs were found. In *Arabidopsis* overexpressing *ARF17*, there was a corresponding increase in 3′ cleavage product. Overexpression of *ARF17* with a mutated miR160 binding site resulted in an increase in *ARF17* mRNA levels, as compared to wild-type plants, but without a corresponding accumulation of 3′ cleavage product. Thus, the appearance of the 3′ cleavage

products was dependent on miR160 (Mallory *et al.*, 2005).

While little is known about how this miRNA-mediated degradation is effected, some insight has emerged from the study of XRN, an exonuclease that has a preference for 5′ monophosphates. *Arabidopsis* XRN4 is a homologue of human Xrn1p. In studies by Souret *et al* examining the decay rate of mRNAs, XRN4 did not have a significant effect on the half-lives of known unstable transcripts. In *xrn4* mutants, the 3′ end RNAs of *SCARECROW-LIKE* RNAs were much more stable than in wild-type plants and mir171 levels were unchanged. 5′ RACE was used to map the 5′ ends of these 3′ RNAs, revealing that they started in the middle of miR171 miRNA binding sites. Thus, XRN4 plays a role in the removal of the 3′ cleavage product. They had similar findings for the miR160 targets *ARF10* and *ARF17*, the miR159 targets *MYB33* and *MYB65*, and the miR165 target *PHV*. This was not the case for the miR167 target *ARF8*, the miR172 target *AP2-like*, the mir-JAW targets *TCP2* and *TCP4*, the miR156 target *SPL10*, and the miR168 target *AGO* (Souret *et al.*, 2004). Therefore, it seems that XRN4 is involved in the removal of some miRNA-regulated target cleavage products in plants, but other mechanisms for this process must also exist. Additional studies are required to further elucidate how target cleavage products are cleared way.

Plants do not always undergo miRNA-mediated mRNA cleavage

While target cleavage is the most common way for miRNAs to regulate gene expression in plants, research shows that this may not be the only method. *APETALA2* (*AP2*) is a transcription factor that functions within the floral meristem. It has an almost perfectly complementary miR172 binding site within the coding region of its mRNA (Park *et al.*, 2002). There are also miR172 sites in other genes with APETALA2 domains. In plants overexpressing miR172a-2, Aukerman and Sakai found a notable decrease in AP2 protein levels as compared to wild-type plants. *AP2* and *AP2*-like target gene RNA levels remained normal in these plants. While *AP2-*

like target gene cleavage products were detected in wild-type and miR172-overexpressing plants using reverse transcriptase-mediated (RT') PCR, these 3' end cleavage products could not be detected by Northern blots (Aukerman and Sakai, 2003). This suggests that mRNA cleavage is not the main mechanism for the miRNA-mediated regulation of *AP2*-like target genes. Likewise, Chen saw that in plants with low miR172 levels, *AP2* RNA levels did not significantly differ from RNA levels in wild-type backgrounds; however, protein levels increased two- to three-fold. Plants overexpressing miR172 had *ap2*-like phenotypes even though *AP2* RNA levels were similar to those of wild-type plants. Furthermore, in *AP2* mutants with a disrupted miR172 binding site, AP2 protein levels were much higher than in plants with the miR172 site of *AP2* intact (Chen, 2004).

Thus, *AP2* mRNA cleavage is not the main method of controlling *AP2* gene expression. There appears to be two ways to regulate *AP2* expression; therefore, plants do not necessarily undergo miRNA-directed mRNA cleavage.

MiRNA-directed cleavage of mRNAs in animals

In 2004, the idea that miRNA-directed target mRNA cleavage is a process exclusive to plants was challenged with the discovery of the first instance of miRNA-regulated cleavage of a target mRNA in animals. Yekta *et al* found that miR-196 regulates its target, *HOXB8*, by cleaving its mRNA (Yekta *et al.*, 2004). *HOXB8* is one of the 39 homeobox genes found in humans. MiR-196 has complementary sites in several of the HOX genes; however, miR-196 almost perfectly base-pairs with the 3' UTR of *HOXB8*, with only one G:U base pair involving the fifth nucleotide of the miRNA. Based on the very high complementarity between miR-196 and the miR-196 binding site in *HOXB8*, it was thought that the transcript would likely be cleaved. 5' RACE using RNA extracted from mice at a time when miR-196 was expressed revealed that miR-196 did, in fact, cleave the *HOXB8* mRNA between the tenth and eleventh nucleotides of the miRNA. This interaction was further confirmed by constructing a luciferase reporter containing the 3' UTR of *HOXB8*. When this reporter was co-transfected with miR-196 into HeLa cells, there was a decrease in both luciferase expression and level of reporter transcript (Yekta *et al.*, 2004).

In plants, a miRNA binds with near perfect complementarity within the coding region of its target to cause cleavage of the mRNA. This case of animal miRNA-mediated cleavage differs from plant miRNA-mediated cleavage because miR-196 binds within the 3' UTR of *HOXB8* and not the coding region.

MiRNA-mediated translational repression without loss of message in animals

Research on *lin-4* and *let-7* suggested that miRNAs bind to the 3' UTRs of their target mRNAs and block protein accumulation. The initial work done on animal miRNA-mediated gene regulation revealed that miRNAs inhibit translation without a significant reduction in target mRNA levels (Wightman *et al.*, 1993).

The first miRNA target identified was *C. elegans lin-14*, a target of the *lin-4* miRNA (Lee *et al.*, 1993). The work of Wightman *et al* showed that LIN-14 protein levels were reduced ten-fold in late larval stages (L2 and L3), as compared to early stages (L1), of wild-type worms, while there was only about a two-fold reduction of LIN-14 protein levels in *lin-4* mutant animals. However, *lin-14* mRNA levels from the same samples remained relatively constant in both wild-type and *lin-4* mutant worms. Thus, *lin-14* mRNA levels were constant while LIN-14 protein was reduced between L1 and L2, corresponding with the accumulation of *lin-4*. Expression of a *lacZ* reporter gene with the *lin-14* 3' UTR was also temporally regulated in a *lin-4*-dependent manner without its mRNA levels being altered (Wightman *et al.*, 1993). Furthermore, Olsen and Ambros found that there was no change in poly(A) tail length of *lin-14* mRNA in the L1 and L2 stages. *lin-14* mRNA was also found to be associated with polyribosomes in the L1 stage, remaining unchanged in the L2 stage, while LIN-14 protein levels at L1 were about fifteen-fold greater than at L2. *lin-4* RNA also co-sedimented with polysomes at the L2 stage. Thus, *lin-4* was bound to *lin-14* mRNA simultaneously with ribosomes being complexed with the mRNA transcript (Olsen and Ambros, 1999).

Another target of *lin-4* is *lin-28*, which controls the L2 to L3 transition in *C. elegans*. According to Western blots performed by Seggerson *et al*, wild-type worms have high levels of LIN-28 protein in the L1 stage and about fifteen-fold less LIN-28 in the L3 stage. Data from RNase protection assays, however, indicated that *lin-28* mRNA levels remain the same from the L1 to mid-L3 stages. *lin-28* mRNA at both the L1 and L3 stages was found to be associated with polysomes in polysome profiles, suggesting that *lin-28* is regulated after the initiation of translation since polysomes were already loaded onto the mRNA. Furthermore, in a translational run-off assay, when *lin-28*-associated polysomes were incubated without an inhibitor of translation elongation, *lin-28* mRNAs shifted to lighter, non-polysome-associated fractions in a polysome profile (Seggerson *et al*., 2002). Therefore, the ribosomes bound to *lin-28* are capable of making proteins, which implies that *lin-28* is likely to be in the process of translation in the L3 stage. This supports the idea that *lin-4*-mediated regulation occurs after the formation of the translational complex.

Reduced target protein levels without loss of mRNA have also been observed for the targets of the *Drosophila* miRNA *bantam*, which is highly active in proliferating cells, suggesting a role in the regulation of proliferation (Brennecke *et al*., 2003). A computational screen by Brennecke *et al* identified *hid*, a pro-apoptotic gene, as a potential target of *bantam*, with the *hid* 3′ UTR harbouring five potential *bantam* binding sites. In support of this prediction, *bantam* expression suppressed *hid*-induced apoptosis in eye imaginal discs, while a GFP reporter containing the *hid* 3′ UTR was down-regulated by *bantam*. Endogenous Hid protein levels were also reduced in the presence of *bantam*, while *hid* mRNA levels remained relatively constant (Brennecke *et al*., 2003).

In 2005, Felli *et al* used miRNA microarrays to examine miRNA levels in erythropoietic (E) differentiation and maturation of cultured cord blood (CB) DC34+ haematopoietic progenitor cells (HPCs). It was discovered that miR-221 and -222 levels decreased in E differentiation and maturation, while *kit*, a predicted target of miR-221 and -222, was highly expressed during E differentiation. Co-transfection of miR-221 and/or miR-222 with a luciferase reporter that included the *kit* 3′ UTR resulted in the repression of *luciferase* gene expression. When miR-221 and/or -222 were added to a cell line with high *kit* expression and low miR-221 and -222 levels, *kit* protein levels declined. However, real-time PCR revealed that *kit* mRNA levels were only slightly affected (Felli *et al*., 2005).

In summary, *lin-4*, *bantam*, miR-221, and miR-222 can regulate target gene expression by inhibiting translation without causing a significant, correlative reduction in target mRNA level. In cases where polysome profiles have been examined, the mRNAs are associated with ribosomes, suggesting the target genes have begun translation, but the miRNAs are acting to prevent proteins from being made.

MiRNA-regulated mRNA degradation in animals

Around a year after miRNA-directed mRNA cleavage was discovered in animals, miRNA-mediated target degradation was seen in *C. elegans*. In worms, *let-7* is known to target *lin-41*. Unlike miR-196 and *HOXB8*, *let-7* complementary sites in the *lin-41* 3′ UTR are not nearly perfect matches (Reinhart *et al*., 2000). Bagga *et al* found that RNA from wild-type *C. elegans* at various developmental stages showed that *lin-41* mRNA levels were greatly reduced at the L3 and L4 stages, the time when *let-7* begins accumulating. However, in *let-7*-mutant worms, three times as much *lin-41* mRNA accumulated at the L4 stage, as compared to wild-type worms. In animals carrying a *lacZ* reporter fused to the *lin-41* 3′ UTR with or without *let-7* complementary sites (LCSs), there was a 3-fold reduction of reporter mRNA levels from L2 to L3, which was dependent on the presence of the LCSs, and mRNA levels of the reporter without LCSs were similar at the L2 and L4 stages. Bagga *et al* also observed a ~500 base pair band was found at the L4 stage in Northern blots from wild-type animals using probes specific to sequences downstream of the LCSs. It turns out that the existence of this ~500 nucleotide band depended on the presence of *let-7* and 5′–3′ exonucleases. This suggested that this band was a result of *let-7*-mediated mRNA degradation by exonucleases. Sequencing of these ~500 base pair RNAs revealed that the majority of RNAs were found to end at the 5′ end of LCS1. 5′ fragments also mapped to a location about 40 base pairs downstream of LCS2. Similar results were also seen for *lin-4* and its targets, *lin-14* and *lin-28* (Bagga *et al.*, 2005). Therefore, it appears that miRNA-targeted mRNAs are being degraded around the miRNA binding sites. It is remarkable that these binding sites are not disrupted, and it is tempting to speculate that they may be protected from degradation as a result of the binding of the RISC-complexed miRNA.

Members of the *lin-4* miRNA family, including human miR-125a and miR-125b, are also involved in target degradation. The *lin-28* 3′ UTR is targeted by *miR-125*, as determined by studies from Wu and Belasco using a firefly luciferase reporter with the 3′ UTR of human *lin-28*. Based on homology to the mouse *lin-28* 3′ UTR, two imperfectly complementary sites, miRE1 and miRE2, were identified. In a miR-125a- or miR-125b-dependent manner, luciferase reporters with six miRE1 or miRE2 sites had a significant reduction in protein synthesis, as well as a significant, but smaller, reduction in mRNA levels. The smaller decrease in mRNA levels, as compared to the decrease in protein levels, may signify that not all of the mRNA is being degraded and that some are just being repressed in a manner to prevent translation, such as by localization to P-bodies (see below). Furthermore, the addition of a stem–loop structure in the 5′ UTR of a luciferase reporter inhibited translation, revealing that miR-125b-mediated mRNA degradation was not dependent upon translation (Wu and Belasco, 2005).

It is important to note that cleavage of target mRNAs did not occur within the miRNA binding sites, as had been observed in plants. Therefore, it is possible that miRNAs are causing mRNA degradation at sites outside of their binding sites, possibly by directing exonucleases to the mRNAs. Research in human cells has also shown that degradation is taking place independently from translation (Wu and Belasco, 2005). This differs from earlier work on *lin-4* targets in *C. elegans*, where ribosomes were found loaded onto the *lin-14* and *lin-28* mRNAs (Olsen and Ambros, 1999; Seggerson *et al.*, 2002).

Localization to P-bodies

Recent research has suggested that miRNAs are directing target mRNAs to processing bodies, or P-bodies, also known as cytoplasmic bodies, GW bodies, or Dcp bodies. These cytoplasmic structures are sites for mRNA degradation (reviewed in Fillman and Lykke-Andersen, 2005) (Fillman and Lykke-Andersen, 2005).

Numerous groups found that tagged and endogenous human Argonaute2 (Ago2) were found throughout the cytoplasm and also in cytoplasmic foci (Chu and Rana, 2006; Liu *et al.*, 2005a; Liu *et al.*, 2005b; Sen and Blau, 2005). This localization pattern was also the case for Ago1, -3, and -4 (Liu *et al.*, 2005a). Furthermore, the co-localization of Ago1 and Ago2 was insensitive to RNaseA treatment, which degrades mRNAs and causes P-bodies to disassemble (Liu *et al.*, 2005a; Sen and Blau, 2005). This interaction, therefore, was not dependent upon RNA being present and occurs outside of these foci. *C. elegans* ALG-1 and *Drosophila* AGO1, the Argonaute proteins involved in miRNA-mediated mRNA repression, also localize to cytoplasmic foci (Behm-Ansmant *et al.*, 2006; Ding *et al.*, 2005). Therefore, proteins involved in miRNA-mediated regulation are localized to cytoplasmic foci, and this seems to be conserved across phyla.

Experiments by Liu *et al* established the importance of miRNAs in the localization of mRNAs to these cytoplasmic foci. Luciferase reporters with or without the *let-7* binding sites of the *C. elegans lin-41* 3′ UTR and multiple MS2 coat binding sites in the 3′ UTR were expressed in human cell cultures with fluorescently tagged MS2 to visualize reporter mRNA localization. In the presence of *let-7*, there was half as much luciferase protein generated by the *lin-41*-containing reporter. Moreover, only these mRNAs were co-localized with Myc-Ago2 in distinct cytoplasmic foci. These findings were confirmed with similar results from reporters that had several imperfect binding sites for CXCR4, a miRNA mimic. Reporter mRNAs with these CXCR4 sites only localized to discrete foci with Argonaute proteins when CXCR4 was co-transfected with the reporter (Liu *et al.*, 2005b). Therefore, only mRNAs being targeted by miRNAs localize to Argonaute-containing cytoplasmic foci.

These foci were determined to be P-bodies based on the co-localization of Argonaute proteins with other P-body associated proteins. For example, human Ago1 and Ago2 co-localized with GW182, phosphorylated autoimmune antigen (Liu *et al.*, 2005a; Sen and Blau, 2005). Moreover, Ago1 and Ago2 association with GW182 was not sensitive to RNaseA treatment, so their interaction does not only occur in P-bodies (Liu *et al.*, 2005a). RNaseA treatment, however, resulted in a reduction of Ago-2-containing P-bodies (Sen and Blau, 2005). Cells treated with siRNA against GW182 altered Ago1 and Ago2 localization in the cytoplasm (Liu *et al.*, 2005a). When Liu *et al* co-transfected CXCR4 and a *Renilla* luciferase reporter with six imperfect CXCR4 sites, reporter expression increased when GW182 levels were reduced. Reduction of GW182 also affected the ability of CXCR4 to repress a reporter with a perfect CXCR4 site. Similar results were seen with endogenous *let-7*. Furthermore, a *Renilla* luciferase reporter with Ago2 tethered to it resulted in about a two-fold reduction of reporter expression. However, when the PAZ domain of Ago2 was mutated to inhibit siRNA binding, Ago2 did not localize to P-bodies or reduce target expression (Liu *et al.*, 2005a; Liu *et al.*, 2005b), although it was still found in GW182 immunoprecipitates (Liu *et al.*, 2005a). Therefore, Argonaute-GW182 is insensitive to RNaseA, but Argonaute needs GW182 to be able to localize to P-bodies. Ago2 also needs to interact with miRNAs to localize to P-bodies, and it appears that the localization to GW182-containing P-bodies plays an important role in target repression.

C. *elegans* AIN-1 is homologous to human GW182, and Ding *et al* saw that AIN-1 is expressed in cytoplasmic foci. MiRNAs are associated with these complexes because sequencing of RNA isolated from the immunoprecipitates identified almost thirty miRNAs. Furthermore, ARGONAUTE-LIKE GENE-1 (ALG-1) localization in cytoplasmic foci was dependent on AIN-1 co-expression (Ding *et al.*, 2005). Thus, it appears that AIN-1 may play a role in miRISC and its localization to P-bodies.

In *Drosophila*, Argonaute1 (AGO1) is involved in miRNA-mediated mRNA repression, and Argonaute2 (AGO2) is involved in siRNA-mediated mRNA degradation (Okamura *et al.*, 2004). Behm-Ansmant *et al* compared mRNA profiles gathered from microarray experiments using *Drosophila* Schneider cells (S2 cells) with reduced levels of either GW182, AGO1, or AGO2 and determined that these three proteins were all found to be involved in the regulation of a shared group of mRNAs. This group includes a significant amount of potential and proven miRNA targets. AGO1 forms a complex with GW182 in S2 cells, and it is the N-terminus of GW182 and the PIWI domain of AGO1 that are important for this interaction. Moreover, AGO1 was only expressed in cytoplasmic foci in the presence of GW182. This interaction was only true for AGO1 and not for AGO2, suggesting that this interaction is specific to miRNA-mediated gene regulation. When GW182 was tethered to a firefly luciferase reporter, there was a sixteen-fold decrease in protein expression, as compared to controls, and a four-fold reduction of reporter mRNA levels. Therefore, it appears that GW182 affects both translation and mRNA degradation, although it seems that the latter is not its primary mechanism. In addition, experiments examining mRNA half-lives revealed that GW182-tethered mRNAs had a much shorter half-life. GW182-bound mRNAs were also deadenylated, and this shortening of the poly(A) tail leads to mRNA degradation. Reducing expression levels of the CCR4:NOT deadenylase complex components resulted in some restoration of reporter mRNA levels but without much increase in protein levels. Knocking down the DCP1:DCP2 decapping complex resulted in restored mRNA levels but not in restored *luciferase* expression (Behm-Ansmant *et al.*, 2006). These findings suggest that mRNA reduction may not necessarily be the cause of the decreased protein levels, or that in order to restore protein levels, these mRNAs must be removed from the P-bodies so that translation can occur.

Behm-Ansmant *et al* went on to show that different miRNAs affected target mRNA levels differently. In comparing *luciferase* activity to mRNA levels for a firefly luciferase reporter with the 3' UTR of a miR-9b target, Nerfin, miR-9b mainly had an effect on protein levels. On the other hand, in the case of a reporter with the 3′ UTR of a miR-12 target, CG10011, mRNA degradation accounted for the suppressed levels of protein expression. The repression that took place for each reporter was reversed when the cells lacked GW182 or AGO1, suggesting that this miRNA-mediated repression depended on P-body localization. Cells lacking the DCP1:DCP2 decapping complex had restored mRNA levels, but GW182- and miRNA-targeted mRNAs were still being deadenylated, indicating that deadenylation may not be driving degradation in this instance. Lastly, the reduction of reporter activity and mRNA levels as a result of tethering GW182 to the mRNA was not dependent upon AGO1 expression (Behm-Ansmant et al., 2006). This suggests that mRNA localization to P-bodies is enough to cause translational repression and mRNA degradation. It also appears that different methods are employed by different miRNAs, although the mechanisms for this remain unclear.

Human Ago2 also co-localized with Dcp1a and Dcp2, decapping enzymes found in cytoplasmic bodies (Chu and Rana, 2006; Liu et al., 2005a; Liu et al., 2005b; Sen and Blau, 2005). Immunoprecipitation assays showed that Ago1 and Ago2 were found in a complex with Dcp1a and Dcp2, and these interactions were insensitive to RNaseA (Chu and Rana, 2006; Liu et al., 2005b). Moreover, in C. elegans AIN-1::GFP co-localized with DCAP-2, the C. elegans homologue of human Dcp2, and FLAG-tagged DCAP-1, the C. elegans homologue of human Dcp1 (Ding et al., 2005).

According to co-immunoprecipitation and co-localization experiments by Chu and Rana, human Ago2 is found to associate with additional proteins found in P-bodies, RCK/p54 and eIF4E. The RCK/p54 association with Ago2 was insensitive to RNaseA treatment, while eIF4E association with Ago2 was sensitive to RNaseA treatment. Ago1, Ago2, and RCK/p54 were found in the RISC complex that was primed with biotin-labelled guide strands. Since eIF4E was not found within this complex, but does associate with Ago2, this association likely takes place in the P-bodies. HeLa cells treated with siRNA against RCK/p54 resulted in the disruption of P-bodies and diffuse Ago2 expression. Therefore, it appears that RCK/p54 is needed for Ago2 to localize to P-bodies. In the case of a Renilla luciferase reporter with CXCR4 sites in the 3′ UTR, reduced levels of RCK/p54 affected miRNA-dependent target regulation and not siRNA-mediated target cleavage (Chu and Rana, 2006). Therefore, RCK/p54 appears to be a part of the miRISC. This interaction takes place before the miRISC enters into P-bodies, and RCK/p54 is needed for Ago2 localization to P-bodies.

While Cy3-labeled pre-let-7 localizes to or next to P-bodies (Pillai et al., 2005), mature let-7 labelled with Cy3 co-localized with P-bodies (Pauley et al., 2006). Pauley et al also observed that let-7 was found in immunoprecipitated P-bodies. This suggested that miRNAs play a role in P-body formation, as seen by the decrease in the number and size of P-bodies in cells treated with Drosha RNAi. When lamin A/C siRNA was transfected into Drosha-reduced cells, P-bodies reappeared, to which the siRNA localized, and lamin A/C mRNA and protein levels declined (Pauley et al., 2006). Therefore, the introduction of an siRNA rescued the phenotype of cells with reduced Drosha expression.

In conclusion, mRNAs being repressed by miRNAs are localized to P-bodies, common sites of mRNA degradation. It appears that the formation of some P-bodies may require miRNAs. Furthermore, RISC-associated-Argonaute proteins interact with proteins found in P-bodies. Ago2 interactions with GW182, Dcp1, Dcp2, and RCK/p54 appear to be insensitive to RNaseA treatment. The interaction between Ago2 and eIF4E, however, is sensitive to RNaseA, and this interaction likely takes place within P-bodies. Thus, since animal miRNAs do not usually cause target cleavage at the miRNA binding sites, mRNA localization to P-bodies appears to be a likely way to bring about target gene repression through mRNA sequestration and potential degradation.

MiRNAs inhibit target gene translation

Studies looking to address which stage of translation is inhibited by miRNAs have been performed. In human cells, Pillai et al observed that *let-7* inhibited the initiation of translation. *Renilla* and firefly luciferase reporter constructs with 3′ UTRs containing either one or three sites with imperfect complementarity to *let-7* or one site with perfect complementarity were transfected into HeLa cells, which produce *let-7* in high quantities. The reporter mRNA with the perfectly complementary site was cleaved, and the presence of more than one imperfect miRNA binding site resulted in more efficient down-regulation. A *Renilla* luciferase reporter with an endoplasmic reticulum (ER) signal sequence was used to show that post-translation initiation steps were not targeted by miRNAs, since targeting the reporter to the ER prevented degradation of the nascent polypeptide. Neither the endogenous *let-7* miRNP nor human Ago2 tethered to the reporter mRNA affected the ER-targeted reporter expression as compared to a non-ER-targeted reporter. Furthermore, polysome profiles revealed that the reporter mRNAs that were either tethered with Ago2 or down-regulated by *let-7* shifted to the top of the gradient, signifying that polysomes were not loaded onto them. This finding suggested that miRNAs could be inhibiting ribosome loading, and thus, translation initiation. In order to take a closer look at translation initiation, m^7G-capped luciferase reporters with either a perfectly complementary or three imperfectly complementary *let-7* sites were made with or without a poly(A) tail and transfected into HeLa cells. The results suggested that repression by *let-7* was not dependent on the poly(A) tail. Next, an internal ribosome entry site (IRES) from the encephalomocarditis virus (EMCV) or the hepatitis C virus (HCV) was used to examine whether or not *let-7* repression was cap dependent, since IRESs bypass the normal translation initiation steps. The IRES-containing reporter constructs were not down-regulated by *let-7*. Furthermore, in the case of a dicistronic reporter consisting of firefly *luciferase* followed by *Renilla luciferase*, whose translation was driven by tethered eIF4E or eIF4G, parts of the protein complex that bind to the mRNA cap, and three imperfect *let-7* binding sites, *let-7* repressed the expression of the firefly *luciferase* but not the *Renilla luciferase* genes. Therefore, *let-7* appears to be acting at the level of translation initiation, before eIF4E

recruits eIF4G. *In situ* hybridizations and cellular fractionations revealed that luciferase constructs with three imperfect *let-7* sites also localized to P-bodies in a *let-7*-dependent manner. Upon closer inspection, it was determined that RNA was localized adjacently to the Dcp1 foci. This finding suggested that there might be compartmentalization within P-bodies. The reporter with a perfectly complementary *let-7* site, however, was not found in P-bodies (Pillai *et al.*, 2005). Therefore, *let-7* appears to be blocking the initiation of translation and localizing target mRNAs with imperfect binding sites to P-bodies.

Additional evidence that miRNAs were acting at the level of translation initiation came from studies suggesting that when the m^7G 5′ capped- and poly(A) tailed-mRNA of a *Renilla* luciferase reporter consisting of a 3′ UTR with four imperfectly complementary binding sites for CXCR4 was transfected with CXCR4 into HeLa cells, *luciferase* expression was repressed. CXCR4 did not regulate *luciferase* expression of a reporter construct that bypassed typical translation initiation as a result of placing an IRES from cricket paralysis virus (CrPV) in front of the *Renilla luciferase*-coding region; this construct also lacked a poly(A) tail. These results suggested that miRNAs are regulating translation initiation. Reporter constructs with or without a poly(A) tail were created with a m^7G cap or an A-cap, which is not functional in translation, to examine whether or not the poly(A) tail is important for miRNA regulation of translation initiation. Constructs with only the m^7G cap or poly(A) tail were less sensitive to the CXCR4 miRNA than the construct with both the m^7G cap and poly(A) tail. The construct with the A-cap and no poly(A) tail was insensitive to CXCR4. Furthermore, the down-regulation of the mRNA constructs in the presence of CXCR4 was not the result of message degradation. Lastly, luciferase reporters with an A-cap and EMCV IRES were built with or without a poly(A) tail. The EMCV IRES-driven translation initiation involves typical initiation factors with the exception of eIF4E. The construct with only an EMCV IRES was not down-regulated by the CXCR4 miRNA, while the construct that also contained a poly(A) tail was down-regulated by CXCR4 at levels like the luciferase construct with an A-cap and poly(A) tail (Humphreys *et al.*, 2005). Thus, CXCR4 appears to be involved in inhibiting translation initiation through eIF4E, confirming the findings of Pillai *et al* involving *let-7* (Pillai *et al.*, 2005).

Based on these findings, miRNAs seem to inhibit the initiation of translation. MiRNAs did not repress IRES-driven reporters. Thus, it looks as if that miRNAs can inhibit protein production by acting at the level of eIF4E recruitment to the m^{7}G cap. Furthermore, target down-regulation may not depend on the poly(A) tail. These studies also support the idea that miRNA-directed target degradation could occur in P-bodies, since reporters with perfectly complementary *let-7* sites did not localize to P-bodies.

MiRNAs can also repress target genes by inhibiting elongation

In contrast to the aforementioned findings regarding miRNA-mediated inhibition of translation initiation, research has also shown that miRNAs can target translation after initiation (Maroney *et al.*, 2006; Nottrott *et al.*, 2006; Petersen *et al.*, 2006). Petersen *et al* co-transfected CXCR4 and a luciferase construct with six imperfect CXCR4 sites in its 3′ UTR and observed down-regulation of reporter gene expression, but this occurred without a decrease in mRNA levels sufficient enough to account for the down-regulation. However, the reported mRNAs, whether being down-regulated or not, associated with polyribosomes in a polyribosome profile. After treating cells with puromycin, a drug that interferes with protein synthesis, down-regulated and non-down-regulated mRNAs shifted in their polysome profiles to fewer polysomes as a result of ribosome removal from the transcript (Petersen *et al.*, 2006). This implied that these miRNA-targeted mRNAs were being actively translated, as was the case for *C. elegans lin-14* and *lin-28* (Olsen and Ambros, 1999; Seggerson *et al.*, 2002). When a bicistronic reporter where either an HCV or CrPV IRES separated the firefly and *Renilla luciferase* coding regions was co-transfected with CXCR4, Petersen *et al* saw that both luciferase gene expressions were simi-

larly down-regulated (Petersen *et al.*, 2006). This finding differed from other work that had shown that miRNAs do not repress IRES-driven reporter expression (Humphreys *et al.*, 2005; Pillai *et al.*, 2005). A pulse-chase experiment using labelled methionine and a firefly reporter with six imperfect CXCR4 sites and a FLAG epitope was then performed by Petersen *et al* to examine protein synthesis. In the case of the CXCR4-regulated reporter, neither full-length proteins nor nascent polypeptides were found. This suggests that CXCR4 is acting before the completion of protein synthesis. They then examined miRNAs and termination of translation through a bicistronic reporter consisting of the firefly and *Renilla luciferase* coding regions separated by three stop codons and six imperfect CXCR4 sites follow the Renilla *luciferase* coding region. CXCR4 caused a slight decrease in translation read-through, suggesting that it may promote premature termination or premature removal of ribosomes. Further examination of this idea revealed that a *luciferase* reporter targeted by CXCR4 and treated with hippuristanol, an eIF4A inhibitor, was more quickly lost from the polysome region of a polysome profile than a reporter targeted by a control small RNA (Petersen *et al.*, 2006). These results suggest that ribosomes are removed faster from short RNA-repressed mRNAs than non-repressed mRNAs.

In HeLa cells, Maroney *et al* saw that miR-21, miR-16, and *let-7* co-sedimented with polysomes. Exposed regions of mRNAs were then digested with a mircococcal nuclease, leaving ribosomes intact, and miR-21 no longer sedimented with ribosomes. Therefore, miRNA-association with polysomes appears dependent upon mRNAs. Next, cells were treated with pactamycin, which blocked translation initiation, resulting in an increase in monosomes and a decrease in polysomes. Just as mRNAs shifted to fractions with fewer ribosomes, miR-21 did so as well, suggesting that miR-21 was bound to actively translating mRNAs. Similar results were seen in cells treated with puromycin or with a hypertonic media, which halts protein synthesis. The effects of the hypertonic media can be reversed using isotonic media, resulting in the recovery of protein synthesis. mRNAs and miRNAs were then found in heavy polysome fractions, suggesting that proteins can be synthesized from miRNA-bound mRNAs (Maroney *et al.*, 2006). They then went on to show that *KRAS* mRNA, which is targeted by *let-7* (Johnson *et al.*, 2005), also followed similar patterns and was associated with polysomes capable of translation (Maroney *et al.*, 2006).

Nottrott *et al* transfected HeLa cells with a Myc-tagged *Renilla* luciferase reporter containing the *C. elegans lin-41* 3′ UTR. Luciferase activity was dependent upon the presence *let-7*; there was a ~3.5-fold reduction of protein levels and a ~1.4-fold reduction in RNA levels when the LCSs of the *lin-41* 3′ UTR were present. Like wild-type controls, reporter mRNAs with LCSs were found to be associated with polysomes. Furthermore, assays with puromycin revealed that these RNAs were being actively translated. They also found that AGO, a part of the *let-7*-miRISC, was associated with the actively translating polysomes. Next, they utilized an iron response element (IRE) stem–loop structure and its cognate RNA-binding protein, iron regulatory protein-1 (IRP-1), as a translational-switch system. In the absence of iron, IRP-1 prevented the cap-dependent translation initiation of reporters. However, in the case of an IRE reporter with the *lin-41* 3′ UTR, there was also a nearly undetectable protein expression level in the presence of iron. Moreover, mRNA levels of this construct in the absence or presence of iron were ~1.5-fold less than the control construct. The IRE reporter constructs were only found to be associated with actively translating polysomes in the presence of iron. Lastly, they used Myc-tagged luciferase reporters and found that immunoprecipitation with Myc antibodies pulled down luciferase RNA only for reporters lacking the *lin-41* 3′ UTR. Thus, *let-7* may be destroying the newly translated peptides or blocking these peptides for later degradation (Nottrott *et al.*, 2006).

These studies suggest that miRNA-targeted mRNAs are loaded with polysomes, which contradict other findings that show that polysomes are not found on transcripts being targeted by miRNAs (Pillai *et al.*, 2005). Therefore, miRNAs may prevent protein production by preventing ribosomes from completely reading the mRNA in only some cases.

MiRNA-mediated deadenylation

Another possible mechanism for miRNA-mediated repression is through the deadenylation of target mRNAs. Giraldez *et al* have shown that miRNAs are involved in deadenylation and removal of maternal mRNAs in zebrafish. A GFP reporter with three imperfect miR-430 sites in the 3′ UTR was degraded in wild-type animals after miR-430 accumulation. The degradation was delayed in MZ*dicer* mutants that lack miR-430, but degradation was restored when miR-430 was injected into these mutants. This result was also true for miR-1 and a corresponding reporter construct. Based on a microarray performed to identify potential targets of miR-430, about 40% of the miR-430 target set consists of maternally deposited mRNAs. Likewise, about 40% of maternal mRNAs have miR-430 sites in their 3′ UTRs, suggesting that miR-430 could be involved in the removal of maternal mRNAs from embryos. Furthermore, target deadenylation was accelerated by miR-430 and miR-1, as mutations in the miRNA binding sites in the 3′ UTRs of targets slowed down target deadenylation. MiRNA-induced deadenylation did not depend on translation because when translation was blocked by morpholinos against the translational start site of a GFP reporter, deadenylation still occurred (Giraldez *et al.*, 2006).

miR-125b, a *lin-4* homolog, has been shown to target and reduce mRNA levels of *lin-28* and luciferase reporter constructs with 3′ UTRs containing copies of miRE1 or miRE2 (imperfect miR125-b complementary elements found in the *lin-28* 3′ UTR) in P19 mouse embryonic carcinoma cells and 293T human embryonic kidney cells, respectively (Wu and Belasco, 2005). Wu *et al* found that when a β-globin reporter driven by the *c-fos* promoter contained two miRE1 sites in its 3′ UTR, degradation of the reporter was accelerated. Adding additional copies of miRE1 further accelerated degradation. There were reduced *luciferase* protein and mRNA levels for the miRE1-containing reporter, but only the mRNA with a perfectly complementary miR-125b binding site was cleaved. Furthermore, the β-globin reporter with two miRE1 sites got smaller over time as a result of shortening of the poly(A) tail, ultimately leading to mRNA degradation. These mRNAs undergoing miR-125b-induced dead-enylation still retained their 5' caps. Without miR-125b present, when a hairpin structure was inserted within the 5' UTR of the β-globin reporter containing two miRE1 sites to block translation initiation, there was only a minimal increase in poly(A) degradation. However, in the presence miR-125b, deadenylation and degradation occurred much more rapidly. On the other hand, when the poly(A) tail of a luci-ferase reporter containing six miRE1 sites was replaced with a 3'-terminal hairpin structure, the translation efficiency was equal for the reporters with either a poly(A) tail or a 3' hairpin, even though the overall amount of repression was higher for the reporter with the poly(A) tail. Wu et al also found that when six copies of the *let-7* binding site in the *lin-28* 3' UTR were put into a luciferase reporter, protein synthesis was down-regulated but mRNA levels only slightly decreased as a result of more rapid deadenylation and degradation in the presence of *let-7* (Wu *et al.*, 2006). Thus, it appears that deadenylation and decay is dependent on miRNA::binding site interaction, while translational repression is not dependent on the presence of a poly(A) tail.

Thus, miRNAs can promote target gene deg-radation through mRNA deadenylation, but the deadenylation is not dependent upon translation. The poly(A) tail, however, is not needed for trans-lational repression. This supports the research in human cells that has found that *let-7*-directed gene repression does not depend on the presence of a poly(A) tail (Pillai *et al.*, 2005). However, in *Drosophila*, it has been shown that mRNAs teth-ered to GW182 are not deadenylated, suggesting that localization to P-bodies does not result in deadenylation (Behm-Ansmant *et al.*, 2006).

MiRNA-mediated mRNA instability

Research also has shown that miRNAs play a role in AU-rich element (ARE)-mediated mRNA instability. AREs are found in the 3' UTRs of short-lived mRNAs. These elements have been known to play a role in mRNA stability.

In a screen to identify genes involved in ARE-RNA degradation in *Drosophila* S2 cells, *Dicer1* was found among their list of candidate genes.

Based on the involvement of *Dicer1* in miRNA processing, it was determined that *Drosophila Ago1* and *Ago2* were also needed for ARE-RNA decay in S2 cells. When HeLa cells were transfected with siRNAs or 2'-O-methyl oligo-nucleotides against miR-16–1, ARE-containing β-globin reporter mRNAs were more stable than the control. Furthermore, overexpressing miR-16–1 increased the overall instability of ARE-RNAs, whereas mutated miR-16–1 did not play a role in reporter ARE-RNA destabili-zation. However, analysis of miR-16–1-mediated instability of reporter constructs with AREs from different genes revealed that not all ARE-RNAs are destabilized by miR-16–1. TTP is a protein that plays a role ARE-RNA degradation. When TTP levels were reduced by siRNA, the ARE-containing reporter was not destabilized by miR-16–1. The opposite was also true; when TTP was overexpressed, reporter stability was only reduced when miR-16–1 was present. While TTP did not bind to miR-16 directly, miR-16–1 was found in the RNA extracted from the TTP-precipitated complex. Human Ago2, a component of the RISC complex, also co-immunoprecipitated with TTP from HeLa cells (Jing *et al.*, 2005). These findings suggest that TTP is found in a complex with Ago2; however, the precise nature of their interaction remains to be discovered.

The exact role *Drosophila* miR-16–1 plays in ARE-RNA instability is still unclear. However, the idea that miRNAs could be involved in such a process is further supported by the finding that human Ago2 is complexed with TTP in HeLa cells.

Conclusions

MiRNAs are small, regulatory RNAs that function post-transcriptionally. MiRNA are encoded in plant and animal genomes. After several processing steps, they bind to their target mRNAs in either the coding regions in plants or the 3' UTRs in animals.

Plant miRNAs bind with perfect comple-mentarity, which generally results in mRNA cleavage within the miRNA::miRNA binding site. However, this is not necessarily the only mechanism of target down-regulation, as plant miRNAs may repress target translation without mRNA degradation.

In animals, however, the roles of miRNAs in target repression are much more varied. The earliest work on the role of *lin-4* found that *lin-4* inhibited protein production without reducing protein levels significantly. However, research since that time suggests that there is not one mechanism that will explain all miRNA-meditated gene regulation. MiRNAs have been shown to mediate target mRNA cleavage and degradation, inhibit the initiation and elongation of target gene translation, promote target mRNA deadenylation, and mediate ARE-mRNA instability. MiRNA-repressed mRNAs may also be localized to cytoplasmic P-bodies, which are sites of mRNA degradation. However, with all of these potential mechanisms for miRNA-mediated gene repression, there have been conflicting findings. For example, research has found that miRNA-targeted mRNAs may or may not have polysomes bound to them. Likewise, there have been conflicting findings as to whether or not miRNA-targeted mRNAs get deadenylated. There are also questions regarding whether or not the initiation of translation even begins on miRNA-targeted mRNAs. Therefore, it is most likely that there is more than one mechanism through which miRNAs regulate gene expression. Determining how and when these mechanisms of suppression are employed by individual miRNAs remains one of the greatest challenges in the field.

References

Achard, P., Herr, A., Baulcombe, D.C., and Harberd, N.P. (2004). Modulation of floral development by a gibberellin-regulated microRNA. Development *131*, 3357–3365.

Aukerman, M.J., and Sakai, H. (2003). Regulation of flowering time and floral organ identity by a micro-RNA and its APETALA2-like target genes. Plant Cell *15*, 2730–2741.

Bagga, S., Bracht, J., Hunter, S., Massirer, K., Holtz, J., Eachus, R., and Pasquinelli, A.E. (2005). Regulation by *let-7* and *lin-4* miRNAs results in target mRNA degradation. Cell *122*, 553–563.

Behm-Ansmant, I., Rehwinkel, J., Doerks, T., Stark, A., Bork, P., and Izaurralde, E. (2006). mRNA degradation by miRNAs and GW182 requires both CCR4:NOT deadenylase and DCP1:DCP2 decapping complexes. Genes Dev. *20*, 1885–1898.

Boehm, M., and Slack, F. (2005). A developmental timing microRNA and its target regulate life span in C. *elegans*. Science *310*, 1954–1957.

Bohnsack, M.T., Czaplinski, K., and Gorlich, D. (2004). Exportin 5 is a RanGTP-dependent dsRNA-binding protein that mediates nuclear export of pre-miRNAs. RNA *10*, 185–191.

Bollman, K.M., Aukerman, M.J., Park, M.Y., Hunter, C., Berardini, T.Z., and Poethig, R.S. (2003). HASTY, the *Arabidopsis* ortholog of exportin 5/MSN5, regulates phase change and morphogenesis. Development *130*, 1493–1504.

Brennecke, J., Hipfner, D.R., Stark, A., Russell, R.B., and Cohen, S.M. (2003). *bantam* encodes a developmentally regulated microRNA that controls cell proliferation and regulates the proapoptotic gene *hid* in *Drosophila*. Cell *113*, 25–36.

Cai, X., Hagedorn, C.H., and Cullen, B.R. (2004). Human microRNAs are processed from capped, polyadenylated transcripts that can also function as mRNAs. RNA *10*, 1957–1966.

Chen, X. (2004). A microRNA as a translational repressor of APETALA2 in *Arabidopsis* flower development. Science *303*, 2022–2025.

Chendrimada, T.P., Gregory, R.I., Kumaraswamy, E., Norman, J., Cooch, N., Nishikura, K., and Shiekhattar, R. (2005). TRBP recruits the Dicer complex to Ago2 for microRNA processing and gene silencing. Nature *436*, 740–744.

Chu, C.Y., and Rana, T.M. (2006). Translation repression in human cells by microRNA-induced gene silencing requires RCK/p54. PLoS Biol. *4*, e210.

Denli, A.M., Tops, B.B., Plasterk, R.H., Ketting, R.F., and Hannon, G.J. (2004). Processing of primary microRNAs by the Microprocessor complex. Nature *432*, 231–235.

Ding, L., Spencer, A., Morita, K., and Han, M. (2005). The developmental timing regulator AIN-1 interacts with miRISCs and may target the argonaute protein ALG-1 to cytoplasmic P bodies in C. *elegans*. Mol. Cell *19*, 437–447.

Doench, J.G., Petersen, C.P., and Sharp, P.A. (2003). siRNAs can function as miRNAs. Genes Dev. *17*, 438–442.

Enright, A.J., John, B., Gaul, U., Tuschl, T., Sander, C., and Marks, D.S. (2003). MicroRNA targets in *Drosophila*. Genome Biol. *5*, R1.

Felli, N., Fontana, L., Pelosi, E., Botta, R., Bonci, D., Facchiano, F., Liuzzi, F., Lulli, V., Morsilli, O., Santoro, S., et al. (2005). MicroRNAs 221 and 222 inhibit normal erythropoiesis and erythroleukemic cell growth via kit receptor down-modulation. Proc. Natl. Acad. Sci. USA *102*, 18081–18086.

Fillman, C., and Lykke-Andersen, J. (2005). RNA decapping inside and outside of processing bodies. Curr. Opin. Cell. Biol. *17*, 326–331.

Giraldez, A.J., Mishima, Y., Rihel, J., Grocock, R.J., Van Dongen, S., Inoue, K., Enright, A.J., and Schier, A.F. (2006). Zebrafish MiR-430 Promotes Deadenylation and Clearance of Maternal mRNAs. Science *312*, 75–79.

Grad, Y., Aach, J., Hayes, G.D., Reinhart, B.J., Church, G.M., Ruvkun, G., and Kim, J. (2003). Computational and experimental identification of C. elegans microRNAs. Mol. Cell 11, 1253–1263.

Gregory, R.I., Yan, K.P., Amuthan, G., Chendrimada, T., Doratotaj, B., Cooch, N., and Shiekhattar, R. (2004). The Microprocessor complex mediates the genesis of microRNAs. Nature 432, 235–240.

Grishok, A., Pasquinelli, A.E., Conte, D., Li, N., Parrish, S., Ha, I., Baillie, D.L., Fire, A., Ruvkun, G., and Mello, C.C. (2001). Genes and mechanisms related to RNA interference regulate expression of the small temporal RNAs that control C. elegans developmental timing. Cell 106, 23–34.

Han, J., Lee, Y., Yeom, K.H., Kim, Y.K., Jin, H., and Kim, V.N. (2004). The Drosha-DGCR8 complex in primary microRNA processing. Genes Dev. 18, 3016–3027.

Humphreys, D.T., Westman, B.J., Martin, D.I., and Preiss, T. (2005). MicroRNAs control translation initiation by inhibiting eukaryotic initiation factor 4E/cap and poly(A) tail function. Proc. Natl. Acad. Sci. USA 102, 16961–16966.

Hutvágner, G., McLachlan, J., Pasquinelli, A.E., Balint, É., Tuschl, T., and Zamore, P.D. (2001). A cellular function for the RNA-interference enzyme Dicer in the maturation of the let-7 small temporal RNA. Science 293, 834–838.

Hutvágner, G., and Zamore, P.D. (2002). A microRNA in a multiple-turnover RNAi enzyme complex. Science 297, 2056–2060.

Jing, Q., Huang, S., Guth, S., Zarubin, T., Motoyama, A., Chen, J., Di Padova, F., Lin, S.C., Gram, H., and Han, J. (2005). Involvement of microRNA in AU-rich element-mediated mRNA instability. Cell 120, 623–634.

John, B., Enright, A.J., Aravin, A., Tuschl, T., Sander, C., and Marks, D.S. (2004). Human MicroRNA targets. PLoS Biol. 2, e363.

Johnson, S.M., Grosshans, H., Shingara, J., Byrom, M., Jarvis, R., Cheng, A., Labourier, E., Reinert, K.L., Brown, D., and Slack, F.J. (2005). RAS is regulated by the let-7 microRNA family. Cell 120, 635–647.

Khvorova, A., Reynolds, A., and Jayasena, S.D. (2003). Functional siRNAs and miRNAs exhibit strand bias. Cell 115, 209–216.

Kiriakidou, M., Nelson, P.T., Kouranov, A., Fitziev, P., Bouyioukos, C., Mourelatos, Z., and Hatzigeorgiou, A. (2004). A combined computational-experimental approach predicts human microRNA targets. Genes Dev. 18, 1165–1178.

Kurihara, Y., and Watanabe, Y. (2004). Arabidopsis micro-RNA biogenesis through Dicer-like 1 protein functions. Proc. Natl. Acad. Sci. USA 101, 12753–12758.

Lagos-Quintana, M., Rauhut, R., Lendeckel, W., and Tuschl, T. (2001). Identification of novel genes coding for small expressed RNAs. Science 294, 853–858.

Lagos-Quintana, M., Rauhut, R., Yalcin, A., Meyer, J., Lendeckel, W., and Tuschl, T. (2002). Identification of tissue-specific microRNAs from mouse. Curr. Biol. 12, 735–739.

Lall, S., Grun, D., Krek, A., Chen, K., Wang, Y.L., Dewey, C.N., Sood, P., Colombo, T., Bray, N., Macmenamin, P., et al. (2006). A Genome-Wide Map of Conserved MicroRNA Targets in C. elegans. Curr. Biol. 16, 460–471.

Landthaler, M., Yalcin, A., and Tuschl, T. (2004). The human DiGeorge syndrome critical region gene 8 and its D. melanogaster homolog are required for miRNA biogenesis. Curr. Biol. 14, 2162–2167.

Lau, N.C., Lim, L.P., Weinstein, E.G., and Bartel, D.P. (2001). An abundant class of tiny RNAs with probable regulatory roles in Caenorhabditis elegans. Science 294, 858–862.

Lee, R.C., Feinbaum, R.L., and Ambros, V. (1993). The C. elegans heterochronic gene lin-4 encodes small RNAs with antisense complementarity to lin-14. Cell 75, 843–854.

Lee, Y., Ahn, C., Han, J., Choi, H., Kim, J., Yim, J., Lee, J., Provost, P., Radmark, O., Kim, S., and Kim, V.N. (2003). The nuclear RNase III Drosha initiates microRNA processing. Nature 425, 415–419.

Lee, Y., Kim, M., Han, J., Yeom, K.H., Lee, S., Baek, S.H., and Kim, V.N. (2004a). MicroRNA genes are transcribed by RNA polymerase II. EMBO J. 23, 4051–4060.

Lee, Y.S., Nakahara, K., Pham, J.W., Kim, K., He, Z., Sontheimer, E.J., and Carthew, R.W. (2004b). Distinct roles for Drosophila Dicer-1 and Dicer-2 in the siRNA/miRNA silencing pathways. Cell 117, 69–81.

Lewis, B.P., Shih, I.H., Jones-Rhoades, M.W., Bartel, D.P., and Burge, C.B. (2003). Prediction of mammalian microRNA targets. Cell 115, 787–798.

Lim, L.P., Lau, N.C., Weinstein, E.G., Abdelhakim, A., Yekta, S., Rhoades, M.W., Burge, C.B., and Bartel, D.P. (2003). The microRNAs of Caenorhabditis elegans. Genes Dev. 17, 991–1008.

Lingel, A., Simon, B., Izaurralde, E., and Sattler, M. (2004). Nucleic acid 3′-end recognition by the Argonaute2 PAZ domain. Nat. Struct. Mol. Biol. 11, 576–577.

Liu, J., Carmell, M.A., Rivas, F.V., Marsden, C.G., Thomson, J.M., Song, J.J., Hammond, S.M., Joshua-Tor, L., and Hannon, G.J. (2004). Argonaute2 is the catalytic engine of mammalian RNAi. Science 305, 1437–1441.

Liu, J., Rivas, F.V., Wohlschlegel, J., Yates, J.R., 3rd, Parker, R., and Hannon, G.J. (2005a). A role for the P-body component GW182 in microRNA function. Nat. Cell. Biol. 7, 1261–1266.

Liu, J., Valencia-Sanchez, M.A., Hannon, G.J., and Parker, R. (2005b). MicroRNA-dependent localization of targeted mRNAs to mammalian P-bodies. Nat. Cell. Biol. 7, 719–723.

Liu, Q., Rand, T.A., Kalidas, S., Du, F., Kim, H.E., Smith, D.P., and Wang, X. (2003). R2D2, a bridge between the initiation and effector steps of the Drosophila RNAi pathway. Science 301, 1921–1925.

Llave, C., Xie, Z., Kasschau, K.D., and Carrington, J.C. (2002). Cleavage of Scarecrow-like mRNA targets directed by a class of Arabidopsis miRNA. Science 297, 2053–2056.

Lund, E., Guttinger, S., Calado, A., Dahlberg, J.E., and Kutay, U. (2004). Nuclear export of microRNA precursors. Science *303*, 95–98.

Ma, J.B., Ye, K., and Patel, D.J. (2004). Structural basis for overhang-specific small interfering RNA recognition by the PAZ domain. Nature *429*, 318–322.

Mallory, A.C., Bartel, D.P., and Bartel, B. (2005). MicroRNA-directed regulation of *Arabidopsis AUXIN RESPONSE FACTOR17* is essential for proper development and modulates expression of early auxin response genes. Plant Cell *17*, 1360–1375.

Maroney, P.A., Yu, Y., Fisher, J., and Nilsen, T.W. (2006). Evidence that microRNAs are associated with translating messenger RNAs in human cells. Nat. Struct. Mol. Biol. *13*, 1102–1107.

Matranga, C., Tomari, Y., Shin, C., Bartel, D.P., and Zamore, P.D. (2005). Passenger-strand cleavage facilitates assembly of siRNA into Ago2-containing RNAi enzyme complexes. Cell *123*, 607–620.

Meister, G., Landthaler, M., Patkaniowska, A., Dorsett, Y., Teng, G., and Tuschl, T. (2004). Human Argonaute2 mediates RNA cleavage targeted by miRNAs and siRNAs. Mol. Cell *15*, 185–197.

Mourelatos, Z., Dostie, J., Paushkin, S., Sharma, A., Charroux, B., Abel, L., Rappsilber, J., Mann, M., and Dreyfuss, G. (2002). miRNPs: a novel class of ribonucleoproteins containing numerous microRNAs. Genes Dev. *16*, 720–728.

Nottrott, S., Simard, M.J., and Richter, J.D. (2006). Human let-7a miRNA blocks protein production on actively translating polyribosomes. Nat. Struct. Mol. Biol. *13*, 1108–1114.

Okamura, K., Ishizuka, A., Siomi, H., and Siomi, M.C. (2004). Distinct roles for Argonaute proteins in small RNA-directed RNA cleavage pathways. Genes Dev. *18*, 1655–1666.

Olsen, P.H., and Ambros, V. (1999). The lin-4 regulatory RNA controls developmental timing in *Caenorhabditis elegans* by blocking LIN-14 protein synthesis after the initiation of translation. Dev. Biol. *216*, 671–680.

Papp, I., Mette, M.F., Aufsatz, W., Daxinger, L., Schauer, S.E., Ray, A., van der Winden, J., Matzke, M., and Matzke, A.J. (2003). Evidence for nuclear processing of plant micro RNA and short interfering RNA precursors. Plant Physiol. *132*, 1382–1390.

Park, W., Li, J., Song, R., Messing, J., and Chen, X. (2002). CARPEL FACTORY, a Dicer homolog, and HEN1, a novel protein, act in microRNA metabolism in *Arabidopsis thaliana*. Curr. Biol. *12*, 1484–1495.

Pasquinelli, A.E., Reinhart, B.J., Slack, F., Martindale, M.Q., Kuroda, M.I., Maller, B., Hayward, D.C., Ball, E.E., Degnan, B., Muller, P., et al. (2000). Conservation of the sequence and temporal expression of *let-7* heterochronic regulatory RNA. Nature *408*, 86–89.

Pauley, K.M., Eystathioy, T., Jakymiw, A., Hamel, J.C., Fritzler, M.J., and Chan, E.K. (2006). Formation of GW bodies is a consequence of microRNA genesis. EMBO Rep. *7*, 904–910.

Petersen, C.P., Bordeleau, M.E., Pelletier, J., and Sharp, P.A. (2006). Short RNAs repress translation after initiation in mammalian cells. Mol. Cell *21*, 533–542.

Pillai, R.S., Bhattacharyya, S.N., Artus, C.G., Zoller, T., Cougot, N., Basyuk, E., Bertrand, E., and Filipowicz, W. (2005). Inhibition of translational initiation by *let-7* microRNA in human cells. Science *309*, 1573–1576.

Rand, T.A., Petersen, S., Du, F., and Wang, X. (2005). Argonaute2 cleaves the anti-guide strand of siRNA during RISC activation. Cell *123*, 621–629.

Reinhart, B.J., Slack, F.J., Basson, M., Pasquinelli, A.E., Bettinger, J.C., Rougvie, A.E., Horvitz, H.R., and Ruvkun, G. (2000). The 21-nucleotide *let-7* RNA regulates developmental timing in *Caenorhabditis elegans*. Nature *403*, 901–906.

Schwarz, D.S., Hutvagner, G., Du, T., Xu, Z., Aronin, N., and Zamore, P.D. (2003). Asymmetry in the assembly of the RNAi enzyme complex. Cell *115*, 199–208.

Seggerson, K., Tang, L., and Moss, E.G. (2002). Two genetic circuits repress the *Caenorhabditis elegans* heterochronic gene *lin-28* after translation initiation. Dev. Biol. *243*, 215–225.

Sen, G.L., and Blau, H.M. (2005). Argonaute 2/RISC resides in sites of mammalian mRNA decay known as cytoplasmic bodies. Nat. Cell. Biol. *7*, 633–636.

Sethupathy, P., Corda, B., and Hatzigeorgiou, A.G. (2006). TarBase: A comprehensive database of experimentally supported animal microRNA targets. RNA *12*, 192–197.

Slack, F.J., Basson, M., Liu, Z., Ambros, V., Horvitz, H.R., and Ruvkun, G. (2000). The *lin-41* RBCC gene acts in the *C. elegans* heterochronic pathway between the *let-7* regulatory RNA and the LIN-29 transcription factor. Mol. Cell *5*, 659–669.

Song, J.J., Smith, S.K., Hannon, G.J., and Joshua-Tor, L. (2004). Crystal structure of Argonaute and its implications for RISC slicer activity. Science *305*, 1434–1437.

Souret, F.F., Kastenmayer, J.P., and Green, P.J. (2004). AtXRN4 degrades mRNA in *Arabidopsis* and its substrates include selected miRNA targets. Mol. Cell *15*, 173–183.

Tomari, Y., Matranga, C., Haley, B., Martinez, N., and Zamore, P.D. (2004). A protein sensor for siRNA asymmetry. Science *306*, 1377–1380.

Wightman, B., Ha, I., and Ruvkun, G. (1993). Posttranscriptional regulation of the heterochronic gene *lin-14* by *lin-4* mediates temporal pattern formation in *C. elegans*. Cell *75*, 855–862.

Wu, L., and Belasco, J.G. (2005). Micro-RNA regulation of the mammalian lin-28 gene during neuronal differentiation of embryonal carcinoma cells. Mol. Cell. Biol. *25*, 9198–9208.

Wu, L., Fan, J., and Belasco, J.G. (2006). MicroRNAs direct rapid deadenylation of mRNA. Proc. Natl. Acad. Sci. USA *103*, 4034–4039.

Xu, P., Vernooy, S.Y., Guo, M., and Hay, B.A. (2003). The *Drosophila* microRNA Mir-14 suppresses cell death and is required for normal fat metabolism. Curr. Biol. *13*, 790–795.

Yan, K.S., Yan, S., Farooq, A., Han, A., Zeng, L., and Zhou, M.M. (2003). Structure and conserved RNA binding of the PAZ domain. Nature *426*, 468–474.

Yekta, S., Shih, I.H., and Bartel, D.P. (2004). MicroRNA-directed cleavage of *HOXB8* mRNA. Science *304*, 594–596.

Yi, R., Qin, Y., Macara, I.G., and Cullen, B.R. (2003). Exportin-5 mediates the nuclear export of pre-microRNAs and short hairpin RNAs. Genes Dev. *17*, 3011–3016.

Zeng, Y., Yi, R., and Cullen, B.R. (2005). Recognition and cleavage of primary microRNA precursors by the nuclear processing enzyme Drosha. EMBO J. *24*, 138–148.

Viral Infection-Related MicroRNAs in Viral and Host Genomic Evolution

7

Yoichi R. Fujii and Nitin K. Saksena

> Valleys in which the fruit blossoms are fragrant pink and white waters in a shallow sea.
>
> John Steinbeck, *The Grapes of Wrath*

Abstract

MicroRNA (miRNA) is a small RNA. The miRNA genes have been discovered in viruses, plants, invertebrates, and vertebrates. These miRNAs can regulate gene expression to inhibit translation of target mRNAs, and sometimes direct many rounds of site-specific mRNA cleavage by RNA interference (RNAi) machinery in mammalian cells. However, role of the viral miRNA for gene regulation is not still well mechanistically understood as compared with host miRNAs. Conversely, the computational methods used to detect and predict the miRNA target genes are common ones in genome informatics. We describe here about the target prediction of a viral miRNA, miR-N367 and the conservation of secondary structure of pre-miR-N367 into two human miRNAs whereas their targets in HIV-1 genome could be related with HIV-1 transcriptional system. Thus, we hypothesize that the orphaned non-selfish miRNAs may be evolved and jumped into other RNAs, and then they had created ancient genome. The miRNAs would be transposably spread out to some plant and vertebrate genomes by feeding of miRNAs-containing foods, viruses, etc. Equivocally, miRNA may finally be picked up into the lentiviral transposon, such as HIV-1. Therefore, the viral miR-N367 would necessarily be a HIV-1 silencer related with the latency to be incorporated into HIV-1 itself.

Introduction

The RNA silencing report has been published as long ago as 1928. Wingard suggested that the upper leaves of tobacco plants initially infected tobacco ringspot virus had somehow become immune to the virus and consequently the plants became asymptomatic and resistant to secondary infection (Wingard, 1928). However, the mechanisms remained to be work out because of before unveiling of inherited DNA. Although Jacob and Monod designed 'the operon' theory to regulate gene by a protein regulator in 1961 (Jacob and Monod, 1961), it is now well known for us that the protein repressor binds to the operator sequence of a gene to suppress the transcription in phage (virus) infected prokaryotes. In 1985, the gene expression has been regulated by complementary micRNA (anti-sense RNA) that can bind to the transcripts of particular genes and make double-stranded RNA (dsRNA), then consequently prevent their translation (Coleman, 1985). This paper may be an appropriate starting point for the current interest in specific gene regulation by dsRNA regulator against viruses. In the case of micRNA (174 nucleotides (nt) in *E. coli*, however, it was very different from RNA interference (RNAi) by short dsRNAs because the longer micRNA in length was more effective inhibitor than the shorter one. Further, since the longer RNAs were locked in another nonspecific interferon (IFN) mechanisms into mammalian cells, during two decades, investigation in mammal has never been allowed to meet the challenge of RNAi therapy as a gene-specific approach. Although Fire *et al.* had been of interest in the sequence-specific post-transcriptional gene silencing (PTGS) by sense, anti-sense and long dsRNAs from coding region, on 1998, consequently they found phenomenon of RNA interference (RNAi) by dsRNA in the nematode *Caenorhabditis elegans* (Fire *et al.*, 1998). However, at one time, any evidence was not shown that RNAi process is different from IFN system especially in mammalian cell even if no IFN in the nematode. We believed that a solution of this problem in RNAi does exist, though it has yet to be proved that human cell researchers will accept it and why not be the RNAi developed into medical approaches. We well knew that expression of the integrated retroviral genome and small RNAs derived from non-coding region in the host genome are far from sequence non-specific IFN system, therefore, by using northern blotting analysis, cloning of microRNAs (miRNAs) from HIV-1-infected cell and an *in vivo* available vector STYLE, which was made from one of retroviruses, apathogenic feline spumavirus cloned by our group (Fujii, 2006), we experimentally showed that a miRNA, miR-N367 derived from the non-coding region LTR can regulate HIV-1 transcription (Omoto and Fujii, 2005; Fujii, 2008). At the same time the *nef* miRNA is refined by a computational method (Bennasser *et al.*, 2004). Activity of the miR-N367 has recently been reproduced as a short interfering RNA (siRNA) agent (Yamamoto *et al.*, 2006). From the aspect of miRNA made incompletely duplex pair, suggesting that RNAi responses with miRNAs in mammalian cells are not completely sequence-specific manner exact. It is different from siRNA, which is pairing with complete complementary pairs. Therefore, to escape from disseminated intravascular coagulation (DIC) by off-target effects and extreme lymphokine secretion with siRNAs, our research done was necessarily enough to make the RNAi to serve as viral-specifically prophylactic and therapeutic arsenal in acquired immunodeficiency syndrome (AIDS) prevention.

Over three hundred of human miRNAs were listed up in miRBase as published miRNA sequence and annotation data (miRBase, http://microrna.sanger.ac.uk). Although experimental identification of miRNA-regulated genes is mainly dependent on computationally predicted miRNA targets (Thadani and Tammi, 2006; Sewer *et al.*, 2005), computational prediction of target sites of miR-N367 in HIV-1 genome

has not yet done. The prediction of miRNA target has been experimentally validated for the miRNAs to be related with development, apoptosis, oncogenesis, and metabolisms (Ouellet *et al.*, 2006). As regulation of the HIV-1 genes by miR-N367 have experimentally been shown (Yamamoto *et al.*, 2002; Omoto *et al.*, 2004), we endeavour here to explain computational investigation of miR-N367 targets that might have unveiled the intracellular interaction between viral miRNAs and the host possibly rendering HIV-1 latent states, and further discuss predictions about the future of RNAi under Darwinism partly elucidated by a challenging therapy of HIV with RNAi-based edible vaccine.

Materials and methods

Viral miRNA cloning and northern blotting assay

Cloning of HIV-1 miRNA and northern blotting assay were described elsewhere (Omoto and Fujii, 2006).

Downloading and preparing software

Computing analysis was performed by the conventional methods as described previously (Rehmsmeier, 2006). RNA hybrid (miRanda v1.0b) with Java™ 2 Platform Standard Edition binary programs were freely downloaded under Windows 2000 (v5.0) environment from the UTR, http://www.microrna.org. The appropriate platform for fasta search program (fasta3) was also downloaded at the same environment. Genetyx v6.0 program was purchased from GENETYX Corp., Osaka, Japan. The human miRNA matured sequences were downloaded from the miRNA Registry 7.1 based on BLAST searching of miRNAs in another species (http://microrna.sanger.ac.uk/sequences/search.shtml).

Prediction of miRNA/target duplex

The HIV-1 genomic sequences were obtained from the UTR, http://ncbi.nlm.nih.gov. The miR-N367 sequence was prepared from Genetyx files to Text file and saved into Fasta format. The filename was assumed as the file name queries. fasta. Bin folder was opened and jmiranda file was opened by Java platform. The target sequence, that is limited into 1×10^3 bp, was prepared with

Fasta format for RNA hybrid search. HIV-1 targets to miR-N367 were initially predicted using miRanda with default parameters (Gap Open Penalty: −2.0; Gap Extend: −8.0; Score Threshold: 80.0; Energy Threshold: −14.0 kcal/mol; Scaling Parameter: 2.0). The miR-N367 and the target sequences queries.fasta were loaded into the miRNAs and the UTRs field, respectively. The 'analyse' button was clicked. After a short time lug, the results were presented into a screenshot. The conserved miRNAs of the viral miR-N367 in the miRNAs database was conventionally investigated with the FIP site (ftp://ftp.sanger.ac.uk/pub/mirbase) for providing with consistent names for novel miRNA gene discoveries before publication. The secondary structures of obtained RNA sequences were analysed by the mfold Web server (http://www.bioinfo.rpi.edu/applications/mfold/pld/rna) and Genetyx v6.0.

Expression of *nef* dsRNAs in *A. thaliana*

To be cloning the HIV-1 SF2 *nef* into the pBI121 plasmids, the *nef* fragment was digested with *Sac*I and *Xba*I (TaKaRa, Kyoto, Japan) from the pCR2.1nef plasmids (Yamamoto *et al.*, 2002). The obtained fragment was ligated into pBI121 cut with *Sac*I and *Xba*I (see Fig. 7.3A below). Three plasmids, *nef* plus-orientated, *nef* minus-orientated and stem–loop of *nef* plasmids were constructed. The constructed plasmids were used for transformation of *Agrobacter tumefaciens* by electroporation (2.5 kV, 25 mF, 400 W) (Gene Purser, Funakosi, Tokyo, Japan). To make the embryonic cell aggregates of *Arabidopsis thaliana*, callus was transformed by incubation with the pBInef plasmid transformed *A. tumefaciens* cells at 22°C, 20 min. After the incubation, in CIM medium (3.2 g/l Gamborg B5, 0.5 g/l MES, 20 g/l glucose, 2.5 g/l gellan gum, 0.5 mg/l 2,4-D, 0.1 mg/l, Kinetin), then transformed aggregates were selected with 50 mg/l carbenicillin and kanamycin. After the selection, the aggregates were cultured for 40 d with SIM medium (3.2 g/l Gamborg B5, 0.5 g/l MES, 20 g/l glucose, 2.5 g/l gellan gum, 0.15 mg/l IAA, 0.5 mg/l 2iP). The cultured callus (100 mg) was homogenated under liquid N_2, total RNA was extracted with TRIzol (Invitrogen). The mi/

siRNAs were detected by northern blotting with probes, sinef#007, 084, 176, 200, 367 and 460 as described previously (Omoto *et al.*, 2004).

Results

HIV-1 miR-N367 targets in the LTR

We have previously cloned and sequenced the miR-N367; however, the predicted stem–loop construct was not obtained by the siRNA Target Finder (http://www.ambion.com/techlib/misc/siRNA_finder.html). Therefore, to investigate a target prediction of pre-miR-N367, we identified targets of the HIV-1 LTR RNA sequences in the HIV-1 genome using miRanda (Table 7.1). Five miR-N367 targets were identified in the HIV-1 LTR. From the mfold analysis, nt 89–112 sequences in the HXBII LTR contained the possibly opposite sequence in another arm of miR-N367. The pre-miR-N367 matured sequences which matched with the miR-N367 were selected corresponding to the lowest of the minimal free energy (mfe) of the predicted secondary structures by the mfold. The selected sequences were again applied for miRanda and mfe of the pre-miR-N367 obtained was −29.73 kcal/mol (Fig. 7.1 and Table 7.1). In order to increase the pair match and the stringency, a cut-off score of alignment length or energy threshold were selected more than 30 and less than −20.0 kcal/mol, respectively in the RNA sequences of HIV-1, HIV-2, HERV-K10, FIV, FeLV, BIV and HTLV LTRs and HIV-1 RRE (Table 7.1), then the miR-N367 targets TAR, RRE of HIV-1, a site of HIV-2 and BIV LTRs, 2 of the HERV-K10 and FeLV LTRs, 3 of HTLV-1 LTRs. The miR-N367 targets 5 sites of the HTLV-2 LTR but does not in the minus-stranded HIV-1 LTR (data not shown).

The miR-N367 family

The miR-N367 was cloned (miRBase ID, hiv1-mir-N367; accession number, MIMA0004478) and its expression was confirmed by northern blot analysis. The predicted pre-miR-N367 has been identified in the HIV-1 LTR U3 region (Fig. 7.1) (accession no. MI0006104). Further, to examine the pre-miR-N367 profiles among vertebrates and plants, miRNA phylogenetic conservation was investigated using BLASTN

Table 7.1 Analysis of the miR-N367 targets

Viruses	Targets	Numbers of nt	Align length	mfe
HIV-1(HXBII)	pre-miR-N367	1–23	26	−29.73
	LTR/*nef*	89–112[*]	26	−24.13
		140–162	24	−15.65
		167–191	25	−19.36
	LTR	361–384	25	−16.14
	LTR/TAR	465–497	33	−15.37
	RRE	21–48	27	−21.19
		109–138	29	−15.93
		160–179	23	−15.13
HIV-2(ROD)	LTR/*nef*	62–86	25	−16.21
	LTR	386–413	27	−16.59
		417–438	23	−15.81
		445–472	27	−15.76
		552–580	30	−15.23
		589–619	31	−19.17
		755–772	22	−16.84
		821–841	23	−15.12
HERV-K10	LTR	223–251	28	−17.64
		280–306	28	−22.41
		327–344	18	−18.06
		360–390	31	−15.22
		410–431	24	−17.44
FIV	LTR	320–346	28	−18.23
FeLV	LTR	127–152	27	−15.02
		216–246	31	−17.93
		318–343	27	−18.29
		457–477	21	−23.26
BIV	LTR	46–72	27	−20.29
		189–210	23	−16.66
HTLV-1	LTR	21–41	22	−18.71
		67–91	24	−20.26
		106–131	25	−21.23
		234–260	28	−15.52
		335–356	24	−15.95
		391–417	26	−25.83
		467–489	24	−15.94
		519–540	24	−18.11
		598–622	26	−16.21

[*]This region involves pre-miR-N367 sequences.

LTR, long terminal repeat; mfe, minimal free energy, kcal/mol; HIV, human immunodeficiency virus; HERV, human endogenous retrovirus; FIV, feline immunodeficiency virus; FeLV, feline leukaemia virus; BIV, bovine immunodeficiency virus; HTLV, human T cell leukaemia virus; shading indicates that alignment length is greater than 30 nts or mfe is less than −20.0 kcal/mol.

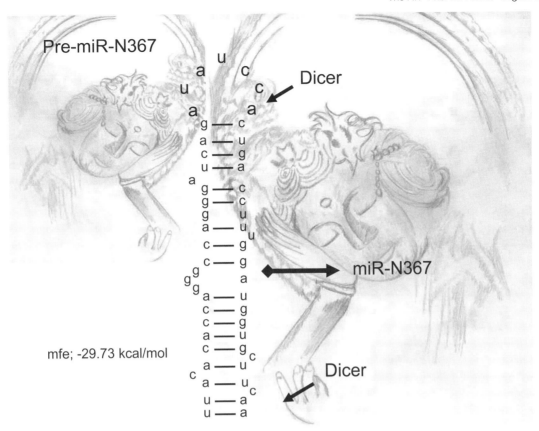

Figure 7.1 Computation of pre-miR-N367 prediction. Prediction of miR-N367 was performed by miRanda and mfold programs. A 54 nt stem–loop was predicted and is represented. The top value of thermodynamic mfe was obtained on –29.7 kcal/mol in the pre-miR-N367 hairpin. The miR-N367 would be digested by Dicer RNase and target to HIV-1 mRNAs. The stem–loop is incomplete pairing symbolized in '25 Bosatu Kannon' statue of 'Zenko' temple in Nagano City of Japan, which shows 10 similar 'kannon' statues in a large panel as miRNA targets.

searching boarding on the stem–loop conformation of pre-miRNAs. Twenty-two candidates of pre-miR-N367 family were selected after deriving scores of 65 to 90 and are summarized in Table 7.2. Of the miRNAs, seven are found in the sense orientation, 15 in the reverse orientation. Two miRNAs hsa-mir-181a-2 and hsa-mir-98 were found within introns of coding transcripts in human chromosome 9 and noncoding transcripts in human chromosome X, respectively. The mir-181a-2 and mir-98 are involved in mir-181 and let-7/mir-98 gene clusters, respectively, therefore, both miRNAs are well conserved in vertebrates. Pre-mir-98 in reverse orientation targeted 18 and 2 sites of HIV-1 LTR and TAR, respectively. However, the seed of mir-98/let-7 hits 3 target sites in the HIV-1 LTR with low value at all. The pre-mir-98 also targeted 9 of TAR. The mir-181a-2 targeted 6

sites of HIV-1 LTR and one site of TAR was also involved, but not top value of energy threshold. MIR-169h, MIR-169X, MIR-394 and mir-12 derived from plants were also involved into pre-miR-N367 family.

Modification of splicing by the miR-N367

The HIV-1 pre-RNA is spliced more than 50 different mRNA species and the mRNAs are grouped into 3 classes (Fig. 7.2). The 9 kb mRNA is the full length transcript encoding the Gag and Pol proteins. The 4 kb encodes the Env, Vpu, Vpr, Vif and Tat, and the 2 kb includes the Nef, Rev and Tat. Though it was speculated by above computational prediction that miR-N367 would be derived from introgenic miRNA as let-7, the introgenic miRNA is generally characterized as a modifier of splicing. To investigate the rela-

Table 7.2 The pre-miR-N367 is conserved in vertebrates and plants

ID	micoRNAs	pre-miR-N367 positions	microRNA positions	Orientation
MI0001218	gga-mir-181a-1	33–67	56–90	+
MI0000269	hsa-mir-181a-2	30–60	70–100	+
MI0002805	ggo-mir-181a-2	30–60	70–100	+
MI0002806	ppa-mir-181a-2	30–60	70–100	+
MI0002807	ptr-mir-181a-2	30–60	70–100	+
MI0005025	bta-mir-98	11–63	57–110	–
MI0000882	rno-mir-98	12–63	51–103	–
MI0000100	hsa-mir-98	12–62	58–109	–
MI0002697	ptr-mir-98	12–62	58–109	–
MI0003468	tni-mir-152	18–68	36–85	–
MI0003467	fru-mir-152	18–68	22–71	–
MI0002017	dre-mir-152	18–45	53–80	–
MI0002258	ptc-MIR169h	1–17	1–17	–
MI0002274	ptc-MIR169X	22–63	13–54	+
MI0001027	osa-MIR394	18–57	46–83	–
MI0004702	mmu-mir-500	15–36	63–84	+
MI0000939	rno-mir-195	1–25	32–56	–
MI0001745	mtr-MIR393	27–66	7–42	–
MI0001644	cfa-mir-429	3–31	35–62	–
MI0001576	ame-mir-12	1–26	25–50	–
MI0004641	mmu-mir-680–2	23–62	28–67	–
MI0005075	mtr-MIR395h	32–53	41–62	–

A cut-off score of 65 was selected and then the data are represented here. Shading represents human pre-miRNA.

tion between the target sites of miR-N367 and splicing donor (SD)/acceptor (SA), I scanned the target every 1,000 bp in the HIV-1 HXB genomic RNA except for the *nef*/3′ LTR region. The clustered target sites were obtained between SD4 and SA7 (Fig. 7.2). Six possible target sites were also predicted into SD1-SA2 and SD2-SA3. Further, recombinational hot spots in the HIV-1 genome were investigated among subtype A, B, C, D, E, F, G and I, and compared with sites of miR-N367 target as well as splicing (Fig. 7.2, asterisks). Six of ten hot spots for HIV-1 recombination were involved into miR-N367 target sites.

Expression of miRNAs from *nef* stem–loop in plants

Since it is well known that plant produces series of siRNAs amplified with RdRP corresponding to a siRNA template and four plant miRNAs,

MIR169h, MIR169X, MIR394 and MIR393 have been predicted as a miR-N367 family, expression of a stem–loop of *nef* containing pre-miR-N367 sequences was extended in the plant to investigate expression of *nef* si/miRNAs (Fig. 7.3). Four transfected calluses with *nef* (+ orientation) plasmid, seven with *nef* (−) and four with stem–loop of *nef* calluses were grown for 94 days in CIM with carbenicillin and kanamycin at 22°C (Fig. 7.3B). The RNA from each one grown callus on the day 215 was subjected for northern blotting with several RNA probes (Fig. 7.3C). The bands around 20 bp of RNAs were detected with #007, #084, #176, #200, #367 and #460 probes corresponding to each position of the LTR by northern blotting (Fig. 7.3C), but no transformed callus showed any bands. It suggests that a model plants, *A. thaliana* may be able to produce similar siRNAs or miRNAs that mammalian cells did as previously described. We

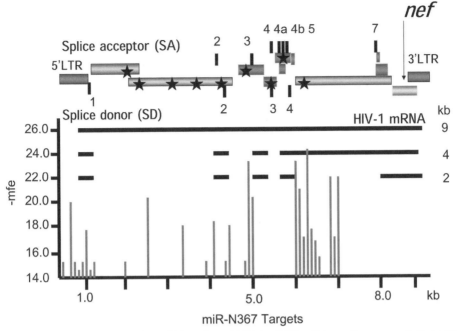

Figure 7.2 The miR-N367 targets for splicing sites of HIV-1 mRNAs. Schematic representation of the 9.7 kb proviral genome specifying the location of open reading frames (orf), splice acceptor (SA) and splice donor (SD) sites (the upper panel). The miR-N367 target sites and -mfe except for nef/3' LTR region were also represented corresponding to the unspliced 9 kb and spliced 4 and 2 kb transcripts (the lower panel). The clusters of targets of miR-N367 are presented on the LTR as well as the down stream of SA4 and SD4. Asterisks represent hot spots of intersubtype recombination among subtype A to I, which were involved in sequence data from strains ZAM184 (A/C), 92RW009.6 (A/C), 93TH253.3 (A/E), 92NG083.2 (A/G), 93BR029.4 (B/F), 94CY032.3 (A/G/I), MAL (A/D/I) and Z32B (A/G/I).

here propose from the data that the mammalian virally miRNA would also be involved in expression system of si/miRNA in the plant.

Discussion

On-target of the miR-N367

As shown in Table 7.3, although virally encoded miRNAs have been discovered in Epstein-Barr virus (EBV) (Pfeffer *et al.*, 2004), the Kaposi sarcoma-associated virus (KSHV), the mouse gammaherpesvirus (MHV), the human cytomegalovirus (HCMV) (Pfeffer *et al.*, 2005), and human immunodeficiency virus type 1 (HIV-1) (Omoto and Fujii, 2005; Fujii, 2008), these miRNAs might also contribute host-virus system interactions. Since 2004, large number of herpesvirus miRNAs was experimentally isolated from latent infected cells (Qi *et al.*, 2006); however, there is no concrete evidence that the encoded miRNAs could influence the pathogenicity. Unexpectedly, the adenovirus VA1-derived miRNA may be a byproduct of

inadvertent processing of VA1 by Dicer (Lu and Cullen, 2004; Andersson *et al.*, 2005). Similarly, the miRNAs of polyomaviruses downregulate the T antigen expression but viral proliferation had less been affected by it (Sullivan *et al.*, 2005). However, from our experiments using HIV-1 encoded miRNA, it is experimentally evidenced that the isolated miR-N367 from *nef* region overlapped the U3/3'-LTR non-coding region inhibited HIV-1 transcription and HIV-1 proliferation (Bennasser *et al.*, 2004; Yeung *et al.*, 2005a). From early time, a negative factor of HIV-1 replication was so named as *nef* because studies found that the *nef* can suppress viral transcription from the LTR and downregulated viral expression (Ahmad *et al.*, 1989; Cheng-Mayer *et al.*, 1989; Niederman *et al.*, 1989). Again, the phenomenon has been well known among retrovirologists that overlap sequences between the 5'- and 3'-LTRs reduced transcription from the 3'-LTR as promoter interference (Cullen *et al.*, 1984), but the mechanisms remained to be clear. Further, the *nef* has been reported to enhance

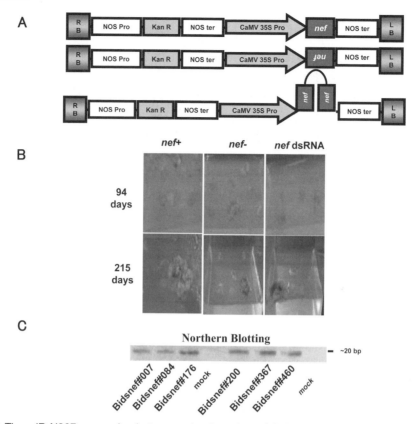

Figure 7.3 The miR-N367 expression in transgenic plant. A model plant, A. thaliana was transfected with three plasmids inserted with HIV-1 *nef* plus orientation, *nef* minus orientation and *nef* stem–loop (A). Calluses were grown on day 215, and RNA were extracted (B). Northern blotting with samples from *nef* stem–loop transgenic aggregates were performed by six probes and bands were detected (C). The plant cells would also produce si/miRNAs similar to mammalian cells.

viral replication (Kestler III *et al.*, 1991; Miller *et al.*, 1994) and the above mentioned *nef* activity as a negative factor have different outcomes, finally, 'controversial *nef* has been documented as a positive factor for HIV-1 proliferation. We hypothesize that since the miR-N367 showed effective reduction of HIV-1 by RNAi machinery, the *nef* might be a negative factor.

Now, the computationally predicted pre-miR-N367 formed a stem–loop and the matured miR-N367 targets TAR as well as RRE. The miRNAs are cognate with transcripts of imperfect base-pairing. Therefore, the pin-point targeting of a miRNA could be responsible for the gene silencing against multiply different mRNAs. Identification of cellular and viral miRNA targets is an important process in elucidating the biological role of the miRNA among viruses and hosts (Table 7.3). Several algorithms currently are available on the web to search for the miRNA seed and to determine the interac-

tion between validated miRNA and mRNA with the mfe (Thadani and Tammi, 2006). The mfe of miR-N367:TAR target sequences is not so high score of thermodynamic parameter but complementarity of align length is broad. Together with experimentally and computationally obtained data, suggesting that character of miRNAs as gene silencer would be a willing workhorse with low thermodynamic perturbation and broad targeting, by which less off-target effect is happened and replication of even mutated viruses can be suppressed. The TAR binding protein (TRBP) not only interacts with HIV-1 TAR RNA but also facilitates mRNA processing to conform the RISC/TRBP/Tat complex (Bennasser *et al.*, 2005). Although not so high pairing ability of miR-N367 has to the TAR region, the RISC/TRBP/miR-N367/Tat complex could recruit the TAR target of the HIV-1 transcriptional region itself (Fig. 7.4) because the miR-N367 eventually suppressed transcriptional process in the HIV-1

Table 7.3 Cellular and viral miRNAs related with viral life cycle

miRNAs		Virus family	Species	Function	Method	References
Cellular	mir-122	Flaviviridae	HCV	Facilitation of viral proliferation	Experimentally	(Jopling et al., 2005)
	mir-23	Retroviridae	PFV-1	Inhibition of viral replication	Experimentally	(Lecellier et al., 2005)
	mir-106a	Retroviridae	MLV SL3-3	Retroviral tumorigenesis	Experimentally	(Lum et al., 2007)
	mir-29, 149, 324, 378	Retroviridae	HIV-1	Latent state of viral infection	Computationally	(Harihara et al., 2005)
	mir-181a, 98/let-7	Retroviridae	HIV-1	Latent state of viral infection	Computationally	In this chapter
	mir-136, 507	Orthomyxoviridae	Flu H5N1	Unknown	Computationally	(Scaria et al., 2006)
Viral	miR-BHRF1-1~3 miR-BART-1~20	Herpesviridae	EBV	Latent state of viral infection	Experimentally	(Pfeffer et al., 2005; Pfeffer et al., 2006)
	miR-K-1~12	Herpesviridae	KSHV	Malignancy	Experimentally	(Cai and Cullen, 2006)
	miR-ULs, USs	Herpesviridae	HCMV	Unknown	Experimentally	(Grey et al., 2005)
	miR-1~8	Herpesviridae	MDV	Transformation	Experimentally	(Burnside et al., 2006)
	miR-LAT	Herpesviridae	HSV-1	Antiapoptosis	Experimentally	(Pfeffer et al., 2005)
	miR-M1-1~9	Herpesviridae	MHV68	Unknown	Experimentally	(Pfeffer et al., 2005)
	miR-rL1-1~16	Herpesviridae	LCV	Unknown	Experimentally	(Cai et al., 2006)
	miR-S1	Papovaviridae	SV40	Silencer of large T antigen	Experimentally	(Sullivan et al., 2005)
	Unnamed	Papovaviridae	Mouse polyomavirus	Silencer of large T antigen	Experimentally	(Sullivan and Ganem, 2005)
	Unnamed	Papovaviridae	SA12 polyomavirus	Silencer of large T antigen	Experimentally	(Cantalupo et al., 2005)
	miR-N367 Unnamed	Retroviridae	HIV-1	Latent state of viral infection HIV-1 virulence	Experimentally and computationally	(Omoto and Fujii, 2005; Bennasser et al., 2004) and in this chapter

HCV, hepatitis C virus; PFV, prototype foamy virus; MLV, mouse leukaemia virus; EBV, Epstein-Barr virus; KSHV, Kaposi's sarcoma-associated herpesvirus; HCMV, human cytomegalovirus; MDV, Marek's disease virus; HSV-1, herpes simplex virus type 1; MHV, mouse gammaherpesvirus; LCV, lymphocryptovirus.

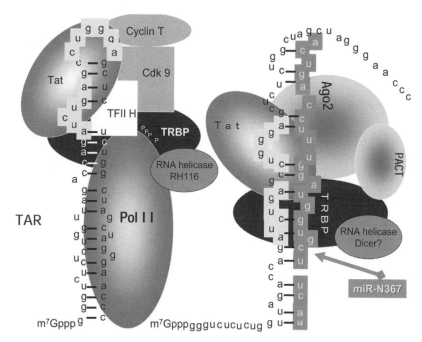

Figure 7.4 The miR-N367 target for TAR and RRE. Model for the silencing action of miR-N367 on transcriptional elongation of HIV-1 mRNAs by Tat. The TRBP and Tat bind to a part of RNA sequences (5′-aucug-3′) in the TAR loop structure. Cyclin T/Cdk9 and TF II H make complex with Tat and TRBP and then the complex promotes phosphorylation (P) of the carboxy terminal domain (CTD) of RNA polymerase II (Pol II) (the left panel). The TRBP could also recruit RISC proteins, such as Ago2 and PACT. Although miR-N367 targeted to the TAR region, the TRBP/RISC/Tat/miR-N367 complex may bind to the TAR RNA sequences and then make secondary structure of TAR change state to transcriptional silencing (the right panel). Some RNA helicase, such as Dicer might have a role for HIV-1 transcription.

infected T cells (Omoto and Fujii, 2005). Further, the RRE and splicing sites of HIV-1 mRNAs would also be the target of miR-N367. RNA helicases, RH116 and DDX3/DDX1 have been related with TAR-dependent HIV-1 transcription and RRE-dependent transport of HIV-1, respectively (Cocude et al., 2003; Dayton, 2004). The Dicer is a RNA helicase (Jeang and Yedavalli, 2006). The complex of Dicer/miR-N367 may work as a scavenger of HIV-1 replication facilitating RNA helicases. Therefore, without degradation of HIV-1 mRNAs (Westerhout et al., 2006), miR-N367 may inhibit maturation of mRNA during splicing into spliceosome with small nuclear RNAs (snRNAs) and then may suppress Rev cis-action since Rev-binding matured HIV-1 RNA is transported as matured HIV-1 genomic RNA from the nucleus to the cytoplasm (Fig. 7.5) (Dayton, 2004). These imperfect and multiple targeting of miR-N367 may render the latent state of HIV-1 even though presence of continuously basal stimulation for re-expression of HIV-1 under the latent states

(Pomerantz, 1990). Furthermore, Morris et al. studied this well-characterized the miR-N367 and HIV-1 mRNA interaction to better understand the TGS (transcriptional gene silencing) (histone deacetylation/methylation and/or DNA methylation) neo-pathway on RNA interference (Weinberg and Morris, 2006). Details of the TGS by miR-N367 are discussed by Kevin V. Morris in Chapter 2.

Latency under the miR-N367

The miRNA genes are encoded in the genome of vertebrates. The encoded miRNAs may play an important role in haematopoiesis, oncogenesis and neurite growth (Abdelrahim et al., 2006; Presutti et al., 2006; Venturini et al., 2006). For instance, mir-1 regulates the balance between differentiation and proliferation of cardiomyocytes in mice (Lim et al., 2005). The mir-15a and -16-1 down-regulate expression of B-cell lymphoma 2 (Bcl2) gene and possibly induce apoptosis (Cimmino et al., 2005). The mir-21 is expressed in the human malignant brain tumour and suppresses

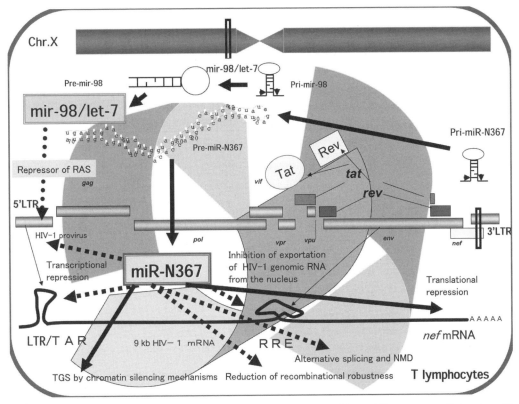

Figure 7.5 The HIV-1 silencer upon viral and host miRNAs. The HIV-1 Tat and Rev are essential and positive regulators of HIV-1 production. Tat and Rev recognize specific high-affinity sites on the LTR/TAR and RRE, respectively. Both TAR and RRE are targeted by the miR-N367 from pre-miR-N367 in the *nef*/LTR sequences of HIV-1 proviral DNA, therefore, binding of Tat and Rev to RNA could be prevented by the miR-N367 similar to the decoy RNA. Human mir-98/let-7 would be involved in the pre-miR-N367 family and may target to the LTR via repression of RAS activation. Taken together, the viral miR-N367 and host mir-98/let-7 could be a silencer of HIV-1 production followed by the latent state of HIV-1 infection in T lymphocytes.

apoptosis in the tumour (Chan *et al.*, 2005). A current view is that has-mir-507 and -136 had potential binding sites in polymerase B2 (PB2) and haemagglutinin (HA) of Influenza A, H5N1, respectively (Scaria, 2006). The mir-136 is specifically expressed in human lung. Both PB2 and HA activities were correlated with flu virulence; however, function of mir-136, -507 is not yet known and less be discussed at all. On the contrary, hsa-mir-122 expressed in liver has been reported to cause facilitation of HCV replication by binding to the 5′ UTR (Jopling *et al.*, 2005). On the other hand, the mir-32 restricted accumulation of prototype foamy virus type 1 (PFV-1) in human cells (Lecellier *et al.*, 2005). Although HCV occasionally induces liver cancer, mir-106 was involved in MLV-mediated tumorigenesis (Lum *et al.*, 2007). Intriguingly, the mir-106 cistron is a homologue of the oncogenic mir-17 cistron including mir-19b-1

and -92–1, which are all implicated in cancer development. We have not yet known whether the mir-122 would be related with oncogenesis in the HCV-infected liver, the analysis of host cellular miRNAs speculates that viral infection may cause significant overexpression of organ specific miRNAs indicating role of the miRNAs in disease development, such as oncogenesis of liver, lung etc.

The different species of viruses were targeted by the different miRNAs; however, mir-29a and mir-29b target to the *nef*, the mir-149, -378 and -324–5p also target to *vpr*, *env* and *vif*, respectively (Harihara *et al.*, 2005). Further, the miR-N367 could target to several LTRs of retroviruses including HERV-K10 of a settled retrotransposon in human genome. Alignment of the pre-miR-N367 sequences was encoded by the HIV-1 LTR and it targeted evolutionarily conserved retroviral LTRs, so it was speculative data that a significant number of miRNA paralogues among the different clusters of registered vertebrate miRNA genes may be implied into an evolution process targeting the potentially conserved LTR in retroviruses. Actually, the phylogenic conservation between viral and cellular miRNAs can also be predicted by computational program. Origin of the viral pre-miR-N367 may be involved in the hsa-mir-98/let-7 and -181 human miRNA clusters. The identified human mir-98 is involved into the cluster of let-7a-1, which is related with haematopoietic lineage, enriched expression of human lung cancers (Takamizawa et al., 2004), and regulation of late developmental transition (Yu et al., 2006). Intriguingly, the clusters of human mir-98/let-7a-1, -7a3, -7b, and -7f-2 and -181a, -181c, and -181d were expressed in Jurkat T highly and also in HL-60, U937 and THP-1 monocyte cell lines (Harihara et al., 2005). Further, the let-7 is a repressor of KRAS (Johnson et al., 2005). HIV-1 replication is reinduced from latency by antigen-driven T-cell activation via TCR. Since TCR activation via RAS/Raf/MEK/ERK pathway is mediated by downstream transcription factors such as NF-κB and the Ras response element binding factor 2 (BRF-2) and then binding of both factors into the BRE III cis-element and NF-?B sites in the HIV-1 LTR renders the state of HIV transcription high (Chen et al., 2005). From our data, the mir-98/let-7 could target even with low score to the element (location of −159 ~ −136) between the USF (−166) and BRE (−129) sites in the U3/LTR. Since removal of the negative response element (NRE) of the U3/5′LTR dramatically reduced the miR-N367 effect against transcriptional activity of the HIV-1 LTR containing NRE (Omoto and Fujii, 2005), both the mir-98/let-7 and miR-N367 may be silencer of HIV-1 transcription in T cell, which is related with RAS activation via TCR pathway. On the other hand, the HIV-1 genomic DNA transfected HeLa cells have recently been reported to completely downregulate the let-7 cluster expression (Yeung et al., 2005b). As primary role of TAT for highly HIV-1 transcription, this phenomenon would possibly be explained that TRBP/Dicer complex may be due to be recruited by Tat and scavenge of TRBP from RISC might inhibit mir-98/let-7 expression in T lymphocytes. Further,

since it has been reported that a miRNA could regulates other miRNAs, we predicted whether miR-N367 can target to pre-miR-98/let-7 and found that with very low thermodynamic value (−14.4 kcal/mol). The mir-98/let-7 was hit as a target of miR-N367. The miR-N367 may weakly inhibit expression of pre-mir-98/let-7. TAT is an activator of RAS (Toschi et al., 2006), therefore, high concentration of TAT in the cell inhibits the activity of Dicer (Bennasser et al., 2005) as well as enhances that of RAS. This may be followed by high production of infectious HIV-1 progeny from provirus DNA.

By thermodynamically changing policy from TAT-protein party to miR-N367-RNA party, RAS activation would be suppressed via miR-N367 silencer and re-overexpression of miR-98/let-7 cluster and could maintain latency of HIV-1 infection. We have previously presented the miRNA mutation model for internal elimination segment in the LTR (Fujii, 2008). The miR-N367 as a scanRNA (scRNAs) can target with Twi and Pdd-like proteins to the proviral DNA LTR sequences and including CpG methylation of DNA. Then, the miR-N367 could degenerate the proviral DNA segment. Actually, nef sequences in the LTR are deleted and nef is also quasispecies (Krichhoff et al., 1995). At least in part, the deletion of nef DNA sequences may induce latency or long-term nonprogression for infected individuals. This silencing model requires RNA-directed RNA polymerase (RdRP) to amplify miRNAs using a template of pre-miR-N367 under RNA wave. The RdRP activity in retroviral reverse-transcriptases (RTs) of HIV-1, HERV-K and foamy virus has been identified that strongly may support this model (Omoto and Fujii, 2005). Thus, it is again here hypothesized as a new dogma that not only RNAi with miRNAs can control the expression of transposable elements but also the miRNA may be necessary to inclusively incorporate the transposable elements as 'a non-selfish RNA' into the mammalian genome as a 'selfish gene' for evolution of its own genome (Fujii, 2007) because the RNAi machinery has been presented as the oldest and most ubiquitous gene silencing system, probably even from the term of RNA world (Fig. 7.6) (Koonin et al., 2006).

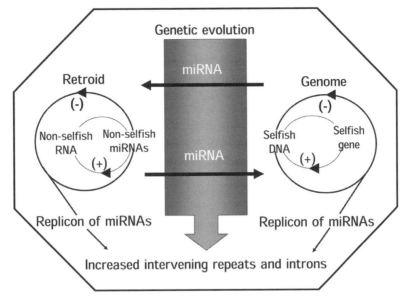

Figure 7.6 Model of genetic evolution based on miRNAs. The panels represented that genome containing selfish genes may be increased intervening repeats and introns on its genome size by miRNAs from extracellular retroid as non-selfish RNAs via incorporation and/or ingestion of miRNAs. The pseudogenes may also be created by miRNAs. Mammalian genomes would encode several hundred miRNAs. Further, virally encoded miRNAs have been discovered. Vaccinia and influenza viral proteins suppress RNA interference machinery called as suppression of the RNA interference system (SRS). The longer RNAs induced to produce non-specific interferon (IFN) into mammalian cells. Recent developments have revealed that in heterochromatin, tandem repeats or multiple copies of transposable elements could be converted to dsRNA through the activity of cellular RdRP in plant. Although one possible pathway upon RNA replication in mammalian cells except for above pol II, RT of human endogenous retrotransposon is a candidate of a cellular RdRP. The miRNAs as non-selfish RNA codes contained in the inserted intervening sequences in foods may be ingested, amplified by RNA PCR wave, then the miRNAs induced PTGS wave and finally could be converted into DNA by reverse transcription with RTs and then integrate into non-homologous RNA codes within selfish genes according to Neo-Darwinism. The DNA codes from the converted miRNAs could be inherited in this model.

Transgenic plants for the miR-N367

The demonstrations of siRNA-mediated repression in mammalian cells showed effective suppression of the GFP gene in mouse embryos (Wianny and Zernicka-Goetz, 2000), *Mos* or *Plat* in mouse oocytes (Svoboda *et al.*, 2000) and *Luc* gene (Ui-Tei *et al.*, 2000; Elbashir *et al.*, 2001) in cultured hamster and human cell lines. Although subsequently siRNA and shRNAs were delivered by liposome, nanoparticle and lentivector based deliveries via nostril, trachea, lobule, muscular, vagina, venous, peritoneal and alternative oral delivery (Fujii, 2006; Xiang *et al.*, 2006). Naked RNA or DNA injection may have potential to be extended beyond mice to human because an announcement of RNAi-based therapeutic examination for phase I in treatment of age-related macular degeneration (ARMD) shows no toxic effects and disease stabilized in all examined patients with intravitreally injected

siRNAs targeting VEGFR (Whelan, 2005). Growing on demand for therapeutic capacity of miRNA, a mass and cost-effective production of si/miRNAs, even vector based si/miRNAs, represents paradigm shift for the manufacturing of pharmaceuticals. Although oral si/miRNA agents that could be effective against HIV infection in intestine *in situ* because of protection against infection of macrophage and dendritic cells (Aquaro *et al.*, 2002) and incorporation of ingested phage DNA via intestinal mucosa to spleen and liver of mice (Schubbert *et al.*, 1997), transgenic vegetables expressing mi/siRNAs targeting viral genes can protect the invading of the real pathogens when they were fed. Further, as a source of manufacturing of si/miRNAs, the plant-based expression of si/miRNAs might provide an efficient, high capacity in pharmaceuticals alternative to traditional cell-culture and bacterial expression. Off-target effects by siRNAs were worried to occur at the protein synthesis levels (Snove Jr. and Holen, 2004), however the recombinant plant could filter the danger of sever off-target effect because of no growing the plants. Further, the application of these kinds of plant-based system could avoid many potential issues such as resource poor, contamination. With hope and promise, pioneering HIV treatment based on RNAi has already begun combining RNAi and gene therapy by Morris and Rossi (Morris and Rossi, 2006). From the aspect of use of siRNA in bacterial system, RNAi oral agents could also avoid opportunistic and infectious pathogens from AIDS pathogenesis as well. Of these, assessing human and ecological risk plus usage of biological weapon from the plant-based RNAi pharmaceuticals offers the best approach for valuing the si/miRNA agents (Smyth and Phillips, 2003). However, creating plant RNAi source hopefully may be sure to benefit pharmaceutical manufacturing for prophylaxis and therapy of AIDS.

Darwinism and HIV-1

Nearly 97% of the human genome consists of non-coding DNA (Venter et al., 2001). The most of non-coding regions transcribed into precursor messenger RNAs (pre-mRNAs). HIV-1 provirus DNA in the human genome contains non-coding region corresponding to the introns. HIV-1 mRNA is matured to separate introns from exons by exonic splicing and intron removal to form a mature mRNA. Given that decrease of genetic robustness is associated with exon shuffling, transgene integration, and transposon activities (Herbert and Berg, 1999; Lewis et al., 2002; Modrek and Lee, 2003), miR-N367: target data is important. Although HIV-1 has activities of integration and transposition using viral particles, there is no idea at all to affect gene arrangements with alternative splicing. In general, alternative splicing and non-sense-mediated mRNA decay (NMD) were shown to increase evolutionary changes of eukaryotic genome resulting from exon shuffling and eliminates intron-containing pre-mRNA in case of splicing failure (Lin, et al., 2006). HIV-1 has been classified into two monophyletic clades termed the major (M) and outlier (O) groups. HIV-1 recombines among inter- and intra-clades, which is plug into

evolutionary change of HIV-1 (Robertson and Gao, 1998). However, the recombination is the complexity of their hybrid pattern indicating that recombination involves multiple crossover events and it has been difficult to identify preferential sites or hot spot of recombination. As concerning to recombination among gag-pol and env genes, nearly 60% of hot spots were closed or identical to miR-N367 target sites. Since introns homologous repressed by RNAi machinery in expression of target gene including the introns and its specific exons in the pre-mRNA, there is some possibility that miR-N367 may target to hot spots for recombination. We have a model of HIV-1 recombination based on the miRNA target that the target region of HIV-1 genomic RNA in one strain would be eliminated by miR-N367 and RdRP activity of RT may repair the target region templating with another genomic RNA from other HIV-1 strains because of existence of double genomic RNA in the HIV-1 particle. Substantially, two strains of HIIV-1 could be shuffled in a viral particle.

The pre-miR-N367 was conserved in vertebrates and plants. Mammalian herpesvirus miRNAs were conserved across over 13 million years of evolution that separate primate EBV and LCV (Cai et al., 2006). However, 53 and 56 evolutionally unconserved miRNAs in human and plants has been identified by computing analysis, respectively, suggesting that the number of miRNAs are specific to small groups of related species (Bentwich et al., 2005; Lindow and Krogh, 2005). The 'non-selfish miRNA' may independently be born and evolved. The intornic miRNAs may also be inclusively incorporated into the ancient genome. Evidently, the unconserved miRNAs encoded by DNA genome and non-functional miRNAs encoded by DNA viruses may be a proof of that and therefore, miRNAs from DNA viruses could be conserved over once upon a time as DNA virus itself. Further, it is possible to reliably speculate that an ancient retroid may specifically become precursor of miRNA itself like viroid causing pathogenicity against plants (Wang et al., 2004). These retroviral miRNAs (precursors of miRNA) were remained as orphan genes and evolved with the host genome. In the virus world concept, it is predicted that in primordial genetic pool, selfish RNA viruses evolve first, followed by retroid element (Koonin *et al.*, 2006); however, our concept is that shows non-selfish miRNA evolution first, may be never out of rule because of existence of over 90% non-coding RNA, involving retroviral miRNAs family. Some incorporated and ingested RNA information as RNA code may be transmitted among species (Fujii, 2007), whereas we might have to decipher 'RNA code' to elucidate the evolution of genome gifted from our God.

References

Abdelrahim, M., Safe, S., Baker, C., and Abudayyeh, A. (2006). RNAi and cancer: implications and applications. J. RNAi Gene Silen. 2, 136–145.

Ahmad, N., Maitra, R.K., and Venkatesan, S. (1989). Rev-induced modulation of Nef protein underlies temporal regulation of human immunodeficiency virus replication. Proc. Natl. Acad. Sci. USA, 86, 6111–6115.

Andersson, M.G., Haasnoot, P.C., Xu, N., Berenjian, S., Berkhout, B., and Akusjarvi, G. (2005). Suppression of RNA interference by adenovirus-associated RNA. J. Virol. 79, 9556–9565.

Aquaro, S., Calio, R., Balzarini, J., Bellocchi, M.C., Garaci, E., and Perno, C.F. (2002). Macrophages and HIV infection: therapeutical approaches toward this strategic virus reservoir. Antiviral Res. 55, 209–225.

Bennasser, Y., Le, S.Y., Yeung, M.L., and Jeang, K.T. (2004) HIV-1 encoded candidate micro-RNAs and their cellular targets. Retrovirology 1, 43.

Bennasser, Y., Le, S.Y., Benkirane, M., and Jeang, K.T. (2005). Evidence that HIV-1 encodes an siRNA and a suppressor of RNA silencing. Immunity 22, 607–619.

Bentwich, I., Avniel, A., Karov, Y., Aharonov, R., Gilad, S., Barad, O., Barzilai, A., Einat, P., Einav, U., Meiri, E., Sharon, E., Spector, Y., and Bentwich, Z. (2005). Identification of hundreds of conserved and nonconserved human microRNAs. Nat. Genet. 37, 766–772.

Burnside, J., Bernberg, E., Anderson, A., Lu, C., Mayers, B.C., Green, P.J., Jain, N., Isaacs, G., and Morgan, R.W. (2006). Marek's disease virus encodes microRNAs that map to meq and the latency-associated transcript. J. Virol. 80, 8778–8786.

Cai, X., and Cullen, B.R. (2006). Transcriptional origin of Kaposi's sarcoma-associated herpesvirus microRNAs. J. Virol. 80, 2234–2242.

Cai, X., Schafer, A., Lu, S., Bilello, J.P., Desrosiers, R.C., Edwards, R., Raab-Traub, N., and Cullen, B.R. (2006). Epstein-Barr virus microRNAs are evolutionarily conserved and differentially expressed. PLOS Pathogens 2, 236–247.

Cantalupo, P., Doering, A., Sullivan, C.S., Pal, A., Peden, K.W., Lewis, A.M., and Pipas, J.M. (2005). Complete nucleotide sequence of polyomavirus SA12. J. Virol. 79, 13094–13104.

Chan, J.A., Krichevsky, A.M., and Kosik, K.S. (2005). MicroRNA-21 is an antiapoptotic factor in human glioblastoma cells. Cancer Res. 65, 6029–6033.

Chen, J., Malcolm, T., Estable, M.C., Roeder, R.G., and Sadowski, I. (2005). TFII-I regulates induction of chromosomally integrated human immunodeficiency virus type 1 long terminal repeat in cooperation with USF. J. Virol. 79, 4396–4406.

Cheng-Mayer, C., Iannello, P., Shaw, K., Luciw, P.A., and Levy, J.A. (1989). Differential effects of nef on HIV replication: implications for viral pathogenesis in the host. Science 246, 1629–1632.

Cimmino, A., Calin, G.A., Fabbri, M., Iorio, M.V., Ferracin, M., Shimizu, M., Wojcik, S.E., Aqeilan, R.I., Zupo, S., Dono, M., Rassenti, L., Alder, H., Volinia, S., Liu, C.G., Kipps, T.J., Negrini, M., and Croce, C.M. (2005). miR-15 and miR-16 induce apoptosis by targeting BCL2. Proc. Natl. Acad. Sci. USA 102, 13944–13949.

Cocude, C., Truong, M.J., Billaut-Mulot, O., Delsart, V., Darcissac, E., Capron, A., Mouton, Y., and Bahr, G.M. (2003). A novel cellular RNA helicase, RH116, differentially regulates cell growth, programmed cell death and human immunodeficiency virus type 1 replication. J. Gen. Virol. 84, 3215–3225.

Coleman, J., Hirashima, A., Inokuchi, Y., Green, P.J., and Inouye, M. (1985). A novel immune system against bacteriophage infection using complementary RNA (micRNA). Nature 315, 601–603.

Cullen, B.R., Lomedico, P.T., and Ju, G. (1984). Transcriptional interference in avian retrovirus – implication for the promoter insertion model of leukaemogenesis. Nature 307, 241–245.

Dayton, A.I. (2004). Within you, without you: HIV-1 Rev and RNA export. Rerovirology 1, 35.

Elbashir, S.M., Harborth, J., Lendeckel, W., Yalcin, A., Weber, K., and Tuschl, T. (2001). Duplexes of 21-nucleotide RNAs mediate RNA interference in cultured mammalian cells. Nature 411, 494–498.

Fire, A., Xu, S., Montgomery, M.K., Kostas, S.A., Driver, S.E., and Mello, C.C. (1998). Potent and specific genetic interference by double-stranded RNA in *Caenorhabditis eleganse*. Nature 391, 806–811.

Fujii, Y.R. (2006). The design of RNAi from HIV-1 before two decades. In RNAi therapeutics, H. Takaku and N.Yamamoto, ed. (Kerala, India: Transworld Research Network), pp. 17–29.

Fujii, Y.R. (2008). Lost in translation: regulation of HIV-1 by microRNAs and a key enzyme of RNA-directed RNA polymerase. In microRNAs: from Basic Science to Disease Biology, K. Appasani, ed. (Cambridge UK: Cambridge University Press), pp. 424–440.

Fujii, Y.R. (2007). Symphony of AIDS. In Regulation of Gene Expression by Small RNAs, J.J. Rossi, and R.K. Gaur, ed. (Boca Raton FL: CRC Press), in press.

Grey, F., Antoniewicz, A., Allen, E., Saugstad, J., McShea, A., Carington, J.C., and Nelson, J. (2005). Identification and characterization of human cytomegalovirus-encoded microRNAs. J. Virol. 79, 12095–12099.

Harihara, M., Scaria, V., Pillai, B., and Brahmachari, S.K. (2005). Targets for human encoded microRNAs in HIV genes. Biochem. Biophys. Res. Commun. *337*, 1214–1218.

Herbert, A., and Berg, P. (1999). RNA processing and the evolution of eukaryotes. Nat. Genet. *21*, 265–269.

Jacob, F., and Monod, J. (1961). Genetic regulatory mechanisms in the synthesis of proteins. J. Mol. Biol. *3*, 318–356.

Jeang, K.–T., and Yedavalli, V. (2006). Role of RNA helicases in HIV-1 replication. Nucleic Acids Res. *34*, 4198–4205.

Johnson, S.M., Grosshans, H., Shingara, J., Byrom, M., Jarvis, R., Cheng, A., Labourier, E., Reinert, K.L., Brown, D., and Slack, F.J. (2005). RAS is regulated by the let-7 microRNA family. Cell *120*, 635–647.

Jopling, C.L., Yi, M., Lancaster, A.M., Lemon, S.M., and Sarnow, P. (2005). Modulation of hepatitis C virus RNA aboundance by a liver-specific microRNA. Scienece *309*, 1577–1580.

Kestler III, H.W., Ringler, D.J., Mori, K., Panicali, D.L., Sehgal, P.K., Daniel, M.D., and Desrosiers, R.C. (1991). Importance of the nef gene for maintenance of high virus loads and for development of AIDS. Cell *65*, 651–662.

Koonin, E.V., Senkevich, T.G., and Dolja, V.V. (2006). The ancient virus world and evolution of cells. Biology Direct *1*, 29.

Krichhoff, F., Greenough, T.G., Brettler, D.B., Sullivan, J.L., and Desrosiers, R.C. (1995). Absence of intact *nef* sequences in a long-term survivor with nonprogressive HIV-1 infection. N. Engl. J. Med. *332*, 228–232.

Lecellier, C.–H., Dunoyer, P., Arar, K., Lehmann-Che, J., Eyquem, S., Himber, C., Saib, A., and Voinnet, O. (2005). A cellular microRNA mediates antiviral defense in human cells. Science *308*, 557–560.

Lewis, B.P., Green, R.E., and Brenner, S.T. (2002). Evidence for the widespread coupling of alternative splicing and non-sense-mediated mRNA decay in humans. Proc. Natl. Acad. Sci. USA, *100*, 189–192.

Lim, L.P., Lau, N.C., Garrett-Engele, P., Grimson, A., Schelter, J.M., Castle, J., Bartel, D.P., Linsley, P.S., and Johnson, J.M. (2005). Microarray analysis shows that some microRNAs downregulate large numbers of target mRNAs. Nature *433*, 594–596.

Lin, S.-L., Miller, J.D., and Ying, S.–Y. (2006). Intronic MicroRNA (miRNA). J. Biomed. Biotech. 26818.

Lindow, M., and Krogh, A. (2005). Computational evidence for hundreds of non-conserved plant microRNAs. BMC Genomics *6*, 119.

Lu, S., and Cullen, B.R. (2004). Adenovirus VA1 noncoding RNA can inhibit small interfering RNA and microRNA biogenesis. J. Viol. *78*, 12868–12876.

Lum, A.M., Wang, B.B., Lauri, L., Channa, N., Bartha, G., and Wabl, M. (2007). Retroviral activation of the mir-106a microRNA cistron in T lymphoma. Retrovirology *4*, 5.

Miller, M.D., Warmerdam, M.T., Gaston, I., Greene, W.C., and Feinberg, M.B. (1994). The human immunodeficiency virus-1 nef gene product: a positive factor for viral infection and replication in primary lymphocytes and macrophages. J. Exp. Med. *179*, 101–113.

Modrek, B., and Lee, C.J. (2003). Alternative splicing in the human, mouse and rat genomes is associated with an increased frequency of exon creation and/or loss. Nat. Genet. *34*, 177–180.

Morris, K.V., and Rossi, J.J. (2006). Lentivirus-mediated RNA interference therapy for human immunodeficiency virus type 1 infection. Hum. Gene Ther. *17*, 479–486.

Niederman, T.M.J., Thielan, B.J., and Ratner, L. (1989). Human immunodeficiency virus type 1 negative factor is a transcriptional silencer. Proc. Natl. Acad. Sci. USA, *86*, 1128–1132.

Omoto, S., Ito, M., Tsutsumi, Y., Ichikawa, Y., Okuyama, H., Brisibe, E.A., Saksena, N.K., and Fujii, Y.R. (2004). HIV-1 nef suppression by virally encoded microRNA. Retrovirology *1*, 44.

Omoto, S., and Fujii, Y.R. (2005). Regulation of human immunodeficiency virus 1 transcription by *nef* microRNA. J. Gen. Virol. *86*, 751–755.

Omoto, S., and Fujii, Y.R. (2006). Cloning and detection of HIV-1-encoded microRNA. Methods. Mol. Biol. *342*, 255–265.

Ouellet, D.L., Perron, M.P., Gobell, L.–A., Plante, P., and Provost, P. (2006). MicroRNAs in gene regulation: when the smallest governs it all. J. Biomed. Biotech. *4*, 69616.

Pfeffer, S., Zavolan, M., Grasser, F.A., Chien, M., Russo, J.J., Ju, J., John, B., Fnright, A.J., Marks, D., Sander, C., and Tuschl, T. (2004). Identification of virus-encoded microRNAs. Science *304*, 734–736.

Pfeffer, S., Sewer, A., Lagos-Quintana, M., Sheridan, R., Sander, C., Grasser, F.A., van Dyk, L.F., Ho, C.K., Shuman, S., Chien, M., Russo, J., Ju, J., Randall, G., Lindenbach, B.D., Rice, C.M., Simon, V., Ho, D.D., Zavolan, M., and Tuschl, T. (2005). Ideantification of microRNAs of the herpesvirus family. Nat. Method. *2*, 269–276.

Pomerantz, R., Trono, D., Feinberg, M.B., and Baltimore, D. (1990). Cells nonproductively infected with HIV-1 exhibit an aberrant pattern of viral RNA expression: a molecular model for latency. Cell *61*, 1271–1276.

Presutti, C., Rosati, J., Vincenti, S., and Nasi, S. (2006). Non-coding RNA and brain. BMC Neuro. *7*, S5.

Qi, P., Han, J., Lu, Y., Wang, C., and Bu, F. (2006). Virus-encoded microRNAs:future therapeutic targets? Cell. Mol. Immunol. *3*, 411–419.

Rehmsmeier, M. (2006). Prediction of microRNA target. Methods. Mol. Biol. *342*, 87–99.

Robertson, D.L., and Gao, F. (1898). Recombination of HIV genomes. In Human Immunodeficiency Virues: Biology, Immunology and Molecular Biology, N.K. Saksena, ed. (Genoa, Italy: Medical Systems Spa), pp. 183–208.

Scaria, V., Hariharan, M., Maiti, S., Pillai, B., and Brahmachari, S.K. (2006). Host-virus interaction: a new role for microRNAs. Retrovirology *3*, 68

Schubbert, R., Renz, D., Schmitz, B., and Doerfler, W. (1997). Foreign (M13) DNA ingested by mice reaches peripheral leukocytes, spleen, and liver via the intestinal wall mucosa and can be covalently linked to mouse DNA. Proc. Natl. Acad. Sci. USA, *94*, 961–966.

Sewer, A., Paul, N., Landgraf, P., Aravin, A., Pfeffer, S., Brownstein, M.J., Tuschl, T., and van Nimwegen, E. (2005). Identification of clustered microRNAs using *ab initio* prediction method. BMC Bioinfo. *6*, 267.

Smyth, S., and Phillips, P.W.B. (2003). Labeling to manage marketing of GM foods. Trends Biotech. *21*, 389–393.

Snove Jr., O., and Holen, T. (2004). Many commonly used siRNAs risk off-target activity. Biochem. Biophys. Res. Commun. *319*, 256–263.

Sullivan, C.S., and Ganem, D. (2005a). MicroRNAs and viral infection. Mol. Cell *20*, 3–7.

Sullivan, C.S., Grundhoff, A.T., Tevethia, S., Pipas, J.M., and Ganem, D. (2005b). SV40-encoded microRNAs regulate viral gene expression and reduce susceptibility to cytotoxic T cells. Nature *435*, 682–686.

Svoboda, P., Stein, P., Hayashi, H., and Schultz, R.M. (2000). Selective reduction of dormant maternal mRNAs in mouse oocytes by RNA interference. Development *127*, 4147–4156.

Takamizawa, J., Konishi, H., Yanagisawa, K., Tomida, S., Osada, H., Endoh, H., Harano, T., Yatabe, Y., Nagino, M., Nimura, Y., Mitsudomi, T., and Takahashi, T. (2004). Reduced expression of the let-7 microRNAs in human lung cancers in association eith shortened postoperative survival. Cancer Res. *64*, 3753–3756.

Thadani, R., and Tammi, M.T. (2006). MicroTar: predicting microRNA targets from RNA duplexes. BMC Bioinfo. *7*, 520.

Toschi, E., Bacigalupo, I., Strippoli, R., Chiozzini, C., Cereseto, A., Falchi, M., Nappi, F., Sgadari, C., Barillari, G., Mainiero, F., and Ensoli, B. (2006). HIV-1 Tat regulates endotherial cell cycle progression via activation of the Ras/ERK MAPK signaling pathway. Mol. Biol. *17*, 1985–1994.

Ui-Tei, K., Zenno, S., Miyata, Y., and Saigo, K. (2000). Sensitive assay of RNA interference in *Drosophila* and Chinease hamster cultured cells using firefly luciferase gene as target. FEBS Lett. *479*, 79–82.

Venter, J.C., Adams, M.D., Myers, E.W., *et al.* (2001). The sequence of the human genome. Science *291*, 1304–1351.

Venturini, L., Eder, M., and Scherr, M. (2006). RNA-mediated gene silencing in hematopoietic cells. J. Biomed. Biotech. 87340

Wang, M.-B., Bian, X.Y., Wu, L.M., Liu, L.X., Smith, N.A., Isenegger, D., Wu, R.M., Masuta, C., Vance, V.B., Watson, J.M., Rezaian, A., Dennis, E.S., and Waterhouse, P.M. (2004). On the role of RNA silencing in the pathogenicity and evolution of viroids and viral satellites. Proc. Natl. Acad. Sci. USA, *101*, 3275–3280.

Westerhout, E.M., ter Brake, O., and Berkhout, B. (2006). The virion-associated incoming HIV-1 RNA genome is not targeted by RNA interference. Retrovirology 3, 57.

Weinberg, M., and Morris, K.V. (2006) Are viral-encoded microRNAs mediating latent HIV-1 infection? DNA Cell Biol. 25, 223–231.

Whelan, J. (2005). First clinical data on RNAi. Drug Discov. Today *10*, 1014–1015.

Wianny, F., and Zernicka-Goetz, M. (2000). Specific interference with gene function by double-stranded RNA in early mouse development. Nat. Cell Biol. 2, 70–75.

Wingard, S.A. (1928). Hosts and symptoms of ring spot, a virus disease of plants. J. Agric. Res. 37, 127–153.

Xiang, S., Fruehauf, J., and Li, C.J. (2006). Short hairpin RNA-expressing bacteria elicit RNA interference in mammals. Nature Biotech. *24*, 697–702.

Yamamoto, T., Omoto, S., Mizuguchi, M., Mizukami, H., Okuyama, H., Okada, N., Saksena, N.K., Brisibe, E.A., Otake, K., and Fujii, Y.R (2002). Double-stranded *nef* RNA interferes with human immunodeficiency virus type 1 replication. MicroBiol. Immunol. *46*, 809–817.

Yamamoto, T., Miyoshi, H., Yamamoto, N., Yamamoto, N., Inoue, J.–I., and Tsunetsugu-Yokota, Y. (2006). Lentivectors expressing short hairpin RNAs against the U3-overlapping region of HIV *nef* inhibit HIV replication and infectivity in primary macrophages. Blood 108, 3305–3312.

Yeung, M.L., Bennasser, Y., Le, S.Y., and Jeang, K.T. (2005a). siRNA, miRNA and HIV: promises and challenges. Cell Res. *15*, 935–946.

Yeung, M.L., Bennasser, Y., Myers, T.G., Jiang, G., Benkirane, M., and Jeang, K.-T. (2005b). Changes in microRNA expression profiles in HIV-1-transfected human cells. Retrovirology 2, 81.

Yu, J., Wang, F., Yang, G.-H., Wang, F.-L., Ma, Y.-N., Du, Z.-W., and Zhang, J.-W. (2006). Human microRNA clusters: genomic organization and expression profile in leukemia cell lines. Biochem. Biophys. Res. Commun. *349*, 59–68.

Regulation of Mammalian Mobile DNA by RNA-based Silencing Pathways

Harris S. Soifer

Abstract

The field of RNA silencing has grown quickly in less than a decade from the unexpected observation of colour variation in transgenic flowers to a conserved eukaryotic mechanism for regulating gene expression at the post transcriptional level. Genetic and biochemical dissection of RNA silencing processes in many eukaryotes indicate that multiple pathways operate to control the expression of target genes, often in a developmentally or tissue-specific manner. What has also become clear is that specialization of RNA silencing components and trigger molecules that occurred during speciation has lead to differences in the regulation of target RNAs across species. In lower eukaryotes, such as *Drosophila* and *C. elegans*, the activity of mobile genetic elements within these genomes is under the direct control of one or more RNA-based silencing mechanisms. Mammalian genomes are dominated by retrotransposons, mobile genetic elements that move through an RNA intermediate and thus would seem to be obvious targets for an RNA-based silencing mechanism that relies on complementarity with the transcript. The idea that mammalian RNA silencing pathways function, in part, to regulate the activity of mobile genetic elements is supported by an increasing number of empirical studies. In this chapter, an overview of RNA silencing mechanisms in mammalian cells is presented along with an outline of the different mobile genetic elements that inhabit the mouse and human genomes. A significant portion of the chapter is devoted to empirical evidence supporting a direct role for mammalian RNA silencing pathways in the regulation of mobile DNA. Furthermore, recent bioinformatic data suggesting that ancient mobile elements once regarded as 'junk' have evolved into functional, non-coding RNA that serve as triggers for RNA-based gene regulation in mammalian cells is also discussed.

An overview of RNA silencing in the animal kingdom

RNA silencing is an evolutionarily conserved mechanism(s) in eukaryotes by which double-stranded RNA (dsRNA) triggers the inactivation of gene expression of homologous sequences. RNA silencing phenomena have been observed in most, if not all, eukaryotes in one form or another. In plants and animals, three RNA silencing processes have been well characterized and play critical roles in development and survival (Meister and Tuschl, 2004). An important link to each of these pathways is that each is triggered by 19–31 nt small RNAs, which themselves are usually processed from a larger dsRNA or hairpin RNA precursors. Gene silencing by small RNAs often occurs post- transcription, through the action of short interfering RNA (siRNA) duplexes that degrade target mRNAs through target site cleavage. Degradation of cognate mRNAs by an siRNA duplex describes a process most often referred to as RNA interference (RNAi). A second mode of post-transcriptional gene silencing is mediated by the large population of plant and animal microRNAs (miRNA), which bind to target sites in the 3′ untranslated region (3′ UTR) of mRNAs and cause a block translation (Pillai *et al.*, 2005). A third form of RNA silencing termed transcriptional gene silencing (TGS) occurs through the action of siRNA duplexes that recruit negative chromatin factors, and in some cases cause heritable epigenetic changes in chromatin structure at DNA target sequences (Castanotto *et al.*, 2005; Morris *et al.*, 2004) (Fig. 8.1).

The origin of the dsRNA silencing trigger can be endogenous, as is the case for *Schizosaccaromyces pombe* centromere repeats and plant/animal pre-miRNA hairpin RNAs. Alternatively, external forces can introduce the dsRNA silencing trigger into the organism, as often happens when the eukaryotic cell is infected by an RNA virus or through uptake of ectopic dsRNA intended for experimental

purposes (Meister and Tuschl, 2004). Whatever the mode of entry, the dsRNA molecule is processed quickly into smaller 19–31 nt effector units through the action of cellular enzymes, especially members of the RNAse III family of endoribonuclease enzymes (Hammond, 2005). Following cleavage and processing, 19–31 nt effector molecules are incorporated into one or more multiprotein complexes collectively referred to as the RNA-induced silencing complex or RISC, which completes the gene silencing process. Members of the Argonaute family of proteins are core components of RISC, with some family members functioning as the endonuclease that cleaves target mRNAs during an RNAi response (Hammond, 2005; Liu *et al.*, 2004) (Fig. 8.1).

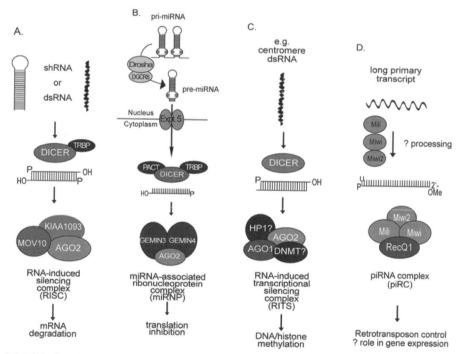

Figure 8.1 RNA silencing pathways in mammals. (A) Dicer cleaves long double-stranded RNA (dsRNA) or short-hairpin RNA (shRNA) with the help of Tar RNA binding protein (TRBP) into functional siRNA with the characteristic 3′ overhangs. siRNAs are incorporated into RISC and cleavage is performed by AGO2. (B) Primary miRNA transcripts, some of which are polycistronic (diagrammed), are processed in the nucleus by the Microprocessor complex (DROSHA and DGCR8) into precursor microRNAs (pre-miRNA) and then exported into the cytoplasm through Exportin 5 (Expt. 5). Dicer and its binding partners TRBP and PACT (protein activator of interferon-induced protein kinase) cleaves the pre-miRNA hairpin in the cytoplasm, with the mature functional miRNA associating with AGO2 in a miRNA ribonucleoprotein (miRNP). miRNPs recognize their target mRNAs resulting in translation inhibition. (C) Transcriptional gene silencing is initiated by Dicer-mediated cleavage of long dsRNA (e.g. centromere dsRNA) into siRNA that associate with the RITS complex. Putative components of *H. sapiens* RITS are depicted: AGO1-argonaute 1; AGO2-argonaute 2; DNMT-DNA methyltransferase; HP1-heterochromatin protein 1. (D) Individual piRNAs are generated through processing of long primary transcripts, perhaps through cleavage directed by the PIWI proteins themselves. The piRNA molecule is larger than si- or miRNAs and associates with germline piwi homologues MILI, MIWI, MIWI2 and the DNA helicase RecQ1. P, 5′ phosphate; 2′-OM, 2′-O-Methyl.

A physiological role for RNA silencing in innate immunity

The genomes of all eukaryotes contain high copy numbers of mobile genetic elements, which many consider parasitic DNA sequences that increase the mutational burden placed upon eukaryotes (Kazazian, 1998). Because of this high copy number, combined with their potential for autonomous movement, one might speculate that eukaryotic genomes are prone to vast instabilities due to retro- and DNA transposon activity. For the most part, however, transposons and retrotransposons move and recombine in wildtype organisms at rates much lower than their copy number indicates, suggesting that eukaryotes handle the mutational burden of mobile genetic elements quite effectively (Fedoroff, 2002). The notion that eukaryotic genomes are relatively stable despite the large number of mobile genetic elements that inhabit them lead many to consider that active cellular processes keep mobile genetic elements in check. In fact, extensive work in a wide variety of eukaryotes over the past decade has characterized several RNA-based silencing pathways that function to restrict the activity of eukaryotic mobile genetic elements (Soifer, 2006; Vastenhouw and Plasterk, 2004).

The first evidence that RNA silencing pathways are important regulators of transposon activity came from EMS mutagenesis screens of *C. elegans* performed to identify mutant worms deficient in RNAi (Ketting *et al.*, 1999; Ketting and Plasterk, 2000; Tabara *et al.*, 1999). From these designed screens, two independent groups identified several genes with essential roles in C. *elegans* germline RNAi and the post-transcriptional regulation of the Tc1 DNA transposon (Hammond, 2005; Ketting *et al.*, 1999; Liu *et al.*, 2004; Tabara *et al.*, 1999). Further screens in C. *elegans* demonstrated that while not all genes necessary for RNA silencing are also required for transposon silencing in worms, there is substantial cross-talk between several related RNA silencing pathways (Ketting and Plasterk, 2000; Vastenhouw *et al.*, 2003). Additional evidence supporting a role for RNA silencing in the management of transposable element (TE) activity has been demonstrated through genetic analysis in a number of eukaryotes including *Drosophila* (*piwi*) (Sarot *et al.*, 2004), *Trypanosoma* (*Ago1*) (Shi *et al.*, 2004), *Neurospora* (*qde-2, dcl1/dcl2*)

(Nolan *et al.*, 2005), and *Arabidopsis* (*Ago4*) (Zilberman *et al.*, 2003) (Table 8.1).

Although certain generalities apply to RNA silencing mechanisms, none of these pathways function exclusively to restrict mobile genetic elements in their respective genome, with RNA silencing pathways serving distinct and sometimes overlapping roles in eukaryotic development. One problem has been translating the results obtained in these model organisms to the more complex mammalian genome. Moreover, mammalian genomes are populated by a multitude of distinct mobile elements, many of which are non-functional due to point mutations and structural rearrangments (Deininger and Batzer, 2002). Adding to the complexity of RNA silencing in the animal kingdom is a new class of germline-specific small RNAs that may not be RNAse III ribonuclease products, but do associate with argonaute-piwi proteins in developing germ cells (Kim, 2006). This chapter will provide an overview of mobile/transposable genetic elements that inhabit mammalian genomes and then present a thorough discussion of RNA silencing mechanisms that may play an active role in suppressing mobile genetic activity in mammals.

Mobile genetic elements in mammalian genomes

The genomes of higher eukaryotes are complex, with protein-coding genes dispersed among individual chromosomes whose sequence is comprised mostly non-coding DNA sequence. Data from the both the Human and Mouse Genome Projects confirmed what was long inferred from Cot curve analysis: mammalian genomes are comprised mostly of repetitive DNA (Flamm, 1972), with the ~25,000 unique genes occupying only a small percentage (3–5%) of genome sequence. In fact, 46% of the human genome (Lander *et al.*, 2001) and 39% of the mouse genome (Waterston *et al.*, 2002) are represented by interspersed repeats (Fig. 8.2). This repetitive DNA can have a functional role in genome physiology, such as the importance of pericentromeric satellite repeat sequence in the formation of heterochromatin that is essential for proper chromosome separation during mitosis. On the other hand, other repeat sequences like the large number of intact mobile genetic elements that

Table 8.1 Evidence for RNA silencing of eukaryotic mobile genetic elements. Mammalian orthologues of RNA silencing genes, if present in the Homologene database, are indicated. ND, not determined; UTR, untranslated region; ORF, open reading frame; *AtSN1 siRNA was determined by Northern blot with a full-length 159-nt sense AtSN1 RNA probe.

Organism	RNA silencing genes implicated in silencing TEs	Human homolog(s)	Transposable element silenced	siRNAs detected?
Caenorhabditis elegans	mut-7 rde-2 mut-16 mut-14 ppw-2	—— —— —— —— PIWI family	Tc1, Tc3, Tc5 DNA transposons	Terminal inverted repeat (TIR) of Tc1
Drosophila melanogaster	piwi	PIWI family	Gypsy ERV copia retrotransposon mdg1 retrotransposon	5' UTR N.D. N.D.
	aubergine	——	copia, I element Het-A, TART	ORFs Yes
Trypanosma bruceii	Ago1	AGO2	INGI retroposon SLACS retrotransposon	ORF 1 ORF 1 and 3' UTR
Neurosporra crassa	qde-2 dcl1/dcl2	AGO2 DICER1	Tad retrotransposon	ORF 1 and ORF 2
Mus musculus	Mili Miwi2 Dicer	HILI HIWI2 Dicer	IAP and L1 IAP and L1 SINE B1, MT	(piRNAs) L1, IAP SINE
Arabidopsis thaliana	Ago4	——	AtSN1 retroelement	AtSN1*

remain capable of autonomous movement, may be provide no benefit to the host and remain a genome parasite. At present, mobile genetic elements are the major factors influencing the health and evolution mammalian genomes (Kazazian, 2004).

Mammalian mobile DNA can be divided into two main categories: DNA transposons and retrotransposons. DNA transposons are discreet DNA sequences bound on either end by terminal inverted repeats (TIRs) (Plasterk et al., 1999). Retrotransposons includes endogenous retroviruses-like elements and retrotransposons that do not contain long terminal repeats, so-called non-LTR retrotransposons (Deininger and Batzer, 2002). DNA transposons are distinct from retrotransposons in their genome structure, as well as their mechanism for moving within the genome (Fig. 8.2).

DNA transposons

DNA transposons, such as the well-characterized Tc1/mariner family, are DNA sequences bound on either end by terminal inverted repeats

(TIRs) that serve as recognition sites for the element-encoded transposase enzyme (Fig. 8.2). Mammalian DNA transposons have a similar structure to the *C. elegans* Tc1 transposon that is under germline RNAi surveillance. DNA transposons move throughout the genome using a 'cut-and-paste' mechanism, whereby the transposase protein recognizes the TIRs and catalyses the excision of the DNA element leaving behind an empty parental site with a double-strand break that must be repaired by host-cell DNA repair proteins (Plasterk et al., 1999). Upon encountering the new genome target site, for which there might be some sequence preference, the ends of the DNA transposons attack the target DNA to generate staggered nicks allowing for integration of the transposon DNA and ligation by host proteins (Plasterk, 1996). In contrast with the continuing DNA transposon activity in the genomes of *C. elegans* and *Drosophila*, mammalian genomes are largely devoid of active DNA transposons, with only 3% of the human genome and 1% of the mouse genome represented by sequences related DNA transposons (Smit, 1999).

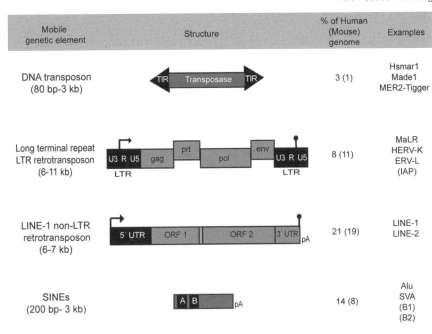

Figure 8.2 Classes of interspersed repeats in mammalian genomes. TIRs, terminal inverted repeats; LTR, long terminal repeat; *gag*, group specific antigen; *prt*, protease; *pol*, polymerase; *env*, envelope; UTR, untranslated region; ORF, open reading frame;, polyadenylation signal; pA, polyA tail; A and B boxes represent conserved regions of the RNA Pol III promoter in Alu sequences.

One can attribute the lack of DNA transposon activity in mammalian genomes to the absence of a functional transposase enzyme encoded by mammalian DNA transposons (Robertson and Zumpano, 1997). Mammalian cells are capable, however, of undergoing DNA transposition following introduction of a eukaryotic DNA transposon vector, such as the reconstituted Sleeping beauty transposon from teleost fish (Ivics *et al.*, 1997).

In the human genome, DNA transposon sequences have been characterized through bioinformatics efforts and grouped into a large category of repetitive DNA sequences called medium reiterated repeats (MERs) (Jurka *et al.*, 1996; Jurka *et al.*, 1993; Kaplan *et al.*, 1991). These human DNA transposon elements are also named miniature inverted-repeat transposable elements (MITEs) (Feschotte, 2002). More than 200,000 MER sequences populate the human genome, although there is no evidence that this family of repeat sequences continues to be active today (Oosumi *et al.*, 1995; Robertson and Zumpano, 1997). Further derivation of an MER consensus sequence indicated that some of these elements contain terminal inverted repeats typical of DNA transposons (Smit and Riggs,

1996). One of the MER sequences, designated Hsmar1, is a human DNA transposon related to the *Tc1/mariner* family of DNA transposons in *C. elegans*. Approximately 200 copies of Hsmar1 occur within the human genome and consist of 30 bp TIRs flanking an internal region ~1200 bp long where the transposase gene would normally reside (Robertson and Zumpano, 1997; Smit and Riggs, 1996). For all but one Hsmar1 sequence, which contains an apparently intact transposase ORF fused in-frame with a cellular gene, the internal region of every other Hsmar1 elements harbours insertions and other rearrangements that disrupt the transposase gene (Robertson and Zumpano, 1997). Another related *mariner*-like transposon element in the human genome is the 80 bp *mariner*-derived element 1 or Made1. Made1 sequences, which are represented in the genome by ~2,400 copies, possess 37 bp TIRs flanking a 6 bp intervening region (Smit and Riggs, 1996). Thermodynamic stability profiles suggest the Made1 palindromic sequence would form a stable hairpin structure if transcription occurred across the Made1 TIRs (Piriyapongsa and Jordan, 2007). For another MER group (MER2), which possess different TIRs than *Tc1/mariner*-like elements, several thousand copies of ~2.4 kb elements known as *Tigger1* and *Tigger2* are predicted by sequence analysis to encode an open reading frame with homology to the *Drosophila pogo* transposase enzyme (Smit and Riggs, 1996). An even larger number of MER2 sequences occupying the human genome appear to be internal deletion products of *Tigger* elements (Smit and Riggs, 1996).

Retrotransposons

The retrotransposons, which represent the majority of mobile genetic elements in mammalian genomes, move by way of an RNA intermediate that is converted into cDNA by the element-encoded reverse transcriptase. The integration process is completed using both retrotransposon and host proteins. Retrotransposons that can carry out this process of replicative integration through an RNA intermediate using their own element-encoded proteins are known as autonomous retrotransposons (Deininger and Batzer, 2002; Wei *et al.*, 2001). Autonomous retrotransposons can be further subdivided into long terminal repeat (LTR)-containing retrotransposons and non-LTR retrotransposons (Fig. 8.2). Within each class of retrotransposons (LTR-containing and non-LTR) exist a greater number of sequences that are non-autonomous retrotransposons because they harbour deletions or mutations that render them defective and incapable of carrying out retrotransposition on their own. An additional class of non-autonomous retrotransposons is represented by the short interspersed nuclear elements (SINEs) (Fig. 8.2). Many SINEs are derived from cellular RNAs that become integrated into the genome through the action of reverse transcriptase and integration proteins supplied by autonomous retrotransposons (Deininger and Batzer, 2002; Dewannieux and Heidmann, 2005a; Dewannieux and Heidmann, 2005b; Wei *et al.*, 2001).

Integration at the new genomic site takes place differently depending on whether the retrotransposon contains long terminal repeats (LTRs). For LTR retrotransposons like the mouse intracisternal A particle (IAP) (Kuff and Lueders, 1988), integration occurs much like it does for extracellular retroviruses (e.g. MMLV, HIV) and is carried out by the retrotransposon-encoded integrase *pol* complex. Non-LTR retrotransposons like the human LINE-1 integrate at new genomic locations using an entirely different mechanism, instead coupling reverse transcription to integration in a combined step that takes place following exposure of a single-strand nick at the target site mediated by the element's endonuclease protein (Luan *et al.*, 1993). This nick serves to prime reverse transcription of the RNA template to generate a DNA copy of the element, which becomes inserted into the new genomic location with the help of host proteins. Because SINEs integrate using the proteins supplied by non-LTR retrotransposons, SINE integration is analogous to the mechanism of LINE-1 retrotransposition (Dewannieux *et al.*, 2003; Dewannieux and Heidmann, 2005a; Esnault *et al.*, 2000).

Non-LTR retrotransposons: The Long INterspersed Elements

Long INterspersed Elements (LINE-1 or L1) remain an active autonomous non-LTR retrotransposon in both the mouse and human genomes. Approximately 20% of the mouse genome (~500,000 copies) comprises L1 sequence, separated into three active sub-families (T_F, A, and G_F) comprising ~3000 full-length L1s with intact ORFs that are potentially capable of autonomous retrotransposition (Goodier *et al.*, 2001; Naas *et al.*, 1998). Full-length mouse L1s (mL1) contain a bipartite 5′ untranslated region (5′ UTR) with promoter activity (DeBerardinis and Kazazian, 1999), two overlapping open reading frames (ORF1 and ORF2), and a 3′ UTR that ends in a poly(A) tail (Ostertag and Kazazian, 2001) (Fig. 8.2). ORF1 encodes a 40 kDa RNA binding protein that displays nucleic acid chaperone activity *in vitro* (Kolosha and Martin, 2003; Martin and Bushman, 2001; Martin *et al.*, 2005). ORF2 encodes a protein with an N-terminal endonuclease domain and a C-terminal reverse transcriptase domain (Goodier *et al.*, 2001; Naas *et al.*, 1998). Full-length mL1 transcripts and ORF1 protein (mORF1p) are present in the cytoplasmic fraction of F9 and C44 embryonal carcinoma cells, as well as early stage spermatocytes, suggesting that mouse L1s are both temporally and developmentally regulated (Branciforte and Martin, 1994; Martin and Branciforte, 1993; Trelogan and Martin, 1995). Genetic inactivation of DNMT3L, an important component of the DNA methyltransferase machinery, causes a dramatic increase in L1 expression in developing spermatocytes indicating that DNA methylation of the CpG-rich 5' UTR is an important mechanism controlling the expression of mouse L1s (Bourc'his and Bestor, 2004).

In the human genome, L1s represent about one fifth of genomic DNA sequence (Lander *et al.*, 2001). Human L1s (L1Hs) have a similar structure to their mouse counterparts in that L1Hs contains a 5' UTR with an internal promoter, two non-overlapping open reading frames, and a 3' UTR with a weak polyadenylation signal that terminates in a polyA tail (Ostertag and Kazazian, 2001) (Fig. 8.2). The L1Hs 5' UTR is ~900 bp long and is capable of transcribing 5' UTR sequence in both directions (Speek, 2001; Swergold, 1990). That is, the L1Hs 5' UTR contains a sense promoter (L1 SP) that transcribes the L1 ORFs beginning around the first nucleotide of the 5'UTR. The L1 also contains an antisense promoter that transcribes minus strand L1 sequence beginning at around position +380 with respect to the 5' end of the L1 (Yang and Kazazian, 2006). Recent evidence suggests that the mL1 5' UTR also possesses an antisense promoter (Muotri *et al.*, 2005). As observed for mL1s, L1Hs ORF1 encodes a 40 kDa protein that forms ribonucleoprotein particles with L1 RNA in the cytoplasm and is an important *cis*-acting factor required for retrotransposition (Hohjoh and Singer, 1996; Hohjoh and Singer, 1997; Kulpa and Moran, 2006). ORF2 encodes a protein with an N-terminal endonuclease domain (Feng *et al.*, 1996) and a C-terminal reverse transcriptase domain (Mathias *et al.*, 1991). For both human and mouse L1s, there is a strong preference for ORF1 and ORF2 to act on the RNA that encodes them, a characteristic known as *cis*-preference that limits the movement of defective L1s and other non-autonomous retrotransposons (Wei *et al.*, 2001).

Although two other distantly related L1 families, LINE-2 and LINE-3, have been uncovered through analysis of the human genome sequence, these elements show great divergence

and are not active (Deininger and Batzer, 2002). In fact, most of the L1 sequences in the human genome are defective, either through 5' end truncations, deletions, or mutations. Of the 3000 full-length L1s that inhabit the human genome, less than 100 are considered to have the capacity for autonomous retrotransposition (Brouha *et al.*, 2003). L1Hs expression has been examined almost exclusively in transformed tissue or cultured human cell lines. Extensive study of human germ cell tumours indicates that L1Hs mRNA and protein expression is very high in these transformed tissues associated with a loss of DNA methylation in the CpG-rich 5' UTR (Bratthauer and Fanning, 1992; Bratthauer and Fanning, 1993; Leibold *et al.*, 1990; Skowronski *et al.*, 1988; Skowronski and Singer, 1985; Su *et al.*, 2007; Thayer *et al.*, 1993). L1Hs expression has also been observed in transformed somatic tissues, suggesting that processes promoting cellular transformation bring on increases in L1 expression (Alves *et al.*, 1996; Asch *et al.*, 1996).

LTR retrotransposons

LTR retrotransposons are flanked by long terminal repeats and have a structure similar to that of retroviruses. Because of this retrovirus-like structure, LTR retrotransposons are often called endogenous retroviruses (ERVs). The ERV LTRs contain the necessary *cis*-regulatory sequences for efficient transcription of the ERV ORFs. For ERV sequences with intact ORFs, the first protein translated is the structural polyprotein *gag* that forms virus-like particles (VLPs) after binding the ERV RNA transcript (Maksakova *et al.*, 2006). Other ERV genes encode the protease (*pro*) and integrase (*pol*) proteins required for polypeptide cleavage and integration, respectively. Some ERV sequences contain a third ORF encoding the envelope protein (*env*) known to play a role in the extracellular life cycle of retroviruses (de Parseval and Heidmann, 2005; Maksakova *et al.*, 2006) (Fig. 8.2).

Mouse LTR retrotransposons

LTR retrotransposons represent the most active retrotransposons in the mouse genome. These murine endogenous retroviruses (muERVs) are divided into three classes based on their similarity to infectious retrovirus clones. Class

I muERVs are endogenous gammaretroviruses that entered the mouse genome through germline infection and are typified by murine leukaemia virus (MLV) (Maksakova *et al.*, 2006). Class II muERVs are endogenous betaretroviruses of which mouse mammary tumour virus (MMTV) is a member. Several class II members, such as the intracisternal A particle (IAP) and Early Transposon (ETn) element, remain active and are responsible for most of the spontaneous mutations observed in mouse strains (Maksakova and Mager, 2005). IAP retrotransposons, present in 1000 copies per haploid genome, are expressed at low levels through embryonic development and into adulthood (Dewannieux *et al.*, 2004). IAP expression is up-regulated in some mouse tumours, especially plasmacytomas and teratocarcinomas (Kuff and Lueders, 1988). IAP expression also increases in Dnmt3L-null mice, similar to the effect on the expression of mL1, implicating DNA methylation in the suppression of IAP expression as well (Bourc'his and Bestor, 2004).

The group of ETn (early transposon) elements, another type of active LTR retrotransposon in the mouse genome, is made up of two closely related elements ETnI (200 copies/haploid genome) and ETnII (40 copies/haploid genome) that differ slightly in their DNA sequences (Baillie *et al.*, 2004; Baust *et al.*, 2003; Maksakova and Mager, 2005). ETns possess LTRs, but instead of containing the retrovirus gag and pol genes, the internal sequence of both ETnI and ETnII elements is mostly non-viral sequence of unknown origin (Mager and Freeman, 2000). In contrast with the broad expression of IAP retrotransposons, the expression of ETn elements is limited to specific cell types at defined time points during embryogenesis (Baust *et al.*, 2003). Movement of ETn elements occurs through *trans*-acting retroviral proteins supplied by a highly related class II ERV called MusD (Ribet *et al.*, 2004). The MusD family (100 copies/haploid genome) is a ~7.5 kb-long element that contain LTRs flanking the genes for *gag*, *pro*, and *pol* (Baust *et al.*, 2003; Ribet *et al.*, 2004). RT-PCR analysis with primers that can distinguish between ETn and MusD LTRs showed that both of these elements are expressed in embryonic tissues, with relative ETn RNA levels higher than that of MusD (Baust *et al.*, 2003).

LTR retrotransposons that fall into class III include the MuERV-L family and other subgroups of LTR-driven mammalian-like retrotransposons called MaLR sequences (Benit *et al.*, 1999; Maksakova *et al.*, 2006; Smit, 1993). The class III ERV sequences account for ~5% of mouse genome sequence, but most of its members are considered inactive due to point mutations and other rearrangements. Despite the apparent absence of coding-competent class III ERVs in the mouse genome, transcripts from MaLR and MuERV-L elements are detected in oocytes indicating that a basal level of transcription of these LTR retrotransposons is maintained (Smit, 1993). Moreover, the LTR sequences of several class III ERV elements, including MuERV-L, can serve as alternative promoters and regulate the expression of several genes in full grown oocytes and cleavage-stage embryos (Maksakova *et al.*, 2006). Thus, even though class III ERVs are unlikely to impact the genome through insertional mutagenesis, their physiological role in regulating host cell gene expression is a novel function of retroelements than one must consider when defining the true role of endogenous small RNAs that may originate from these LTR retrotransposons.

Human LTR retrotransposons

The human genome contains an estimated 8% of DNA sequence derived from human endogenous retroviruses (HERVs) (de Parseval and Heidmann, 2005). HERV members are transcribed from their LTRs, but in contrast to MuERVs, the vast majority of HERV provirus sequences harbour defective *gag, pro* and *pol* genes, yet contain an intact *env* gene (de Parseval and Heidmann, 2005). Although the majority of HERVs lack intact structural genes, HERV virus-like particles formed through the action of the gag protein have been observed in placenta and teratocarcinoma cell lines, indicating that some HERV members express coding-competent structural proteins. Real-time PCR analysis using primers capable of discriminating between full-length *env* transcripts from different HERV members indicates that *env* expression is highest in placenta tissue and much lower in other human tissues. The high level expression of the HERV *env* gene product in placenta, combined with the fusogenic properties of the *env* protein, suggests that these proteins plays an important role in cell–cell fusion during formation of the syncytiotrophoblast layer in pregnancy (Mi *et al.*, 2000). Formal proof that HERV *env* proteins are required for normal placenta formation awaits the demonstration that inhibiting specific *env* gene function, perhaps through RNAi directed against *env* transcripts, significantly disrupts syncitia formation.

In addition to HERV sequences, approximately 5% of the human genome comprises class III ERV-L sequences. Most of these are represented by the MaLR subfamily whose structure contains LTRs that flank a ~1,300 bp ORF of unknown identity (Smit, 1993). Many of these elements exist as solitary LTRs and show a strong bias against existing within an intron in the same orientation as transcription occurs, possibly owing to the polyadenylation signals that reside in LTRs that could terminate transcription prematurely (Smit, 1993). A consequence of this bias is that MaLR sequences are unlikely to be expressed through the large amount of read-through transcription occurring in the human genome. In spite of this bias, however, a databank screen using a consensus sequence derived from the MaLR ORF identified 48 MaLR-containing ESTs in a brain cDNA library (Smit, 1993).

SINEs in the mammalian genome

SINEs, or Short INterspersed Elements, are 80–400 bp non-coding DNA sequences separated into different families, some of which are primate or rodent specific. Some SINEs contain their own mode of transcription from an internal RNA polymerase III promoter (Fig. 8.2). Other SINE sequences, such as processed pseudogenes, do not contain a promoter sequence and are unlikely to be highly expressed in mammalian tissues. In the human genome, the most abundant SINE element is the Alu retrotransposon, which is represented in at least 1.5 million copies per haploid genome and comes from the 7SL RNA. Full-length Alus are ~ 280 bp consisting

of two related monomer sequences, which lack any discernible ORF, and a terminal polyA tail (Moran, 2002). Alu transcription is initiated by an RNA pol III promoter that may be induced by genotoxic stress (Hagan *et al.*, 2003; Rudin and Thompson, 2001). Despite the fact that Alus lack proteins required for their retrotransposition, empirical evidence indicates that Alus remain the most promiscuous non-LTR retrotransposon in the human genome by co-opting for their own movement the retrotransposon proteins of other non-LTR retrotransposons (i.e. LINEs) (Dewannieux *et al.*, 2003). Indeed, the ability of Alus to efficiently use LINE proteins for their own movement, combined with their high copy number and propensity to serve as substrates for unequal recombination, make Alus the most mutagenic of all human mobile genetic elements (Deininger and Batzer, 2002). In addition to Alus, the human genome contains ~2800 copies of a composite element named SVA, for its main components: SINE, VNTR, and Alu. Although the expression of SVA elements awaits confirmation, at least four *de novo* mutations are attributed to insertional mutagenesis by SVA indicating that at least a small level of SVA transcription occurs (Ostertag and Kazazian, 2001).

While Alus are primate-specific SINEs, the rodent genome is populated by another SINE derived from the 7SL RNA, the B1 and related B2 elements. B1s are present in ~50,000 copies per haploid genome and are comprised of a ~135 bp monomer also derived from the 7SL RNA (Moran, 2002). Similar to Alu, B1 contains an internal RNA Pol III promoter and a terminal polyA tail, possibly added post-transcriptionally following termination of Pol III transcription at T-rich termination sequence. B1 elements can use the retrotransposition machinery of full-length L1s to integrate, although no spontaneous mouse mutations have been attributed to B1 integration (Dewannieux and Heidmann, 2005a). Various other SINEs are known to exist in both the mouse and human genomes and derive from tRNAs referred to as mammalian-wide interspersed repeats (MIR) (Smit and Riggs, 1995). None of these MIRs appear capable of expression and likely represent fossils of ancient mobile elements.

RNA silencing of mobile genetic elements in the mouse and human genomes

The rich bioinformatics resources developed over the last decade, combined with large-scale biochemical purification from human cells, demonstrated that mammalian RNA silencing pathways are largely analogous to similar mechanisms in other eukaryotes. The area of mammalian RNA silencing that has lagged behind its mechanistic counterpart in other eukaryotes is a greater understanding of the biological role for these RNA silencing pathways beyond the action of mammalian microRNAs (Soifer, 2006). In model eukaryotes, the high rate of homologous recombination (HR) and ability to perform large-scale genetic screens, permits the study of mutant phenotypes through insertion and/or inactivation of specific genes. Moreover, the recent application of RNAi technology to selectively inhibit gene function in mammalian cells both in culture and *in vivo* has made it less necessary to rigorously pursue methods that enhance the efficiency of HR in mammalian cells. While it is possible to achieve transient knockdown of RNAi gene components by transfecting mammalian cells with large quantities (> 50 nM) of siRNA, functional inhibition of the RNAi pathway is not stable and varies in direct proportion to the transfection efficiency (Hutvagner *et al.*, 2001; Meister *et al.*, 2005). A recent report, however, demonstrates that stable knockdown of Drosha, a nuclear RNAse III enzyme that cleaves primary miRNA transcripts, was possible in a transformed human cell line by expressing an anti-Drosha short hairpin RNA (shRNA) under the control of a tetracycline-inducible RNA Pol III promoter (Aagaard *et al.*, 2007). In addition, some animal and plant viruses contain gene products that are natural inhibitors of mammalian RNA silencing pathways. For example, over expression of the adenovirus VA1 non-coding RNA decreases the function of some cellular miRNAs by successfully competing with Drosha-processed pre-miRNA molecules for the Exportin 5 transport pathway out of the nucleus (Lu and Cullen, 2004). Other plant virus proteins, such as the tombusvirus P19 and tenuivirus NS3, bind and sequester siRNA triggers in inactive complexes (Bucher *et al.*, 2003; Lakatos *et al.*, 2004; Vargason *et al.*, 2003).

Recognizing the difficulties in obtaining genetic evidence that mammalian mobile genetic elements are under the control of specific components of the RNA silencing machinery, some investigators have focused on determining whether mammalian cells contain small RNAs derived from DNA transposons and retrotransposons. In lower eukaryotes where classical genetics has established a direct link between RNAi and the control of mobile genetic elements, siRNAs have been detected for both DNA transposons and retrotransposons. In many cases, stability of the transposable element-derived (TE-derived) siRNA is dependent on RNA silencing components (Hamilton *et al.*, 2002; Kuhlmann *et al.*, 2005; Nolan *et al.*, 2005; Sijen and Plasterk, 2003). Surprisingly, few groups have used mobile genetic element sequences as probes for Northern blot analysis of small RNAs in mammalian cells. In one published report, 21 nt siRNAs were detected in the human embryonic kidney cell line HEK293 using strand-specific RNA probes spanning nts. 1–500 of the L1Hs 5′ UTR (Yang and Kazazian, 2006). Probes from other regions of the human L1, including the two ORFs, failed to detect any additional 21 nt L1 RNAs, indicating that the region of convergent transcription within the L1 5′ UTR is required for siRNA production. It remains possible that this 21mer hybridization signal represents a pool of L1 5′ UTR-derived siRNA processed from a larger precursor. Cloning and additional characterization of this L1 siRNA population in HEK293 cells is necessary before it can be confirmed as a functional siRNA duplex.

Additional efforts to characterize small RNAs in mammalian cells have combined biochemical fractionation of protein-bound small RNAs with small RNA cloning methods. In one small-scale study performed in murine ES cells, investigators cloned small RNAs that bound to the tombusvirus P19 protein, a viral inhibitor of RNAi that preferentially binds dsRNAs from 18 to 20 nt (Calabrese and Sharp, 2006; Vargason *et al.*, 2003). More than 1300 clones were sequenced from murine ES cells following immunoprecipitation with an anti-P19 antibody and analysed by bioinformatics for alignment to the mouse genome. No obvious 18–20 nt dsRNA that pulled down in the P19 complex showed homology to murine mobile genetic elements. In fact, a majority of P19-bound small RNAs came from ribosomal RNA genes and are of unknown significance. Several clones sequenced from the supernatant control fractions that did not bind P19, however, did show sequence homology to murine retroelements (Calabrese and Sharp, 2006). TE-derived small RNAs in the supernatant fractions mapped to a number of different retrotransposons including B1 SINEs, ETn and class III ERVs, and LINE-1 non-LTR retrotransposons. The fact that these retroelement-derived small RNAs do not bind P19, which has a much lower affinity for ssRNA and dsRNA longer than 20 nucleotides (Vargason et al., 2003), suggests that retroelement-derived small RNAs in murine ES cells are either single-stranded or, if they are duplexes, their individual RNA strands are larger than 20 nt and do not bind to P19. It should be possible to determine whether these cloned small RNAs can inhibit transposable elements by PTGS through the use of retrotransposon-indicator plasmids that exhibit measurable autonomous movement in cultured murine cells (Dewannieux and Heidmann, 2005a; Heidmann and Heidmann, 1991; Moran et al., 1996; Soifer, 2006).

The notion that siRNAs and small RNAs derive from mammalian mobile genetic elements suggests that they are processed from larger dsRNA precursors by cellular RNAse enzymes. In mammalian cells, the RNAse III enzyme Drosha works with its partner DGCR8 in a microprocessor complex to cleave primary miRNA hairpin transcripts in the nucleus into smaller precursor miRNAs (Gregory et al., 2004; Han et al., 2004; Han et al., 2006; Landthaler et al., 2004; Lee et al., 2003; Wang et al., 2007). Drosha is an intriguing player for nuclear processing of TE-derived dsRNA in mammalian cells because Drosha cleavage of TE-derived dsRNA into smaller units could prevent cellular responses that occur in the cytoplasm when specific proteins such as PKR (protein kinase-dsRNA) recognize dsRNA over 30 nt (de Veer et al., 2005; Karpala et al., 2005). The two published reports demonstrating functional inhibition of Drosha in human cell lines did not examine the up-regulation of mobile genetic elements (Aagaard et al., 2007; Lee et al., 2003).

Drosha action on TE-derived dsRNA could produce smaller dsRNAs for further processing by Dicer into TE siRNA. The human and mouse genomes encode one Dicer (DCR) protein (Nicholson and Nicholson, 2002; Provost et al., 2002), an enzyme with two RNase III domains that forms an intramolecular dimer to cleave dsRNA in a processive manner producing 21–25 nucleotide siRNAs (Macrae et al., 2006; Zhang et al., 2004). While the embryonic lethality observed in mice with a homozygous Dicer knockout genotype (Dcr-1 −/−) confirms an essential role for Dicer in mammalian development, the early death (E7.5) also prevents further biochemical study of knockout embryos (Bernstein et al., 2003). To examine Dicer function in cell culture, Dicer-deficient mouse embryonic stem (ES) cells were created by gene targeting and homologous recombination (Kanellopoulou et al., 2005). Increased transcription of murine L1 and IAP elements was observed in the absence of Dicer, but not wild-type ES cells, providing the first genetic evidence that Dicer controls the expression of murine retrotransposons (Kanellopoulou et al., 2005). The expression of both L1 and IAP retrotransposons was measured by quantitative RT-PCR and one cannot exclude the contribution of non-functional L1 and IAP transcripts to the increased level of expression in the absence of Dicer. This work does support an earlier report in which IAP and MuERV-L transcripts were upregulated following injection of anti-Dicer dsRNA into two- and eight-cell stage mouse embryos (Svoboda et al., 2004).

Conditional knockout of Dicer in mouse oocytes produce viable offspring with defects in oogenesis (Murchison et al., 2007). In addition, this model provides some interesting observations concerning Dicer-mediated control of mobile genetic elements in the female germline. First, transcription of L1 and IAP retrotransposons was not significantly changed in Dicer-null oocytes, suggesting that these active retrotransposons are under control of a Dicer-independent silencing process. On the other hand, the transcription of several non-functional retrotransposons such as class III ERV (MT and MaLR) retrotransposons and SINE B elements was significantly increased in Dicer-null oocytes. The increase in class III ERV transcripts, which come from inactive ERV copies implies that mobile genetic elements not considered a threat for autonomous movement are under the control of Dicer-mediated RNA silencing. The second interesting observation was that SINE and MaLR retrotransposon sequence motifs were significantly enriched in the 3′ UTR of mRNA transcripts that are upregulated in the absence of Dicer. In contrast, sequences from LINE retrotransposons were not enriched within the 3′ UTR of these transcripts. Thus, oocyte transcripts with TE-derived sequences in their 3′ UTRs could be under Dicer regulation. It is intriguing to speculate that the trigger molecules that binds these TE-derived 3′ UTR motifs are produced through Dicer processing of TE-derived dsRNA (Murchison et al., 2007).

Examining the function of Dicer in human cells beyond the important role in pre-miRNA processing has been difficult because of the inability to obtain stable inhibition of Dicer gene function. Several groups have achieved transient knockdown of Dicer mRNA and proteins levels by as much as 60% using anti-Dicer siRNA transfected into human HeLa cells (Hutvagner et al., 2001; Lee et al., 2002; Yang and Kazazian, 2006), however, it is unclear how much functional inhibition of Dicer activity is achieved with siRNA. One published report measured a 15–20% increase in L1 expression as determined by quantitative RT-PCR in HeLa cells following siRNA treatment that reduced Dicer gene expression by 50% (Yang and Kazazian, 2006). While this small increase in L1 expression proved statistically significant, it remains to be determined if full-length, coding-competent L1s constitute the bulk of L1 transcripts induced by Dicer knockdown, or if the increased number of transcripts contain mostly read-through RNAs containing non-functional L1 sequences. This study also showed the presence of L1-derived siRNAs in human cell lines, but did not present data assessing L1 siRNA levels in HeLa cells treated with Dicer siRNA (Yang and Kazazian, 2006). Stable attenuation of Dicer activity has been achieved by deleting exon 5 of DCR1 in the human colon carcinoma cell line HCT116 through homologous recombination. Further study is required, however, to determine whether

human TEs are up regulated in this Dicer hypomorph cell line (Cummins et al., 2006).

For the mammalian RNAi pathway, siRNAs produced by Dicer are handed off to the RNA-induced silencing complex (RISC) (Chendrimada et al., 2005; Hammond, 2005). While the totality of factors that make up mammalian RISC await full characterization, siRNA-mediated knockdown of AGO2 in HeLa cells (Meister et al., 2004), as well as gene targeting experiments in mice (Liu et al., 2004), demonstrate that AGO2 is essential for target mRNA cleavage. AGO2, one of eight argonaute proteins in the human genome, contains the characteristic PAZ RNA binding domain and Piwi catalytic motif where cleavage activity resides (Carmell et al., 2002; Sasaki et al., 2003). Immunoprecipitation of AGO2 complexes in human HeLa cells lead to the identification of additional proteins that co-localize with AGO proteins in cytoplasmic processing bodies and are necessary for siRNA-directed cleavage (Meister et al., 2005). These additional proteins include the putative DExD-box helicase MOV10 and an unknown protein KIAA1093 that contains glycine/tryptophan repeats (GW repeat) and an RNA recognition motif. MOV10 is a candidate homologue of the *Drosophila* RNA helicase *armitage* known to play a role in RNAi in the *Drosophila* germline. Although *Drosophila armitage* mutants fail to assemble an active RISC complex, the induction of *Drosophila* mobile genetic elements in *armitage* mutants awaits determination (Tomari et al., 2004).

Selective inactivation of AGO2 orthologues in lower eukaryotes demonstrates that RISC-associated Ago proteins are required for silencing both DNA transposons and retrotransposons. For example, loss of the AGO2 orthologue *qde-2* in *Neurospora crassa* leads to increased expression of the LINE-like retrotransposon, Tad (Nolan et al., 2005). In addition, increased transcript levels of the Ingi and SLACS retrotransposon elements are observed in cells lacking Ago1, the AGO2 orthologue of *Trypanosoma brucei* RISC (Shi et al., 2004). Several other spontaneous or induced AGO mutants, such as the *Arabidopsis Ago4* (Zilberman et al., 2003) and *Drosophila piwi* mutants (Sarot et al., 2004), also show elevated levels of retrotransposon expression (Table 8.1).

So far, there is no evidence linking any mammalian RISC component to RNA silencing of mobile genetic elements. Increased activity of murine transposable elements was not examined when Ago2 was disrupted by homologous recombination and was likely complicated by the embryonic lethality of Ago2-null mice (Liu *et al.*, 2004). Furthermore, activation of human mobile genetic elements was not assessed in HeLa cells subjected to siRNA-mediated knockdown of AGO2, MOV10, and KIAA1093 (Meister and Tuschl, 2004). Future experiments in cultured human cells using these effective 'anti-RISC' siRNAs will be valuable in assessing whether mammalian mobile genetic elements are subject to PTGS.

Mammalian TE-derived small RNAs associate with germline Argonaute proteins *in vivo*

For eukaryotes where RNA silencing mechanisms have been characterized, members of the Argonaute family are known in all cases to play important roles. Argonaute proteins contain two characteristic domains: a PAZ (Piwi/Argonaute/Zwille) domain that binds nucleic acids and a PIWI domain with a crystal structure similar to that of RNAse H (Hammond, 2005). In human cells, AGO2 is a key component of RISC and acts through the PIWI domain to perform endonucleolytic cleavage of target RNAs (Rivas *et al.*, 2005). In addition, both AGO1 and AGO2 have been shown by chromatin immunoprecipitation to associate with DNA loci targeted for transcriptional gene silencing by ectopic siRNA (Janowski *et al.*, 2006; Kim, 2006). In mice, there are seven members of the Argonaute superfamily: four argonaute members (Ago1-Ago4) and three piwi members (MIWI, MILI, MIWI2) (Carmell *et al.*, 2002). The human genome contains eight members of the Argonaute superfamily: four argonaute members (AGO1-AGO4) and four piwi members (HIWI, HILI, PIWIL3, HIWI2) (Sasaki *et al.*, 2003). For the most part, Ago members are ubiquitously expressed and associate with miRNAs and siRNAs (Mourelatos *et al.*, 2002). The expression of Piwi proteins, on the other hand, is largely restricted to the male germline and stem cells (Carmell *et al.*, 2002). Miwi- and Mili-null mice are viable with unique

defects in spermatogenesis, indicating that these proteins have important but distinct roles in germ cell development (Deng and Lin, 2002; Kuramochi-Miyagawa *et al.*, 2004).

Recent studies reporting the characterization of unique small RNAs from rodent and human germ cells provide direct evidence of an RNA-based silencing pathway that regulates mammalian mobile genetic elements (Aravin *et al.*, 2006; Girard *et al.*, 2006; Grivna *et al.*, 2006a; Grivna *et al.*, 2006b; Lau *et al.*, 2006; Watanabe *et al.*, 2006). For a few of these studies, the cloning of germline small RNAs was prompted by the observation that total RNA from adult mouse testis contained an abundant population of RNAs that migrated at ~30 nt and could be visualized by ethidium bromide staining or through 5' ^{32}P-labelling. Further purification showed that these ~30 nt RNAs associate with germline argonaute homologues of *Drosophila Piwi*, specifically Miwi and Mili (Aravin *et al.*, 2006; Girard *et al.*, 2006; Grivna *et al.*, 2006a; Grivna *et al.*, 2006b; Lau *et al.*, 2006). The *in vivo* association of these 30 nt RNAs with germline homologues of *Piwi* lead to a new classification of endogenous small RNAs known as *Piwi*-interacting RNAs (piRNAs) (Girard *et al.*, 2006). PiRNA expression is markedly reduced in Mili- and Miwi-null indicating that these proteins are required for piRNA biogenesis and/or stability (Aravin *et al.*, 2007; Grivna *et al.*, 2006a).

Detailed analysis of cloned small RNA libraries from the mammalian germline revealed several interesting characteristics of piRNAs. First, piRNAs are single-stranded molecules ~26–30 nt in length with a strong bias for a uridine at the 5'-most position. Second, piRNAs tend to be clustered in shorter genome regions, with some chromosomes harbouring more piRNAs clusters than others. Third, piRNAs originating from the same cluster are usually derived from the same orientation, suggesting that piRNAs are processed from a long primary transcript. Additionally, piRNA expression within some clusters abruptly switches from one DNA strand to the other, implying that a central bi-directional promoter might be responsible for transcribing these bi-directional piRNA clusters (Aravin *et al.*, 2006; Girard *et al.*, 2006; Lau *et al.*, 2006). Northern blot analysis confirmed that piRNAs are single-stranded RNAs expressed in developing spermatids beginning around day 14 post partum and in adult mice (Aravin *et al.*, 2006; Girard *et al.*, 2006; Watanabe *et al.*, 2006). Biochemical probing of mouse piRNAs revealed that the 5′-termini of piRNAs contain a phosphate, whereas the 3′-termini are fully modified by 2′-O-methylation (Kirino and Mourelatos, 2007; Ohara *et al.*, 2007). The 2′-O-methyl ribose group at the 3′ terminus could make the mature piRNA more resistant to degradation by cellular ribonucleases.

Bioinformatics analysis of cloned piRNAs indicated that the great majority map to unique regions of the rodent and human genomes, with only ~17% of piRNAs coming from mammalian mobile genetic elements, mostly retrotransposons. For several of these experiments, an unbiased approach to germline small RNA cloning was taken by constructing small RNA libraries using 18–30 nt RNAs excised from gels of separated total germ cell RNA. In one study, Watanabe *et al.* examined the small RNA profile of mouse oocytes, whose small RNA content differs from that of the mouse testis in that oocytes lack the prominent 30 nt testes-specific piRNA band (Watanabe *et al.*, 2006). From a small library of ~400 clones, 40 different retroelement-derived small RNAs ranging in size from 20–25 nt were identified. Retroelement-derived small RNAs in mouse oocytes arose mostly from L1 and IAP retrotransposons and were cloned with both sense and antisense polarity. Approximately half of retroelement-derived small RNAs cloned from oocytes mapped to ancient LTR retrotransposons such as MaLR and other ERV III retroelements that are likely to be completely inactive in the mouse genome. The authors went on to demonstrate that insertion of either L1, IAP, or MaLR sequences into the 3′ UTR of a GFP reporter gene resulted in reduced GFP levels when injected into oocytes, suggesting that the endogenous retroelement-derived small RNAs in oocytes are active and can suppress their target retrotransposons by post-transcriptional gene regulation (Watanabe *et al.*, 2006). It is important to note that some of the mobile genetic elements for which these oocyte-specific small RNAs map were also found as potential seed sites in the 3′ UTR of transcripts that are upregulated when Dicernull mouse oocytes (Murchison *et al.*, 2007). Therefore, these retrotransposon-derived small RNAs in mammalian oocytes might have a second role in regulating the expression of transcripts that contain retrotransposon seed sites in their 3′ UTR.

The study by Watanabe *et al.* also examined the small RNA profile in mouse testes at three different stages and identified 44 different retroelement-derived small RNAs from mouse testes that ranged in size from 18–36 nucleotides (Watanabe *et al.*, 2006). Most of these clones were derived from IAP and L1 retrotransposons, although a significant portion of clones map to B1 SINEs. No class III ERV retrotransposons were cloned from this small subset of piRNAs from mouse testes. Furthermore, the expression of two retroelement-derived piRNAs (IAP and B1) was confirmed by Northern blot analysis of total testicular RNA in pubescent mice. Some of these cloned retroelement-derived small RNAs, which this paper called repeat-associated RNAs (rasiRNAs), are probably not true piRNAs because they are smaller (18–23 nt) than other piRNAs or they were cloned from pre-pubescent day 8 mice at a time when MIWI and MILI are not expressed (Deng and Lin, 2002; Kuramochi-Miyagawa *et al.*, 2004; Watanabe *et al.*, 2006). The fact that different retroelement-derived small RNAs were cloned from mouse oocytes and testes is not surprising considering that these two germ cell organs possess abundant small RNAs of distinct sizes.

Rather than rely on the preparation of small RNA libraries from mammalian testes by excising different sized RNAs from total RNA fractions, several groups prepared small RNA libraries from RNAs that immunoprecipitated with MIWI or MILI. To examine small RNAs that associate with the germline argonaute protein MIWI, Girard *et al.* performed highly parallel pyrosequencing of small RNA libraries after MIWI immunoprecipitation of mouse testis lysate. The pyrosequencing method permitted the sequence identification and bioinformatics assessment of >50,000 RNA sequences with a mean length of 29–30 nucleotides (Girard *et al.*, 2006; Margulies *et al.*, 2005). For this data set of MIWI-interacting piRNAs, approximately 3% mapped to mouse L1 and LTR retrotransposons, with a much smaller number of MIWI-interacting piRNAs coming from murine SINE or more ancient retroelements. No further detail was given to distinguish piRNAs originating from different types of LTR retrotransposons, prohibiting a direct comparison of this large dataset with the LTR-derived piRNAs cloned in the smaller study by Watanabe et al (Girard et al., 2006; Watanabe et al., 2006). In addition to cloning piRNAs from mouse testis, this study by Girard et al. went further by determining the sequence of candidate piRNA populations in human and rat testes. Human and rat piRNAs have the same characteristics as mouse piRNAs indicating that piRNAs are a conserved class of small RNAs in the testes of mammals. Human and rat piRNA sequences were deposited into the GenBank database, but the number of retroelement-derived piRNAs in the human and rat piRNA data sets was not reported (Girard et al., 2006).

At the same time as Girard et al. and others were reporting MIWI-interacting piRNAs (Girard et al., 2006; Grivna et al., 2006a), Tom Tuschl and colleagues immunoprecipitated MILI ribonucleoprotein complexes from the adult mouse testes and used 5′ ^{32}P-labelling to reveal a population of 26–28 nt RNAs that were absent in newborn mice (Aravin et al., 2006). MILI-interacting piRNAs are smaller than piRNAs shown to bind the related argonaute protein MIWI. The size difference between individual piRNAs populations could be due to separate biogenesis pathways mediated by the interaction of primary piRNA transcripts with either Mili or Miwi. Small RNA libraries were prepared from MILI immunoprecipitated fractions from adult mouse testis, along with two additional populations of small RNAs ranging in size between 18–26 nt and 24–33 nt that did not co-purify with MILI. Over 15,000 sequences were obtained from the three small RNA libraries, with the majority (~65%) of the 18–26 nt library comprised of miRNAs, ribosomal RNA (rRNA), and sequences that map to known mRNAs. In contrast, the 24–33 nt and MILI-interacting RNA libraries contained a majority of clones that mapped to unique positions in the genome or mapped to murine mobile genetic elements (Aravin et al., 2006).

Bioinformatics analysis determined that 13% of MILI-interacting RNAs derive from mouse TEs including some DNA transposons, LTR retrotransposons, and non-LTR retrotransposons. Interestingly, retrotransposon-derived piRNAs were cloned with both sense and antisense polarity, distinguishing these retrotransposon-derived piRNAs from the majority of piRNAs, which map to only one chromosome strand (Aravin et al., 2006). For piRNAs that mapped to non-LTR retrotransposon, ~80% of clones were derived from mL1s, with the remaining non-LTR retrotransposon-derived MILI-interacting piRNAs derived from non-active L2 and RTE elements. The piRNAs that mapped to LTR retrotransposons showed a greater distribution amongst different types of endogenous retrovirus sequences. As with the study by Girard et al., the identity of cloned LTR retrotransposon-derived piRNAs is not sufficiently detailed in the text to distinguish individual retrotransposons, although MaLR and ERVK (e.g. IAP) piRNAs appear to be most abundant (Aravin et al., 2006; Girard et al., 2006).

In a third independent study, Lau et al. fractionated extract from rat testis by ion-exchange chromatography to uncover endogenous small RNAs that were larger than ~22 nt mature miRNAs (Lau et al., 2006). Small RNA libraries constructed from column elute and flow-through material was subjected to high-throughput sequencing to generate almost 100,000 raw sequence reads. In contrast to the RNAs obtained from flow-through material that consisted primarily of miRNAs, the RNA sequences obtained from the eluate fraction showed the characteristics of piRNAs in that they were 29–31 nt long, had a strong bias for a 5′ uridine, and mapped to unique regions of the genome not previously regarded as regions of high transcriptional activity. Multi-step purification of testes lysate lead to the identification of the rat piwi orthologue Riwi as one important protein component of the rat piRNA complex (Lau et al., 2006). This multi-step biochemical purification also identified the rat homologue of RecQ1 (rRecQ1), an important ATP-dependent DNA helicase, as a member of the rat piRNA complex as well. Although the rat small RNA libraries were not cloned directly from immuno-precipitated piRNA complexes, as was done for the previous studies reporting Mili- and Miwi-bound small RNAs, the existence of a piRNA ribonucleoprotein complex was confirmed indirectly by monitoring column fractions for the presence of 29–31 nt RNA by 5′ end-labelling along with Riwi and RecQ1 by immunoblot. This study by Lau *et al.* also examined an equal number of piRNA sequence reads from mouse testes (Lau *et al.*, 2006).

Bioinformatics indicated that only 20% of the rat and mouse testis small RNAs matched mobile genetic elements, a smaller percentage than would be expected by random sampling of the rat genome, but a result consistent with the other large piRNA cloning experiments (Aravin *et al.*, 2006; Girard *et al.*, 2006). In agreement with other studies, the individual small RNAs that mapped to mobile elements possessed sense and antisense polarity (Aravin *et al.*, 2006; Girard *et al.*, 2006; Lau *et al.*, 2006). For both the rat and mouse small RNA data sets, several of the individual small RNAs mapping to mobile genetic elements were cloned more than once, although the majority of small RNAs were identified as single sequence reads. For the set of small RNAs derived from non-LTR retro-transposons, a majority of the clones mapped to murine LINE-2 sequences with a smaller subset represented by piRNAs from LINE-1. For the class of LTR retrotransposons, small RNAs from class III muERV-L and MaLR sequences were identified most often, along with a smaller number of RNAs derived from class II ERVK retrotransposons like IAPs. In addition to the retrotransposon-derived small RNAs, this study by Lau *et al.* also cloned a set of testis-specific RNAs that mapped to DNA transposons of the MER1 family (Lau *et al.*, 2006). A noteworthy observation from this study was the identification of numerous small RNAs that mapped to DNA transposons and retrotransposons, but that were smaller than the expected size for piRNAs. If their *in vivo* association with piwi orthologues, as well as their larger size of 26–31 nt classify piRNAs as distinct RNA silencing triggers, then some of the small RNAs derived from murine mobile genetic elements are clearly not piRNAs

because they are 18–25 nt in length and do not bind piwi proteins *in vivo*.

Evidence is building for a Piwi-driven silencing program that specifically targets mammalian retrotransposons in the germline. In the testes of Mili-null mice, a significant upregulation of L1 and IAP transcription is observed. Elevated levels of L1 retrotransposon transcripts are due in part to loss of DNA methylation at the L1 5′ UTR, implying that Mili, and likely Mili-interacting piRNAs, are required for transcriptional gene silencing of L1 promoter sequences in the mammalian testis (Aravin *et al.*, 2007). High-content sequence analysis of Mili-interacting RNAs indicated that pre-pachytene spermatocytes contain more piRNAs derived from mobile genetic elements, with the vast majority coming from SINE and LTR retrotransposons. A similar up-regulation of L1 and IAP retrotransposons was also observed in the testes of Miwi2 knockout mice. Loss of DNA methylation at L1 5′ UTR and IAP LTR sequences was found in testis genomic DNA but not tail DNA, indicating that the defect in Miwi2 control of L1 and IAP is limited to the male germline (Carmell *et al.*, 2007). These results implicating Piwi proteins in the regulation of L1 and IAP retrotranspons in the male germline, combined with the results in mouse oocytes indicating that MT and SINE retrotransposons but not L1 and IAP are under Dicer regulation, suggest that both Dicer-dependent and Piwi-driven pathways suppress different mobile genetic elements in mammalian genomes.

Important studies in the future, employing both bioinformatics and biochemical approaches, should refine the alignment of piRNAs to active sub-families of mammalian TEs and determine using cell culture assays whether TE-derived piRNAs function by post-transcriptional gene silencing. Furthermore, it is not clear whether the upregulated L1 and IAP transcripts and Mili and Miwi2 knockout testes are retrotransposition-competent and pose a threat the germline integrity through insertional mutagenesis and therefore warrant protection against. Immunohistochemical measurement of L1 and IAP protein expression in these knockout mice using antibodies derived from functional L1 and IAP clones would be valuable. As mobile genetic elements that move in the germline insure passage of their genetic information to the next generation without impacting host fitness, it makes sense that mammals would combat this threat with an active germline-specific pathway against transposable elements. A mammalian RNA silencing pathway to combat germline retrotransposition is also consistent with characterized germline defence mechanisms in *C. elegans* and *Drosophila* (Vagin et al., 2006; Vastenhouw and Plasterk, 2004). Therefore, retroelement-derived piRNAs in mammalian testes, along with the smaller retroelement-derived RNAs characterized in mouse oocytes, represent the most likely triggers for an RNA silencing mechanism against mobile genetics elements and warrant further study to define their possible unique role in developing male and female germ cells.

Mammalian microRNA genes derived from mobile genetic elements

Despite millions of years of evolution in different hosts, the high structural conservation exhibited by eukaryotic mobile genetic elements suggests that some selective pressure is placed upon mammalian transposable elements to maintain certain features of ancient mobile DNA. For example, the human genome contains 200 sequences of the Hsmar1 DNA transposon that retain recognizable terminal inverted repeats even though almost all copies lack a functional transposase gene. Additionally, thousands of solitary ERV LTRs are dispersed throughout the mouse and human genomes, including the globin locus where coordinated gene expression is influenced by RNAi components (Haussecker and Proudfoot, 2005). Therefore, it is possible that RNA silencing in mammalian cells functions not only to suppress mobile genetic activity, but RNA silencing pathways could also interact with structured RNAs produced from mobile DNA sequences and function to coordinate the expression of other genes. This possibility is supported by recent observations in mouse oocytes of Dicer-dependent transcription control of mRNAs that have TE-derived sequences in their 3′ UTRs (Murchison et al., 2007).

As the number of cloned mammalian microRNAs continues to grow, several groups have turned to bioinformatics analysis to link miRNA silencing pathways with mammalian mobile genetic elements. The first genome-wide examinations microRNAs in mammalian cells failed to uncover any relevant homology to mobile genetic elements (Lagos-Quintana et al., 2003). Because many initial bioinformatics efforts focused on predicting the identity of mammalian microRNAs with particular hairpin structures, it is possible that these algorithms overlooked mammalian mobile elements, many of which are retrotransposons that lack terminal repeats. A re-examination of known microRNAs using the RepeatMasker program revealed that the putative miRNA precursor of several highly conserved miRNAs contained sequences of mammalian mobile genetic elements, including LINE-2, LTR, DNA transposon (*mariner*), and SINE (Smalheiser and Torvik, 2005). Precursor miRNAs that derived from LINE-2 retrotransposons were mapped to two discrete locations in the 3′ end of the LINE-2 consensus sequence. For some of these LINE-2-derived miRNAs, the entire hairpin precursor is made up of LINE-2 sequence or is formed by the transcription of adjacent LINE-2 sequences oriented in opposite directions.

Because LINE-2 and related sequences are often incorporated into mRNAs, one might speculate that LINE-2-derived miRNAs can target these transcripts, as well as their seed match-predicted target(s). A query of the EST and Refseq databases identified numerous transcripts with perfect complementarity to two, LINE-2-derived miRNAs (miR-95 and miR-151*) (Smalheiser and Torvik, 2005). Further experimentation is required to determine whether these specific mRNAs are true targets for post-transcriptional regulation by LINE-2 derived miR-95 and miR-151*. It is also possible that LINE-2 derived miRNAs regulate aberrant read-through transcripts that contain LINE-2 sequences. A follow-up study identified 30 miRNAs with seed complementarity to a short consensus sequence of the Alu SINE retrotransposon (Smalheiser and Torvik, 2006). The seed matches were with Alu sequences that are highly conserved in the 3′ UTR of some mRNAs, suggesting that these miRNAs may represent a novel form of regulating gene expression by making Alu-containing mRNAs subject to miRNA-mediated translation control.

It is possible that these microRNAs also target Alu transcripts themselves since Alu sequences are very promiscuous non-autonomous retrotransposons and likely come under some form of active regulation. Alu retrotransposons do not code for an open reading frame, leaving one to predict that post-transcriptional inhibition of Alu retrotransposons by microRNAs would occur through cleavage and degradation of Alu transcripts. Alu-based retrotransposition indicator and expression cassettes have been described (Dewannieux et al., 2003), making it possible to test the whether these microRNAs have the capacity to regulate Alu retrotransposition in human cells. As of yet, there is no evidence that orthologous murine microRNAs target murine mobile genetic elements, although the human LINE-2-derived miRNAs have rodent miRNA counterparts (Smalheiser and Torvik, 2005).

The derivation of microRNAs from mobile genetic elements can help explain the evolution of species-specific miRNA genes. In addition to the primate specific miR-95 derived from a LINE-2 retrotransposon, the precursor of a rodent-specific miRNA (rat miR-333) arises from insertion of a B2 SINE retrotransposon that is only found in the rodent genome (Smalheiser and Torvik, 2005). Furthermore, a recent study demonstrates that the primate-specific miR-548 gene family is derived from the *mariner*-like DNA transposon sequence Made1 (Piriyapongsa and Jordan, 2007). Made1 elements can be considered palindromes, in that they are 80 bp long and contain 37 bp TIRs flanking a 6 bp intervening sequence (Smit and Riggs, 1996). Expression of full-length Made1 sequences is predicted to give rise to a hairpin transcript consisting of a ~37 bp stem and short loop sequence that could be a substrate for Dicer processing into mature miRNA. A query of the EST database identified over 140 human ESTs that contained sequence similarity along the length of Made1 elements, suggesting that full-length Made1 sequences could be transcribed by read-through transcription. Through a combination of miRNA target site prediction and EST database alignments, 29 mRNAs were identified as putative miR-548

targets that also contained Made1 sequence. An examination of gene expression profiles found an inverse correlation between putative target gene mRNA levels and the cloning frequency of miR-548 (Piriyapongsa and Jordan, 2007). These results, combined with the evidence that other mammalian miRNAs arise from LINE-2 and Alu sequences, suggests that these TE-derived miRNAs represent an intermediate between siRNAs and miRNAs. Therefore, TE-derived small RNAs in mammalian cells may provide a novel means of controlling mobile genetic element, while at the same time have the capacity to regulate the expression of cellular transcripts.

Concluding remarks

The high degree of conservation observed for RNA silencing components across the eukaryotic kingdom suggests that these pathways perform similar functions in diverse eukaryotes. There is extensive experimental evidence, through genetic manipulation and biochemical purification, that RNA silencing pathways control the activity of transposable elements in many lower eukaryotes. Progress is being made to assign a functional role for RNA silencing in the surveillance of mammalian mobile genetic elements. The notion that some specific miRNA families derive from mammalian mobile genetic elements and might target cellular mRNAs that contain TE sequences suggests that RNA silencing pathways and mammalian TEs continue to interact, perhaps co-evolving with one another to bring forth a new mode of gene regulation. Continued efforts combining empirical and bioinformatics approaches will be necessary to fully characterize TE-derived small RNAs and assign a functional role for these molecules and the RNA silencing pathways in the suppression of, and possibly interaction with, transcripts containing sequences of mammalian mobile genetic elements.

References

Aagaard, L., Amarzguioui, M., Sun, G., Santos, L.C., Ehsani, A., Prydz, H., and Rossi, J.J. (2007). A facile lentiviral vector system for expression of doxycycline-inducible shRNAs: Knockdown of the pre-miRNA processing enzyme drosha. Mol. Ther. 15, 938–945.

Alves, G., Tatro, A., and Fanning, T. (1996). Differential methylation of human LINE-1 retrotransposons in malignant cells. Gene 176, 39–44.

Aravin, A., Gaidatzis, D., Pfeffer, S., Lagos-Quintana, M., Landgraf, P., Iovino, N., Morris, P., Brownstein, M.J., Kuramochi-Miyagawa, S., Nakano, T., et al. (2006). A novel class of small RNAs bind to MILI protein in mouse testes. Nature 442, 203–207.

Aravin, A.A., Sachidanandam, R., Girard, A., Fejes-Toth, K., and Hannon, G.J. (2007). Developmentally regulated piRNA clusters implicate MILI in transposon control. Science 316, 744–747.

Asch, H.L., Eliacin, E., Fanning, T.G., Connolly, J.L., Bratthauer, G., and Asch, B.B. (1996). Comparative expression of the LINE-1 p40 protein in human breast carcinomas and normal breast tissues. Oncol. Res. 8, 239–247.

Baillie, G.J., van de Lagemaat, L.N., Baust, C., and Mager, D.L. (2004). Multiple groups of endogenous betaretroviruses in mice, rats, and other mammals. J. Virol. 78, 5784–5798.

Baust, C., Gagnier, L., Baillie, G.J., Harris, M.J., Juriloff, D.M., and Mager, D.L. (2003). Structure and expression of mobile ETnII retroelements and their coding-competent MusD relatives in the mouse. J. Virol. 77, 11448–11458.

Benit, L., Lallemand, J.B., Casella, J.F., Philippe, H., and Heidmann, T. (1999). ERV-L elements: a family of endogenous retrovirus-like elements active throughout the evolution of mammals. J. Virol. 73, 3301–3308.

Bernstein, E., Kim, S.Y., Carmell, M.A., Murchison, E.P., Alcorn, H., Li, M.Z., Mills, A.A., Elledge, S.J., Anderson, K.V., and Hannon, G.J. (2003). Dicer is essential for mouse development. Nat. Genet. 35, 215–217.

Bourc'his, D., and Bestor, T.H. (2004). Meiotic catastrophe and retrotransposon reactivation in male germ cells lacking Dnmt3L. Nature 431, 96–99.

Branciforte, D., and Martin, S.L. (1994). Developmental and cell type specificity of LINE-1 expression in mouse testis: implications for transposition. Mol. Cell. Biol. 14, 2584–2592.

Bratthauer, G.L., and Fanning, T.G. (1992). Active LINE-1 retrotransposons in human testicular cancer. Oncogene 7, 507–510.

Bratthauer, G.L., and Fanning, T.G. (1993). LINE-1 retrotransposon expression in pediatric germ cell tumors. Cancer 71, 2383–2386.

Brouha, B., Schustak, J., Badge, R.M., Lutz-Prigge, S., Farley, A.H., Moran, J.V., and Kazazian, H.H., Jr. (2003). Hot L1s account for the bulk of retrotransposition in the human population. Proc. Natl. Acad. Sci. USA 100, 5280–5285.

Bucher, E., Sijen, T., De Haan, P., Goldbach, R., and Prins, M. (2003). Negative-strand tospoviruses and tenuiviruses carry a gene for a suppressor of gene silencing at analogous genomic positions. J. Virol. 77, 1329–1336.

Calabrese, J.M., and Sharp, P.A. (2006). Characterization of the short RNAs bound by the P19 suppressor of RNA silencing in mouse embryonic stem cells. RNA 12, 2092–2102.

Carmell, M.A., Xuan, Z., Zhang, M.Q., and Hannon, G.J. (2002). The Argonaute family: tentacles that reach into RNAi, developmental control, stem cell maintenance, and tumorigenesis. Genes Dev. 16, 2733–2742.

Carmell, M.A., Girard, A., van de Kant, H.J., Bourc'his, D., Bestor, T.H., de Rooij, D.G., and Hannon, G.J. (2007). MIWI2 is essential for spermatogenesis and repression of transposons in the mouse male germline. Dev. Cell 12, 503–514.

Castanotto, D., Tommasi, S., Li, M., Li, H., Yanow, S., Pfeifer, G.P., and Rossi, J.J. (2005). Short hairpin RNA-directed cytosine (CpG) methylation of the RASSF1A gene promoter in HeLa cells. Mol. Ther. 12, 179–183.

Chendrimada, T.P., Gregory, R.I., Kumaraswamy, E., Norman, J., Cooch, N., Nishikura, K., and Shiekhattar, R. (2005). TRBP recruits the Dicer complex to Ago2 for microRNA processing and gene silencing. Nature 436, 740–744.

Cummins, J.M., He, Y., Leary, R.J., Pagliarini, R., Diaz, L.A., Jr., Sjoblom, T., Barad, O., Bentwich, Z., Szafranska, A.E., Labourier, E., et al. (2006). The colorectal microRNAome. Proc. Natl. Acad. Sci. USA 103, 3687–3692.

de Parseval, N., and Heidmann, T. (2005). Human endogenous retroviruses: from infectious elements to human genes. CytoGenet. Genome Res. 110, 318–332.

de Veer, M.J., Sledz, C.A., and Williams, B.R. (2005). Detection of foreign RNA: implications for RNAi. Immunol Cell. Biol. 83, 224–228.

DeBerardinis, R.J., and Kazazian, H.H., Jr. (1999). Analysis of the promoter from an expanding mouse retrotransposon subfamily. Genomics 56, 317–323.

Deininger, P.L., and Batzer, M.A. (2002). Mammalian retroelements. Genome Res. 12, 1455–1465.

Deng, W., and Lin, H. (2002). miwi, a murine homolog of piwi, encodes a cytoplasmic protein essential for spermatogenesis. Dev. Cell 2, 819–830.

Dewannieux, M., Dupressoir, A., Harper, F., Pierron, G., and Heidmann, T. (2004). Identification of autonomous IAP LTR retrotransposons mobile in mammalian cells. Nat. Genet. 36, 534–539.

Dewannieux, M., Esnault, C., and Heidmann, T. (2003). LINE-mediated retrotransposition of marked Alu sequences. Nat. Genet. 35, 41–48.

Dewannieux, M., and Heidmann, T. (2005a). L1-mediated retrotransposition of murine B1 and B2 SINEs recapitulated in cultured cells. J. Mol. Biol. 349, 241–247.

Dewannieux, M., and Heidmann, T. (2005b). LINEs, SINEs and processed pseudogenes: parasitic strategies for genome modeling. Cytogenet. Genome Res. 110, 35–48.

Esnault, C., Maestre, J., and Heidmann, T. (2000). Human LINE retrotransposons generate processed pseudogenes. Nat. Genet. 24, 363–367.

Fedoroff, N. (2002). Control of mobile DNA. In Mobile DNA II, N.L. Craig, R. Craigie, M. Gellert, and A.M. Lambowitz, eds. (Washington, DC: ASM Press), pp. 997–1007.

Feng, Q., Moran, J.V., Kazazian, H.H., Jr., and Boeke, J.D. (1996). Human L1 retrotransposon encodes a conserved endonuclease required for retrotransposition. Cell 87, 905–916.

Feschotte, C., Zhang, X., Wesser, S.R. (2002). Miniature inverted-repeat transposable elements and their relationship to established DNA transposons. In Mobile DNA II, N.L. Craig, R. Craigie, M. Gellert, and A.M. Lambowitz, eds. (Washington, DC: ASM Press), pp. 997–1007.

Flamm, W.G. (1972). Highly repetitive sequences of DNA in chromosomes. Int Rev Cytol 32, 1–51.

Girard, A., Sachidanandam, R., Hannon, G.J., and Carmell, M.A. (2006). A germline-specific class of small RNAs binds mammalian Piwi proteins. Nature 442, 199–202.

Goodier, J.L., Ostertag, E.M., Du, K., and Kazazian, H.H., Jr. (2001). A novel active L1 retrotransposon subfamily in the mouse. Genome Res. 11, 1677–1685.

Gregory, R.I., Yan, K.P., Amuthan, G., Chendrimada, T., Doratotaj, B., Cooch, N., and Shiekhattar, R. (2004). The Microprocessor complex mediates the genesis of microRNAs. Nature 432, 235–240.

Grivna, S.T., Beyret, E., Wang, Z., and Lin, H. (2006a). A novel class of small RNAs in mouse spermatogenic cells. Genes Dev. 20, 1709–1714.

Grivna, S.T., Pyhtila, B., and Lin, H. (2006b). MIWI associates with translational machinery and PIWI-interacting RNAs (piRNAs) in regulating spermatogenesis. Proc. Natl. Acad. Sci. USA 103, 13415–13420.

Hagan, C.R., Sheffield, R.F., and Rudin, C.M. (2003). Human Alu element retrotransposition induced by genotoxic stress. Nat. Genet. 35, 219–220.

Hamilton, A., Voinnet, O., Chappell, L., and Baulcombe, D. (2002). Two classes of short interfering RNA in RNA silencing. EMBO J. 21, 4671–4679.

Hammond, S.M. (2005). Dicing and slicing: the core machinery of the RNA interference pathway. FEBS Lett. 579, 5822–5829.

Han, J., Lee, Y., Yeom, K.H., Kim, Y.K., Jin, H., and Kim, V.N. (2004). The Drosha-DGCR8 complex in primary microRNA processing. Genes Dev. 18, 3016–3027.

Han, J., Lee, Y., Yeom, K.H., Nam, J.W., Heo, I., Rhee, J.K., Sohn, S.Y., Cho, Y., Zhang, B.T., and Kim, V.N. (2006). Molecular basis for the recognition of primary microRNAs by the Drosha-DGCR8 complex. Cell 125, 887–901.

Haussecker, D., and Proudfoot, N.J. (2005). Dicer-dependent turnover of intergenic transcripts from the human beta-globin gene cluster. Mol. Cell. Biol. 25, 9724–9733.

Heidmann, O., and Heidmann, T. (1991). Retrotransposition of a mouse IAP sequence tagged with an indicator gene. Cell 64, 159–170.

Hohjoh, H., and Singer, M.F. (1996). Cytoplasmic ribonucleoprotein complexes containing human LINE-1 protein and RNA. EMBO J. 15, 630–639.

Hohjoh, H., and Singer, M.F. (1997). Sequence-specific single-strand RNA binding protein encoded by the human LINE-1 retrotransposon. EMBO J. 16, 6034–6043.

Hutvagner, G., McLachlan, J., Pasquinelli, A.E., Balint, E., Tuschl, T., and Zamore, P.D. (2001). A cellular function for the RNA-interference enzyme Dicer in the maturation of the let-7 small temporal RNA. Science 293, 834–838.

Ivics, Z., Hackett, P.B., Plasterk, R.H., and Izsvak, Z. (1997). Molecular reconstruction of Sleeping Beauty, a Tc1-like transposon from fish, and its transposition in human cells. Cell 91, 501–510.

Janowski, B.A., Huffman, K.E., Schwartz, J.C., Ram, R., Nordsell, R., Shames, D.S., Minna, J.D., and Corey, D.R. (2006). Involvement of AGO1 and AGO2 in mammalian transcriptional silencing. Nat. Struct. Mol. Biol. 13, 787–792.

Jurka, J., Kapitonov, V.V., Klonowski, P., Walichiewicz, J., and Smit, A.F. (1996). Identification of new medium reiteration frequency repeats in the genomes of Primates, Rodentia and Lagomorpha. Genetica 98, 235–247.

Jurka, J., Kaplan, D.J., Duncan, C.H., Walichiewicz, J., Milosavljevic, A., Murali, G., and Solus, J.F. (1993). Identification and characterization of new human medium reiteration frequency repeats. Nucleic Acids Res. 21, 1273–1279.

Kanellopoulou, C., Muljo, S.A., Kung, A.L., Ganesan, S., Drapkin, R., Jenuwein, T., Livingston, D.M., and Rajewsky, K. (2005). Dicer-deficient mouse embryonic stem cells are defective in differentiation and centromeric silencing. Genes Dev. 19, 489–501.

Kaplan, D.J., Jurka, J., Solus, J.F., and Duncan, C.H. (1991). Medium reiteration frequency repetitive sequences in the human genome. Nucleic Acids Res. 19, 4731–4738.

Karpala, A.J., Doran, T.J., and Bean, A.G. (2005). Immune responses to dsRNA: implications for gene silencing technologies. Immunol Cell. Biol. 83, 211–216.

Kazazian, H.H., Jr. (1998). Mobile elements and disease. Curr. Opin. Genet. Dev. 8, 343–350.

Kazazian, H.H., Jr. (2004). Mobile elements: drivers of genome evolution. Science 303, 1626–1632.

Ketting, R.F., Haverkamp, T.H., van Luenen, H.G., and Plasterk, R.H. (1999). Mut-7 of C. elegans, required for transposon silencing and RNA interference, is a homolog of Werner syndrome helicase and RNaseD. Cell 99, 133–141.

Ketting, R.F., and Plasterk, R.H. (2000). A genetic link between co-suppression and RNA interference in C. elegans. Nature 404, 296–298.

Kim, V.N. (2006). Small RNAs just got bigger: Piwi-interacting RNAs (piRNAs) in mammalian testes. Genes Dev. 20, 1993–1997.

Kolosha, V.O., and Martin, S.L. (2003). High-affinity, non-sequence-specific RNA binding by the open reading frame 1 (ORF1) protein from long interspersed nuclear element 1 (LINE-1). J. Biol. Chem. 278, 8112–8117.

Kuff, E.L., and Lueders, K.K. (1988). The intracisternal A-particle gene family: structure and functional aspects. Adv Cancer Res. 51, 183–276.

Kuhlmann, M., Borisova, B.E., Kaller, M., Larsson, P., Stach, D., Na, J., Eichinger, L., Lyko, F., Ambros, V., Soderbom, F., et al. (2005). Silencing of retrotransposons in Dictyostelium by DNA methylation and RNAi. Nucleic Acids Res. 33, 6405–6417.

Kulpa, D.A., and Moran, J.V. (2006). *Cis*-preferential LINE-1 reverse transcriptase activity in ribonucleoprotein particles. Nat. Struct. Mol. Biol. *13*, 655–660.

Kuramochi-Miyagawa, S., Kimura, T., Ijiri, T.W., Isobe, T., Asada, N., Fujita, Y., Ikawa, M., Iwai, N., Okabe, M., Deng, W., et al. (2004). Mili, a mammalian member of piwi family gene, is essential for spermatogenesis. Development *131*, 839–849.

Lagos-Quintana, M., Rauhut, R., Meyer, J., Borkhardt, A., and Tuschl, T. (2003). New microRNAs from mouse and human. RNA *9*, 175–179.

Lakatos, L., Szittya, G., Silhavy, D., and Burgyan, J. (2004). Molecular mechanism of RNA silencing suppression mediated by p19 protein of tombusviruses. EMBO J. *23*, 876–884.

Lander, E.S., Linton, L.M., Birren, B., Nusbaum, C., Zody, M.C., Baldwin, J., Devon, K., Dewar, K., Doyle, M., FitzHugh, W., et al. (2001). Initial sequencing and analysis of the human genome. Nature *409*, 860–921.

Landthaler, M., Yalcin, A., and Tuschl, T. (2004). The human DiGeorge syndrome critical region gene 8 and Its D. melanogaster homolog are required for miRNA biogenesis. Curr. Biol. *14*, 2162–2167.

Lau, N.C., Seto, A.G., Kim, J., Kuramochi-Miyagawa, S., Nakano, T., Bartel, D.P., and Kingston, R.E. (2006). Characterization of the piRNA Complex from Rat Testes. Science *313*, 263–367.

Lee, Y., Ahn, C., Han, J., Choi, H., Kim, J., Yim, J., Lee, J., Provost, P., Radmark, O., Kim, S., and Kim, V.N. (2003). The nuclear RNase III Drosha initiates microRNA processing. Nature *425*, 415–419.

Lee, Y., Jeon, K., Lee, J.T., Kim, S., and Kim, V.N. (2002). MicroRNA maturation: stepwise processing and subcellular localization. EMBO J. *21*, 4663–4670.

Leibold, D.M., Swergold, G.D., Singer, M.F., Thayer, R.E., Dombroski, B.A., and Fanning, T.G. (1990). Translation of LINE-1 DNA elements *in vitro* and in human cells. Proc. Natl. Acad. Sci. USA *87*, 6990–6994.

Liu, J., Carmell, M.A., Rivas, F.V., Marsden, C.G., Thomson, J.M., Song, J.J., Hammond, S.M., Joshua-Tor, L., and Hannon, G.J. (2004). Argonaute2 is the catalytic engine of mammalian RNAi. Science *305*, 1437–1441.

Lu, S., and Cullen, B.R. (2004). Adenovirus VA1 noncoding RNA can inhibit small interfering RNA and MicroRNA biogenesis. J. Virol. *78*, 12868–12876.

Luan, D.D., Korman, M.H., Jakubczak, J.L., and Eickbush, T.H. (1993). Reverse transcription of R2Bm RNA is primed by a nick at the chromosomal target site: a mechanism for non-LTR retrotransposition. Cell *72*, 595–605.

Macrae, I.J., Zhou, K., Li, F., Repic, A., Brooks, A.N., Cande, W.Z., Adams, P.D., and Doudna, J.A. (2006). Structural basis for double-stranded RNA processing by Dicer. Science *311*, 195–198.

Mager, D.L., and Freeman, J.D. (2000). Novel mouse type D endogenous proviruses and ETn elements share long terminal repeat and internal sequences. J. Virol. *74*, 7221–7229.

Maksakova, I.A., and Mager, D.L. (2005). Transcriptional regulation of early transposon elements, an active family of mouse long terminal repeat retrotransposons. J. Virol. *79*, 13865–13874.

Maksakova, I.A., Romanish, M.T., Gagnier, L., Dunn, C.A., van de Lagemaat, L.N., and Mager, D.L. (2006). Retroviral elements and their hosts: insertional mutagenesis in the mouse germ line. PLoS Genet. *2*, e2.

Margulies, M., Egholm, M., Altman, W.E., Attiya, S., Bader, J.S., Bemben, L.A., Berka, J., Braverman, M.S., Chen, Y.J., Chen, Z., et al. (2005). Genome sequencing in microfabricated high-density picolitre reactors. Nature *437*, 376–380.

Martin, S.L., and Branciforte, D. (1993). Synchronous expression of LINE-1 RNA and protein in mouse embryonal carcinoma cells. Mol. Cell. Biol. *13*, 5383–5392.

Martin, S.L., and Bushman, F.D. (2001). Nucleic acid chaperone activity of the ORF1 protein from the mouse LINE-1 retrotransposon. Mol. Cell. Biol. *21*, 467–475.

Martin, S.L., Cruceanu, M., Branciforte, D., Wai-Lun Li, P., Kwok, S.C., Hodges, R.S., and Williams, M.C. (2005). LINE-1 retrotransposition requires the nucleic acid chaperone activity of the ORF1 protein. J. Mol. Biol. *348*, 549–561.

Mathias, S.L., Scott, A.F., Kazazian, H.H., Jr., Boeke, J.D., and Gabriel, A. (1991). Reverse transcriptase encoded by a human transposable element. Science *254*, 1808–1810.

Meister, G., Landthaler, M., Patkaniowska, A., Dorsett, Y., Teng, G., and Tuschl, T. (2004). Human Argonaute2 mediates RNA cleavage targeted by miRNAs and siRNAs. Mol. Cell *15*, 185–197.

Meister, G., Landthaler, M., Peters, L., Chen, P.Y., Urlaub, H., Luhrmann, R., and Tuschl, T. (2005). Identification of novel argonaute-associated proteins. Curr. Biol. *15*, 2149–2155.

Meister, G., and Tuschl, T. (2004). Mechanisms of gene silencing by double-stranded RNA. Nature *431*, 343–349.

Mi, S., Lee, X., Li, X., Veldman, G.M., Finnerty, H., Racie, L., LaVallie, E., Tang, X.Y., Edouard, P., Howes, S., et al. (2000). Syncytin is a captive retroviral envelope protein involved in human placental morphogenesis. Nature *403*, 785–789.

Moran, J.V., Holmes, S.E., Naas, T.P., DeBerardinis, R.J., Boeke, J.D., and Kazazian, H.H., Jr. (1996). High frequency retrotransposition in cultured mammalian cells. Cell *87*, 917–927.

Moran, J.V., and Gilbert, N. (2002). Mammalian LINE-1 retrotransposons and related elements. In Mobile DNA II, N.L. Craig, R. Craigie, M. Gellert, and A.M. Lambowitz, eds. (Washington, DC: ASM Press), pp. 997–1007.

Morris, K.V., Chan, S.W., Jacobsen, S.E., and Looney, D.J. (2004). Small interfering RNA-induced transcriptional gene silencing in human cells. Science *305*, 1289–1292.

Mourelatos, Z., Dostie, J., Paushkin, S., Sharma, A., Charroux, B., Abel, L., Rappsilber, J., Mann, M., and Dreyfuss, G. (2002). miRNPs: a novel class of ribonucleoproteins containing numerous microRNAs. Genes Dev. *16*, 720–728.

Muotri, A.R., Chu, V.T., Marchetto, M.C., Deng, W., Moran, J.V., and Gage, F.H. (2005). Somatic mosaicism in neuronal precursor cells mediated by L1 retrotransposition. Nature 435, 903–910.

Murchison, E.P., Partridge, J.F., Tam, O.H., Cheloufi, S., and Hannon, G.J. (2005). Characterization of Dicer-deficient murine embryonic stem cells. Proc. Natl. Acad. Sci. USA 102, 12135–12140.

Murchison, E.P., Stein, P., Xuan, Z., Pan, H., Zhang, M.Q., Schultz, R.M., and Hannon, G.J. (2007). Critical roles for Dicer in the female germline. Genes Dev. 21, 682–693.

Naas, T.P., DeBerardinis, R.J., Moran, J.V., Ostertag, E.M., Kingsmore, S.F., Seldin, M.F., Hayashizaki, Y., Martin, S.L., and Kazazian, H.H. (1998). An actively retrotransposing, novel subfamily of mouse L1 elements. EMBO J. 17, 590–597.

Nicholson, R.H., and Nicholson, A.W. (2002). Molecular characterization of a mouse cDNA encoding Dicer, a ribonuclease III ortholog involved in RNA interference. Mamm Genome 13, 67–73.

Nolan, T., Braccini, L., Azzalin, G., De Toni, A., Macino, G., and Cogoni, C. (2005). The post-transcriptional gene silencing machinery functions independently of DNA methylation to repress a LINE1-like retrotransposon in Neurospora crassa. Nucleic Acids Res. 33, 1564–1573.

Oosumi, T., Belknap, W.R., and Garlick, B. (1995). Mariner transposons in humans. Nature 378, 672.

Ostertag, E.M., and Kazazian, H.H., Jr. (2001). Biology of mammalian L1 retrotransposons. Annu. Rev. Genet. 35, 501–538.

Pillai, R.S., Bhattacharyya, S.N., Artus, C.G., Zoller, T., Cougot, N., Basyuk, E., Bertrand, E., and Filipowicz, W. (2005). Inhibition of translational initiation by Let-7 MicroRNA in human cells. Science 309, 1573–1576.

Piriyapongsa, J., and Jordan, I.K. (2007). A family of human microRNA Genes from miniature inverted-repeat transposable elements. PLoS ONE 2, e203.

Plasterk, R.H., Izsvak, Z., and Ivics, Z. (1999). Resident aliens: the Tc1/mariner superfamily of transposable elements. Trends Genet. 15, 326–332.

Provost, P., Dishart, D., Doucet, J., Frendewey, D., Samuelsson, B., and Radmark, O. (2002). Ribonuclease activity and RNA binding of recombinant human Dicer. EMBO J. 21, 5864–5874.

Ribet, D., Dewannieux, M., and Heidmann, T. (2004). An active murine transposon family pair: retrotransposition of 'master' MusD copies and ETn transmobilization. Genome Res. 14, 2261–2267.

Rivas, F.V., Tolia, N.H., Song, J.J., Aragon, J.P., Liu, J., Hannon, G.J., and Joshua-Tor, L. (2005). Purified Argonaute2 and an siRNA form recombinant human RISC. Nat. Struct. Mol. Biol. 12, 340–349.

Robertson, H.M., and Zumpano, K.L. (1997). Molecular evolution of an ancient mariner transposon, Hsmar1, in the human genome. Gene 205, 203–217.

Rudin, C.M., and Thompson, C.B. (2001). Transcriptional activation of short interspersed elements by DNA-damaging agents. Genes Chromosomes Cancer 30, 64–71.

Sarot, E., Payen-Groschene, G., Bucheton, A., and Pelisson, A. (2004). Evidence for a piwi-dependent RNA silencing of the gypsy endogenous retrovirus by the Drosophila melanogaster flamenco gene. Genetics 166, 1313–1321.

Sasaki, T., Shiohama, A., Minoshima, S., and Shimizu, N. (2003). Identification of eight members of the Argonaute family in the human genome small star, filled. Genomics 82, 323–330.

Shi, H., Djikeng, A., Tschudi, C., and Ullu, E. (2004). Argonaute protein in the early divergent eukaryote Trypanosoma brucei: control of small interfering RNA accumulation and retroposon transcript abundance. Mol. Cell. Biol. 24, 420–427.

Sijen, T., and Plasterk, R.H. (2003). Transposon silencing in the Caenorhabditis elegans germ line by natural RNAi. Nature 426, 310–314.

Skowronski, J., Fanning, T.G., and Singer, M.F. (1988). Unit-length line-1 transcripts in human teratocarcinoma cells. Mol. Cell. Biol. 8, 1385–1397.

Skowronski, J., and Singer, M.F. (1985). Expression of a cytoplasmic LINE-1 transcript is regulated in a human teratocarcinoma cell line. Proc. Natl. Acad. Sci. USA 82, 6050–6054.

Smalheiser, N.R., and Torvik, V.I. (2005). Mammalian microRNAs derived from genomic repeats. Trends Genet. 21, 322–326.

Smalheiser, N.R., and Torvik, V.I. (2006). Alu elements within human mRNAs are probable microRNA targets. Trends Genet. 22, 532–536.

Smit, A.F. (1993). Identification of a new, abundant superfamily of mammalian LTR-transposons. Nucleic Acids Res. 21, 1863–1872.

Smit, A.F. (1999). Interspersed repeats and other mementos of transposable elements in mammalian genomes. Curr. Opin. Genet. Dev. 9, 657–663.

Smit, A.F., and Riggs, A.D. (1995). MIRs are classic, tRNA-derived SINEs that amplified before the mammalian radiation. Nucleic Acids Res. 23, 98–102.

Smit, A.F., and Riggs, A.D. (1996). Tiggers and DNA transposon fossils in the human genome. Proc. Natl. Acad. Sci. USA 93, 1443–1448.

Soifer, H.S. (2006). Do small RNAs interfere with LINE-1? J. Biomed. Biotechnol. 2006, 29049.

Speek, M. (2001). Antisense promoter of human L1 retrotransposon drives transcription of adjacent cellular genes. Mol. Cell. Biol. 21, 1973–1985.

Svoboda, P., Stein, P., Anger, M., Bernstein, E., Hannon, G.J., and Schultz, R.M. (2004). RNAi and expression of retrotransposons MuERV-L and IAP in preimplantation mouse embryos. Dev. Biol. 269, 276–285.

Swergold, G.D. (1990). Identification, characterization, and cell specificity of a human LINE-1 promoter. Mol. Cell. Biol. 10, 6718–6729.

Tabara, H., Sarkissian, M., Kelly, W.G., Fleenor, J., Grishok, A., Timmons, L., Fire, A., and Mello, C.C. (1999). The rde-1 gene, RNA interference, and transposon silencing in C. elegans. Cell 99, 123–132.

Thayer, R.E., Singer, M.F., and Fanning, T.G. (1993). Undermethylation of specific LINE-1 sequences in human cells producing a LINE-1-encoded protein. Gene 133, 273–277.

Tomari, Y., Du, T., Haley, B., Schwarz, D.S., Bennett, R., Cook, H.A., Koppetsch, B.S., Theurkauf, W.E., and Zamore, P.D. (2004). RISC assembly defects in the *Drosophila* RNAi mutant armitage. Cell *116*, 831–841.

Trelogan, S.A., and Martin, S.L. (1995). Tightly regulated, developmentally specific expression of the first open reading frame from LINE-1 during mouse embryogenesis. Proc. Natl. Acad. Sci. USA *92*, 1520–1524.

Vagin, V.V., Sigova, A., Li, C., Seitz, H., Gvozdev, V., and Zamore, P.D. (2006). A distinct small RNA pathway silences selfish genetic elements in the germline. Science *313*, 320–324.

Vargason, J.M., Szittya, G., Burgyan, J., and Tanaka Hall, T.M. (2003). Size selective recognition of siRNA by an RNA silencing suppressor. Cell *115*, 799–811.

Vastenhouw, N.L., Fischer, S.E., Robert, V.J., Thijssen, K.L., Fraser, A.G., Kamath, R.S., Ahringer, J., and Plasterk, R.H. (2003). A genome-wide screen identifies 27 genes involved in transposon silencing in C. *elegans*. Curr. Biol. *13*, 1311–1316.

Vastenhouw, N.L., and Plasterk, R.H. (2004). RNAi protects the *Caenorhabditis elegans* germline against transposition. Trends Genet. *20*, 314–319.

Wang, Y., Medvid, R., Melton, C., Jaenisch, R., and Blelloch, R. (2007). DGCR8 is essential for microRNA biogenesis and silencing of embryonic stem cell self-renewal. Nat. Genet. 39, 380–385.

Watanabe, T., Takeda, A., Tsukiyama, T., Mise, K., Okuno, T., Sasaki, H., Minami, N., and Imai, H. (2006). Identification and characterization of two novel classes of small RNAs in the mouse germline: retrotransposon-derived siRNAs in oocytes and germline small RNAs in testes. Genes Dev. *20*, 1732–1743.

Waterston, R.H., Lindblad-Toh, K., Birney, E., Rogers, J., Abril, J.F., Agarwal, P., Agarwala, R., Ainscough, R., Alexandersson, M., An, P., *et al.* (2002). Initial sequencing and comparative analysis of the mouse genome. Nature *420*, 520–562.

Wei, W., Gilbert, N., Ooi, S.L., Lawler, J.F., Ostertag, E.M., Kazazian, H.H., Boeke, J.D., and Moran, J.V. (2001). Human L1 retrotransposition: *cis* preference versus *trans* complementation. Mol. Cell. Biol. *21*, 1429–1439.

Yang, N., and Kazazian, H.H., Jr. (2006). L1 retrotransposition is suppressed by endogenously encoded small interfering RNAs in human cultured cells. Nat. Struct. Mol. Biol. *13*, 763–771.

Zhang, H., Kolb, F.A., Jaskiewicz, L., Westhof, E., and Filipowicz, W. (2004). Single processing center models for human Dicer and bacterial RNase III. Cell *118*, 57–68.

Zilberman, D., Cao, X., and Jacobsen, S.E. (2003). ARGONAUTE4 control of locus-specific siRNA accumulation and DNA and histone methylation. Science *299*, 716–719.

The Role of Non-coding RNAs in Controlling Mammalian RNA Polymerase II Transcription

9

Stacey D. Wagner, Jennifer F. Kugel and James A. Goodrich

Abstract

Controlling transcription of protein-encoding genes into mRNA is central to all basic biological processes and understanding mechanisms of transcriptional control has been the focus of intense investigation. In the case of mammalian protein-encoding genes, a multitude of protein factors that regulate mRNA transcription have been discovered and characterized over the past four decades. During that time a few RNAs were found to play a role in transcriptional regulation in very specific biological situations. Only recently has the scientific community begun to appreciate that RNAs play a much larger part in regulating mammalian transcription. This appreciation grew out of the discoveries of a number of RNA transcriptional regulators, some of which have the potential to control entire transcriptional programs in response to extracellular stimuli. At the same time, studies have shown that a much larger portion of the genome is transcribed than previously thought, and that many of the transcripts produced in mammalian cells do not encode protein, and hence are called non-coding RNAs (ncRNA) (Bertone et al., 2004; Cawley et al., 2004; Cheng et al., 2005; Mattick, 2005; Okazaki et al., 2002; Washietl et al., 2005). With increased awareness of the regulatory potential of ncRNAs, researchers have begun specifically looking for ncRNA transcriptional regulators. This search will likely result in a dramatic increase in the number and complexity of these novel transcriptional regulators over the next few years.

Here, we review ncRNAs that control mammalian mRNA transcription. The ncRNAs discussed are grouped by the specific stage of transcription that they target. Intriguingly, most stages of the reaction are now known to be targeted by at least one ncRNA, from the mobilization of activators through the termination of transcription. Throughout the chapter we also include our thoughts on the pressing questions that could next be addressed to further our understanding of each of the ncRNAs.

Transcription of protein-encoding genes in mammals

Transcription of mRNA in eukaryotes is catalysed by the enzyme RNA polymerase II (for a comprehensive review of mRNA transcription see Thomas and Chiang, 2006 and references therein). This enzyme is composed of 12 protein subunits, and undergoes a complex series of steps to produce an RNA copy of protein-encoding genes. RNA polymerase II is aided by a multitude of protein factors that assist in the recruitment of RNA polymerase II to genes, control the structure of chromatin, facilitate the steps of transcription, and regulate transcription in a gene-specific fashion. The process of turning on transcription of a gene begins with the binding of activator proteins to the regulatory regions within the promoter of the gene. Promoters contain the DNA regulatory sequences that bind the protein factors necessary for directing accurate transcription. Often the binding of activators to DNA is regulated, either directly or at the level of nuclear localization. For an activator to gain access to its DNA binding site it must negotiate chromatin, which is composed of DNA and histone proteins. Chromatin modifying and remodeling factors control the structure of chromatin and can both be recruited to promoters by activators as well as facilitate the association of activators to DNA.

Once activators bind the promoter of a gene, they can recruit other transcription factors to the gene. These can include additional chromatin modifying factors, transcriptional coactivator proteins, and the RNA polymerase II general transcription machinery. Coactivators are a group of factors that do not bind DNA themselves, but work with DNA bound activator proteins to upregulate transcription. The general transcription machinery includes RNA polymerase II and a host of protein factors required for proper assembly at the promoter and initiation of transcription (TFII-A, B, D, E, F, and H, and mediator) as well as factors that are necessary for stimulating elongation of the RNA transcript.

The synthesis of an RNA transcript includes many steps. The general transcription factors are recruited to the DNA in an ordered assembly to form preinitiation complexes (PICs). This recruitment depends on promoter-bound activators, coactivators, and chromatin modifying factors. After RNA polymerase II initiates transcription it undergoes a series of transformations that allow it to escape the promoter and enter the elongation phase of transcription. During early transcription the C-terminal domain (CTD) of the largest subunit of RNA polymerase II is phosphorylated by the general transcription factor TFIIH. A second kinase, positive transcription elongation factor b (P-TEFb) also phosphorylates the CTD, which promotes efficient elongation by the polymerase. Transcription is coupled to RNA processing events, including addition of the 5′-cap, splicing, and polyadenylation of the 3′ end of the terminated transcript. Many of the steps described above are well-known targets of protein transcriptional regulators; more recently, ncRNAs have been discovered that also function to regulate many of the aforementioned steps of RNA polymerase II transcription. A summary of

these ncRNA transcription factors follows, with an emphasis on detailing current knowledge on their distinct mechanisms of action (also see Table 9.1).

ncRNAs that control the intracellular localization and DNA binding activity of transcriptional activator proteins

To be functional, transcriptional activator proteins must be localized in the nucleus and be active for binding DNA at specific sequences in the promoters of genes. A multitude of mechanisms are utilized to control both the nuclear localization and the DNA binding activity of transcriptional activators. Recent studies have identified two ncRNAs that regulate either localization or DNA binding (Fig. 9.1).

Non-coding RNA repressor of NFAT (NRON)

NRON is a ncRNA that was identified using a targeted RNAi based screen for ncRNAs that affect the activity of a transcriptional activator named NFAT (Nuclear Factor of Activated T cells) (Willingham *et al.*, 2005). NFAT proteins are transcriptional activators that bind DNA at specific sites to regulate the expression of genes involved in diverse processes including heart development and the immune response (for a review see Hogan *et al.*, 2003). The nuclear/cytoplasmic trafficking of NFAT proteins is highly regulated. Dephosphorylation of cytoplasmic NFAT by the calcium sensing protein, calcineurin, results in nuclear localization of NFAT. NRON is hypothesized to repress NFAT activity by preventing its translocation into the nucleus (Willingham *et al.*, 2005). Knockdown of NRON increases the nuclear localization of NFAT but not a control transcription factor that is also trafficked between the cytoplasm and nucleus (Willingham *et al.*, 2005).

Eleven proteins were identified to interact with NRON using an NRON interaction assay; NFAT was not among them (Willingham *et al.*, 2005). Four of the NRON interacting proteins cooperate with NRON to decrease NFAT activity at a reporter gene. These proteins include a

Table 9.1 Summary of ncRNAs that regulate mammalian transcription*

ncRNA	Size	*Trans.* by	Target	Effect on transcription
NRON	0.8–3.7 kb	Pol II	Multiple proteins	Blocks NFAT nuclear localization
HSR-1	~600 nt	?	HSF	Trimerization of HSF, activation of hsp genes
SRA	0.7–1.5 kb	?	Steroid hormone receptors	Coactivates transcription from SREs
NRSE	20 bp	?	NRSF/REST	Derepresses transcription through NRSF/REST
Evf-2	3.8 kb	Pol II	Dlx-2	Activates transcription of Dlx5/6
B2	~178 nt	Pol III	Pol II	Inhibits transcription during heat shock
DHFR	?	Pol II	TFIIB	Inhibits transcription of DHFR mRNA
U1	165 nt	Pol III	TFIIH	Enhances initiation and reinitiation by Pol II
7SK	330 nt	Pol III	P-TEFb	Inhibits elongation
TAR	59 nt[†]	Pol II	Tat (P-TEFb)	Stimulates elongation of HIV genes
CBS	9 nt[†]	Pol II	Pol II CTD	Inhibits mRNA 3′ end formation
CoTC	200 nt[†]	Pol II	β-Globin mRNA	Terminates transcription
Xist	17–19 kb	Pol II	X chromosome	X inactivation – gene silencing
Air	108 kb	Pol II	?	Causes imprinting – gene silencing

*See the text for references.
[†]The functional region of the *cis*-acting ncRNA is listed.

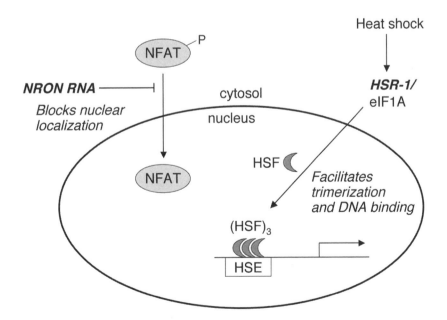

Figure 9.1 ncRNAs that control the intracellular localization and DNA binding activity of transcriptional activator proteins. Non-coding RNA repressor of NFAT (NRON) inhibits the nuclear/cytoplasmic trafficking of the transcriptional activator NFAT. Heat shock RNA (HSR-1) and eIF1A protein function together to facilitate trimerization and DNA binding by the transcriptional activator heat shock factor (HSF), which binds to the heat shock response element (HSE) to activate transcription of heat shock genes. References are in the text.

proteasome component, a phosphatase subunit, a factor involved in the nuclear transport of proteins, and a calcineurin scaffolding protein.

It will be interesting to learn how NRON affects NFAT nuclear trafficking through the proteins identified. For example, NRON could bind to the NFAT transporter and repress its ability to shuttle NFAT from the nucleus to the cytoplasm. NRON may also interfere with factors binding to the calcineurin scaffolding protein. NRON may serve as a scaffold itself given how many proteins were identified in the interaction assay. Future work is also needed to determine how NRON interacts with the identified proteins. Moreover, NRON could play an important role in controlling the immune response, which could lead to the development of new immunoregulatory drugs that target the NRON ncRNA.

Heat shock RNA-1 (HSR-1)

HSR-1 is an ncRNA regulator of the mammalian heat shock response (Shamovsky et al., 2006). The ~600 nt HSR-1 is required for transcription of the inducible heat shock protein 70 (hsp70) gene. A hallmark of the cellular shock response is upregulation of several chaperone proteins called heat shock proteins (for a review see Morimoto, 1998). The transcription of heat shock protein genes is activated by the transcription factor, heat shock factor (HSF). To activate transcription, HSF must trimerize and bind to the heat shock response elements (HSE) located in the promoters of heat shock protein genes, which occurs after cells are stressed by heat shock. The HSR-1 ncRNA and a translation initiation factor, eIF1A, are together necessary for the trimerization and activation of HSF (Shamovsky et al., 2006). Knock down of HSR-1 causes decreased viability during heat shock because cells are unable to cope with stress in the absence of active HSF. It is proposed that the HSR-1 ncRNA senses the increase in temperature and undergoes a temperature induced conformational change, which ultimately allows it to mobilize HSF (Shamovsky et al., 2006).

Exactly how HSR-1 and eIF1A cause HSF to trimerize is not yet known. Moreover, it remains to be determined if HSR-1 and eIF1A bind with HSF to DNA in the promoters of genes activated during heat shock. If so, do these factors have additional functions in regulating transcription beyond facilitating DNA binding by HSF? Investigating the thermosensing function of HSR-1 may provide insight into whether the structure of HSR-1 responds to changes in temperature, and how an RNA conformational change influences HSF trimerization.

ncRNAs that function as transcriptional coactivators

Several ncRNAs have been discovered that have activities normally attributed to proteins that function as transcriptional coactivators. In general, transcriptional coactivators are recruited to specific promoters by DNA-bound transcriptional activator proteins where they mediate transcriptional activation. The ncRNAs that function as coactivators, much like their protein counterparts, are thought to be recruited specifically to promoter DNA where they mediate interactions between protein factors, thereby increasing levels of transcription of target genes (Fig. 9.2).

Steroid receptor RNA activator (SRA)

SRA is a coactivator of some steroid hormone receptors (Lanz et al., 1999). Steroid hormone receptors are transcription factors that are activated by specific ligands, bind their cognate DNA response elements, and activate transcription (for a review see Lonard and O'Malley, 2005). Coregulators play a role in fine-tuning the actions of steroid receptors. Type I receptors (e.g. glucocorticoid, progesterone, and estrogen receptors) possess an AF1 trans-activation domain, which is present at the N-termini of the proteins. SRA was identified in a yeast two-hybrid screen using the AF1 domain of the progesterone receptor as bait (Lanz et al., 1999). SRA is expressed as many isoforms from 0.7–1.5 kb that show cell type specific expression patterns. In human cells, SRA is found associated with the steroid receptor coactivator-1 (SRC-1), which is a protein important for activation by type I steroid receptors. Coactivation by SRA requires at least five predicted structural elements distributed throughout the ncRNA, which complicates the use of minimal domain mutants for mechanistic studies (Lanz et al., 2002).

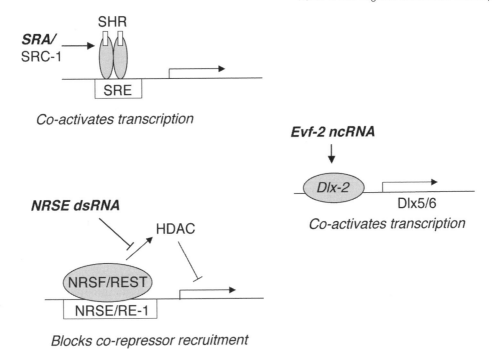

Figure 9.2 ncRNAs that function as transcriptional coactivators. Steroid receptor RNA activator (SRA) functions with steroid receptor coactivator (SRC-1) to coactivate steroid hormone receptors (SHR). The neuronal restricted silencing element (NRSE) ncRNA is double stranded and functions to change the transcription factor neuronal restricted silencing factor/RE-1 silencing factor (NRSF/REST) from a repressor to an activator in neurons by preventing HDAC recruitment. Evf-2 ncRNA and Dlx-2 activate the transcription of genes involved in brain patterning. References are in the text.

The pseudouridine synthase, MPus1p, was recently found to interact with SRA (Zhao et al., 2004). Pseudouridine synthases modify RNAs by isomerizing uridine to pseudouridine. MPus1p interacts with SRA and pseudouridylates SRA in vitro. Protein-protein interaction assays and ChIPs were used to show that Mpus1p also interacts with the retinoic acid receptor (RAR) and is associated with the retinoic acid response element (RARE) consensus site. In reporter assays, both the RAR binding and pseudouridylation activities of Mpus1p are required for SRA to coactivate RAR-driven transcription. These experiments suggest a model where both pseudouridylation of SRA and Mpus1p binding to the RAR are needed for coactivation.

Many questions remain concerning the function of SRA. For example, what is the mechanism by which SRA mediates transcriptional activation? Does it function by helping to recruit SRC-1 to DNA bound steroid receptors? How ubiquitous is SRA as a coactivator; at how many genes does it function?

Neuronal restricted silencer element (NRSE) ncRNA

The NRSE ncRNA consists of an ~20 bp double-stranded (ds)RNA that is thought to be transcribed from the neuronal restricted silencer element (NRSE/RE-1) in neurons (Kuwabara et al., 2004). The NRSE/RE-1 is a 21–23 bp consensus DNA sequence present in many genes involved in neuronal differentiation (Schoenherr and Anderson, 1995a). NRSE/RE-1 is recognized by the neuronal restricted silencing factor/RE-1 silencing transcription factor (NRSF/REST). NRSF/REST represses transcription of the genes controlled by NRSE/RE-1 in non-neuronal cells (Chong et al., 1995; Schoenherr and Anderson, 1995b). It functions by recruiting histone deacetylases (HDAC) and methyl DNA binding proteins that function in transcriptional repression (Huang et al., 1999; Lunyak et al., 2002). In neuronal cells, NRSF/REST occupies the NRSE; however, HDAC and other corepressors are not present, due to the action of the NRSE ncRNA, which interacts directly with NRSF/REST (Kuwabara et al.,

2004). This interaction requires that the NRSE ncRNA is double stranded.

It is thought that NRSE ncRNA converts NRSF/REST from a transcriptional repressor to an activator in neuronal progenitor cells to induce differentiation (Kuwabara *et al.*, 2004). In order to turn on transcription of neuron-specific genes, the cell must first transcribe both strands of the NRSE ncRNA. NRSE ncRNA associates with NRSF/REST to prevent recruitment of co-repressors such as HDACs. The mechanism by which the NRSE ncRNA prevents NRSF/REST from recruiting co-repressors has yet to be determined. Moreover, it is not known how transcription of the NRSE ncRNA is regulated.

evf-2 non-coding RNA

Evf-2 is a 3.8 kb spliced and polyadenylated ncRNA that is transcribed from an intergenic region of the ultraconserved Dlx5/6 enhancer in vertebrates (Feng *et al.*, 2006). Ultraconserved regions of DNA generally show >90% conservation over several million years of evolution (Sandelin *et al.*, 2004). Dlx genes function in neuronal differentiation and brain and limb patterning. Dlx-2 is a transcription factor that upregulates expression of Dlx-5/6 to induce brain patterning (Zerucha *et al.*, 2000; Zhou *et al.*, 2004). The Evf-2 ncRNA cooperates with Dlx-2, to upregulate transcription specifically from the Dlx5/6 enhancer in mouse neuronal cells (Feng *et al.*, 2006). Knocking Evf-2 down with siRNA results in loss of activation of Dlx5/6. Evf-2 coimmunoprecipitates with Dlx-2 from mouse neuron cells and co-localizes with Dlx-2 in mouse neuronal tissue when visualized by fluorescence microscopy. Minimally, a 5′ region of Evf-2 that is approximately 400 nt long binds Dlx-2 to activate transcription.

It is not yet known how Evf-2 coactivates transcription with Dlx-2. Ultraconserved regions are present in areas of the genome that regulate expression of developmental genes in higher vertebrates (Ghanem *et al.*, 2003; Sandelin *et al.*, 2004). It will be interesting to learn whether other ultraconserved regions encode ncRNAs that function to regulate transcription during development.

ncRNAs that target protein subunits of the general transcription machinery

Several ncRNAs have been identified that either stimulate or repress transcription by associating with specific protein components of the RNA polymerase II general transcription machinery (Fig. 9.3). These ncRNAs have the potential to affect transcription of all genes. The predominant outstanding question is how the activities of these ncRNAs are controlled inside cells to achieve specificity in regulation.

B2 RNA

B2 RNAs (~178 nt) are transcribed by RNA polymerase III from short interspersed elements (SINEs) that litter the genome in mouse cells (Kramerov and Vassetzky, 2005). SINEs are conserved mobile DNA elements. Throughout evolution they have duplicated, mobilized, and integrated within the genomic DNA by retrotransposition. Transcription of B2 RNA is upregulated upon heat shock and other types of cell stress (Fornace and Mitchell, 1986; Liu *et al.*, 1995). The heat shock response is characterized by increased transcription of heat shock protein genes and a general decrease in transcription of other mRNA genes (Morimoto, 1998). B2 RNA is required for the general repression of RNA polymerase II transcription in heat shocked mouse cells (Allen *et al.*, 2004). B2 RNA represses transcription by binding directly to RNA polymerase II with high affinity and specificity (Espinoza *et al.*, 2004). Mechanistic studies suggest that B2 RNA inhibits transcription by incorporating into RNA polymerase II preinitiation complexes and rendering them inactive.

A minimal 51 nucleotide region of B2 RNA that contains two stem loops separated by a single stranded region is fully functional for binding RNA polymerase II and repressing transcription *in vitro* (Espinoza *et al.*, 2007). A smaller region of B2 RNA is minimally capable of binding RNA polymerase II, but not inhibiting transcription. Hence, binding of an ncRNA to RNA polymerase II is not sufficient for transcriptional inhibition.

The discovery of B2 RNA and this novel mechanism of transcriptional repression raises

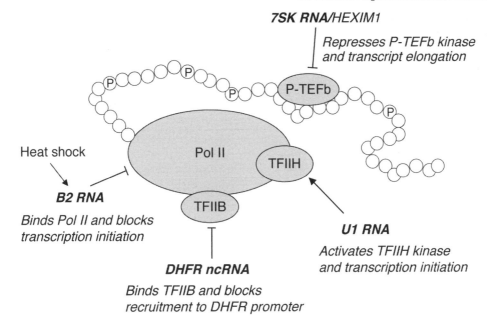

Figure 9.3 ncRNAs that target protein subunits of the general transcription machinery. B2 RNA binds to RNA polymerase II to inhibit the initiation of transcription during the mouse heat shock response. Dihydrofolate reductase (DHFR) ncRNA binds to TFIIB and prevents its recruitment to the DHFR mRNA promoter, thereby repressing transcription. U1 RNA stimulates transcription initiation by enhancing the kinase activity of TFIIH. 7SK RNA and HEXIM1 function together to inhibit the kinase activity of positive transcription elongation factor (P-TEFb). P-TEFb normally phosphorylates the CTD of RNA polymerase II, indicated by the chain of circles, resulting in the stimulation of transcription elongation. References are in the text.

many new questions. Can other ncRNAs in mammalian cells bind RNA polymerase II and modify its activity causing either repression or activation of transcription? Where on RNA polymerase II does B2 RNA bind? Does B2 RNA block DNA from entering its active site or does B2 RNA affect the catalytic activity of RNA polymerase II? How is repression by B2 RNA prevented at heat shock protein genes and when transcription of repressed genes resumes a few hours after heat shock? Does B2 RNA control transcription in response to other cellular stresses?

Dihydrofolate reductase ncRNA (DHFR ncRNA)

The dihydrofolate reductase (DHFR) gene has two promoters (Masters and Attardi, 1985). The DHFR protein is produced from the transcript initiating at the major promoter and is an enzyme involved in the pathway for synthesis of dTTP, which is required for DNA synthesis (Slansky and Farnham, 1996). The transcript produced from the minor, upstream promoter

is DHFR ncRNA, which does not encode a protein (Martianov *et al.*, 2007). The DHFR ncRNA represses transcription of the DHFR mRNA (Martianov *et al.*, 2007). Transcription of DHFR ncRNA appears to be regulated during the cell cycle; it is expressed in quiescent cells whereas the DHFR mRNA is expressed predominantly in cycling cells (Blume *et al.*, 2003; Hendrickson *et al.*, 1980). The repression of DHFR mRNA transcription by DHFR ncRNA can be reconstituted *in vitro* (Martianov *et al.*, 2007). The DHFR ncRNA represses transcription by directly binding TFIIB and decreasing its occupancy at the DHFR major promoter (Martianov *et al.*, 2007).

The working model for the function of DHFR ncRNA is the following (Martianov *et al.*, 2007): under cellular conditions where DHFR is not needed, the DHFR ncRNA is expressed, binds TFIIB, and inhibits preinitiation complex formation at the major promoter. When the cell undergoes DNA replication it requires the DHFR enzyme and DHFR ncRNA is not expressed, allowing transcription of DHFR mRNA from the major promoter. Several of the questions that remain unanswered are: How is the transcription of DHFR ncRNA from the minor promoter controlled during the cell cycle? How does DHFR ncRNA specifically inhibit transcription from the major promoter but not from the minor promoter? Does DHFR ncRNA regulate transcription of other genes?

U1 RNA

U1 RNA (165 nt) is an RNA polymerase III transcript that has diverse roles in the cell. U1 RNA functions in the snRNP splicing complex, which has been studied for many years (for a review see Will and Luhrmann, 2001). More recently, however, U1 RNA was found to also regulate transcription (Kwek et al., 2002). U1 RNA binds to TFIIH, an enzyme that is required for transcription initiation. By functioning through TFIIH, U1 RNA stimulates several activities of RNA polymerase II including initiation and reinitiation *in vitro*.

TFIIH contains several protein subunits including a kinase that phosphorylates the CTD of RNA polymerase II during early transcription. Addition of U1 RNA to reconstituted *in vitro* cyclin H/CDK7 kinase assays results in significant enhancement of kinase activity (O'Gorman et al., 2005). U1 RNA forms a stable complex with the cyclin H subunit of TFIIH and coimmunoprecipitates with cyclin H from human cells. Two regions of cyclin H were identified as being important for interaction with U1 RNA (O'Gorman et al., 2005). Enhancement of TFIIH kinase activity by U1 RNA can be abrogated with cyclin H peptides that block U1 RNA from binding to TFIIH. U1 RNA contains four stem loops, one of which is important for binding and enhancement of cyclin H kinase activity *in vitro* (O'Gorman et al., 2005).

The generality of U1 RNA as a regulator of transcription in cells remains to be tested. It will be interesting to determine the number and identities of genes whose transcription is dependent on U1 RNA. In principal, U1 RNA could itself be considered a general transcription factor if it affects transcription of most genes. In addition, future work will likely determine whether recruitment of U1 RNA to sites of transcription affects splicing of transcripts.

7SK RNA

7SK RNA is a 330 nt RNA polymerase III transcript that associates with the CDK9/cyclin T1 heterodimer, which is also known as positive transcription elongation factor b (P-TEFb) (Nguyen et al., 2001; Yang et al., 2001). The CDK9 subunit of P-TEFb phosphorylates the CTD of RNA polymerase II to stimulate transcription elongation (for a review see Peterlin and Price, 2006). P-TEFb is required for efficient elongation at many genes including those of HIV-1 (Mancebo et al., 1997; Zhu et al., 1997). HIV-1 transcription increases upon UV irradiation; this increase results from 7SK RNA dissociation from P-TEFb (Yang et al., 2001).

The effect of 7SK RNA on transcription appears to be general; it is capable of inhibiting transcription from several promoters (Chen et al., 2004; Yang et al., 2001; Yik et al., 2003). 7SK RNA bridges P-TEFb to another protein, HEXIM1 (Yik et al., 2003). Upon binding to 7SK RNA, P-TEFb is sequestered into a kinase-inactive ternary complex: 7SK RNA/P-TEFb/HEXIM1 (Yik et al., 2003). 7SK RNA is required for the interaction between HEXIM1 and P-TEFb. 7SK RNA contains several hairpins; two of these hairpins are sufficient for interaction with P-TEFb and HEXIM1 (Egloff et al., 2006). The 7SK RNA/HEXIM1 complex has decreased tryptophan fluorescence compared with HEXIM1 alone, indicating that the RNA may cause a conformational change in HEXIM1 (Li et al., 2007). This conformational change may activate HEXIM1 binding to P-TEFb.

It will be interesting to understand how 7SK RNA is regulated to maintain levels of transcription in the cell. 7SK RNA appears to be associated with pools of kinase inactive complexes (Nguyen et al., 2001; Yik et al., 2003) and it does not appear to demonstrate specificity for the genes that it inhibits. Does the amount of 7SK RNA determine how much P-TEFb is available for stimulating transcription elongation? Is the level or the intercellular location of 7SK RNA controlled in cells in response to environmental or developmental cues? Answering these questions will lead to a greater understanding of how an ncRNA that represses a critical transcription elongation factor is itself regulated.

ncRNAs that regulate transcription in *cis*, as part of the growing transcript

Cis regulation by ncRNAs occurs when a nascent transcript itself controls the transcription reaction. Thus, this type of regulatory ncRNA can influence its own transcription. The three mammalian ncRNAs described below influence transcript elongation or termination by either recruiting a protein elongation factor, binding RNA polymerase II, or undergoing self-cleavage (Fig. 9.4).

Trans-activation response region (TAR) RNA

The retrovirus HIV-1 has an RNA genome. After viral RNA is reverse transcribed into DNA, the DNA is integrated in the host genome and subsequently transcribed into mRNA by the host RNA polymerase II (for a review see Roebuck and Saifuddin, 1999 and references therein). HIV-1 mRNAs contain a specific ncRNA regulator near their 5′ ends, named TAR RNA. TAR RNA elements function to regulate transcript elongation of viral genes. TAR RNA elements bind directly to the HIV-1 protein, Tat (Dingwall *et al.*, 1989; Roy *et al.*, 1990; Weeks *et al.*, 1990). Mutational analyses show that activation by the Tat protein requires a specific stem loop in the 59 nt TAR RNA (Feng and Holland, 1988).

When a TAR element is transcribed, it forms a hairpin that recruits Tat protein to the transcript. TAR RNA secondary structure has been implicated in stalling RNA polymerase II to facilitate correct binding to the Tat protein (Laspia *et al.*, 1989; Selby *et al.*, 1989). Tat also binds the cyclin T1 subunit of the P-TEFb protein (Garber *et al.*, 1998). TAR RNA is thought to act as a scaffold by cooperatively binding Tat and P-TEFb (Richter *et al.*, 2002). One residue on each side of the TAR RNA loop is involved in the specific recognition of Tat and cyclin T1 (Richter *et al.*, 2002). The recruitment of P-TEFb to TAR RNA is necessary for efficient elongation through HIV-1 genes. Artificial

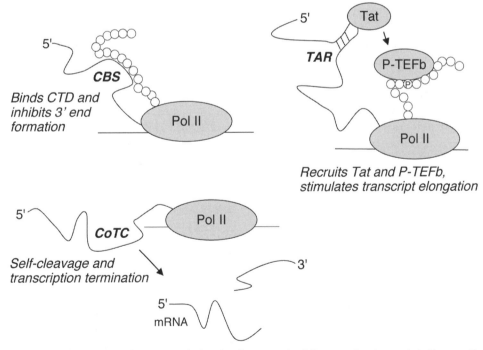

Figure 9.4 ncRNAs that regulate transcription in *cis*, as part of the growing transcript. *Trans*-activating response region (TAR) RNA is located at the 5′ end of HIV-1 mRNAs. TAR forms a hairpin that binds to the HIV-1 protein Tat. Tat recruits the positive elongation factor b (P-TEFb) to stimulate elongation. When present in the 3′ end of a growing transcript, the CTD binding sequence (CBS) binds the CTD of RNA polymerase II and prevents 3′ end formation. An element in the 3′ end of the β-globin transcript acts as a ribozyme to cleave itself, a process referred to as co-transcriptional cleavage (CoTC). Self cleavage results in termination of transcription. References are in the text.

recruitment of P-TEFb to HIV-1 genes results in efficient transcription in the absence of the TAR RNA/Tat protein interaction, indicating that the TAR/Tat complexes function solely to recruit the P-TEFb elongation factor (Bieniasz *et al.*, 1999).

TAR RNA shares some features with transacting transcriptional regulators; it recruits protein factors to the site of transcription to control the reaction. The generality of this mode of regulation for controlling transcription of endogenous mammalian genes remains to be determined. Is this type of regulation limited to viral transcripts that function through viral regulatory proteins? How many genes in mammalian cells are controlled by distinct structural elements within the 5′ ends of the nascent transcripts?

CTD binding sequence (CBS)

The CTD of vertebrate RNA polymerase II was found to interact with SV40 and β-globin pre-mRNAs *in vitro* (Kaneko and Manley, 2005). The interaction repressed polyadenylation *in vitro*. SELEX was used to determine consensus RNA sequences capable of binding the CTD of RNA polymerase II (Kaneko and Manley, 2005). A 9 nt sequence, referred to as the CTD binding sequence (CBS), was identified to preferentially interact with the CTD, although the binding was relatively weak (600 nM). The CTD contains many conserved heptad repeats (YSPTSPS) where serine residues 2 and 5 are differentially phosphorylated to stimulate the activity of RNA polymerase II (for a review see Shilatifard *et al.*, 2003). A hypophosphorylated CTD and specific heptad repeats are important for binding to the CBS (Kaneko and Manley, 2005).

To demonstrate that the RNA/CTD interaction occurs in cells, constructs containing a CBS in a transcript region were transfected, the cells were treated with formaldehyde, immunoprecipitations were performed with an antibody against the CTD, and associated RNAs were detected by RT-PCR (Kaneko and Manley, 2005). The CTD was shown to associate specifically with transcripts containing a CBS. Other transfection studies revealed that constructs encoding a CBS in the 3′ ends of transcripts caused increased read through of termination sites, decreased cleavage of pre-mRNA, and decreased RNA processing. Together these data show that CTD binding to the CBS can repress termination, 3′ end formation, and pre-mRNA processing (Kaneko and Manley, 2005).

The consequences of having a CBS at the 3′ end of pre-mRNAs have been demonstrated using transfected constructs, but the number and identities of endogenous pre-mRNAs containing sequences that can bind the CTD of RNA polymerase II to control transcription in cells remains to be determined. It has been suggested that CTD binding prevents association of factors that are involved in transcription termination and 3′ end processing; however, the mechanism by which binding of the CBS to the CTD affects termination has not yet been investigated.

Co-transcriptional cleavage (CoTC)

The β-globin pre-mRNA acts as a self-cleaving ribozyme to terminate transcription (Teixeira *et al.*, 2004). This phenomenon is termed co-transcriptional cleavage (CoTC). β-Globin pre-mRNA undergoes self-cleavage in the absence of proteins *in vitro* (Teixeira *et al.*, 2004). A minimal 200 nt region in the 3′ end of the RNA is sufficient for *in vitro* self-cleavage. Cleavage requires GTP, which must possess 2′ and 3′ hydroxyl groups. The pre-mRNA, but not the cleavage product, binds GTP. The cleavage site was mapped to a specific region and cleavage results in products containing a 5′ phosphate and a 3′ hydroxyl.

To investigate the effect of the CoTC ribozyme on transcription termination efficiency, minimal ribozyme constructs with upstream β-globin genes were transfected into human cells (Teixeira *et al.*, 2004). The 200 nt minimal ribozyme construct terminated transcription at the predicted site, whereas a catalytic core double point mutant was incapable of terminating transcription. These data along with the *in vitro* characterization of the ribozyme provide evidence that β-globin mRNA utilizes a self-cleaving activity to terminate its own transcription in cells. Conservation of the ribozyme catalytic core in primate β-globin genes suggests that CoTC is generally involved in termination of β-globin gene transcription (Teixeira *et al.*, 2004).

CoTC is the first example of regulation of mRNA transcription by a natural ribozyme present in a pre-mRNA in mammals. Investigation of the structure of the CoTC RNA will provide insight into how this ribozyme functions. The current data provide evidence that GTP acts as a nucleophile to induce cleavage. It remains to be determined whether protein factors influence or regulate self-cleavage in cells. Termination sequences have been identified in the mouse serum albumin pre-mRNA and data suggest that they are also cleaved co-transcriptionally (West et al., 2006), indicating that CoTC may be a more general mechanism of transcription termination in mammals.

ncRNAs involved in dosage compensation and imprinting

ncRNAs play integral roles in epigenetic mechanisms of gene expression in mammals including dosage compensation and genomic imprinting (for a detailed review see Zaratiegui et al., 2007). Dosage compensation equalizes expression from X chromosomes in females and males; in females, an ncRNA functions in silencing the majority of genes from one of the two X chromosomes. Imprinting involves repression of one allele of a gene and depends on whether the gene was inherited from the father or mother (for a review see Reik and Walter, 2001). Imprinting is reprogrammed every generation, which is probably orchestrated by ncRNAs. Gene silencing at the level of chromatin modification is also mediated by microRNAs and short interfering RNAs; however, this topic will not be considered here since it is the focus of another chapter in this book.

X inactive silencing transcript (Xist)

Xist RNA is a large (17–19 kb), spliced, polyadenylated ncRNA expressed from an X chromosome in females (for a detailed review see Chow et al., 2005 and references therein). Expression of Xist RNA is essential to initiate inactivation of one of the two X chromosomes. The inactivation of the X chromosome involves the establishment and maintenance of repressed chromatin, which includes histone modifications and recruitment of repressive factors such as the polycomb group proteins. Upon expression, Xist RNA associates with the X chromosome from the site of transcription and spreads outward to coat the chromosome. RNA polymerase II and the general transcription machinery are excluded from the sites on the chromosome that are coated by Xist RNA. At these same sites histones are deacetylated, polycomb group proteins are found, and histones and DNA are methylated.

Xist RNA is responsible for initiating X chromosome inactivation, but does not play a major role in the maintenance of the inactive state; the repressed chromatin state is maintained through multiple cell divisions even after downregulation of Xist RNA (Brown and Willard, 1994). Xist RNA has several conserved repeats (A-E), which were found to be important for its many functions. For example, mutation of the A repeat results in loss of transcriptional repression even though the mutant Xist RNA coats the X chromosome and polycomb group proteins are recruited (Plath et al., 2004; Wutz et al., 2002). An ncRNA that is antisense to Xist, called Tsix ncRNA (Lee et al., 1999; Lee and Lu, 1999), is important for choosing the active versus inactive X chromosome and establishes the active X chromosome by suppressing the effects of Xist RNA (for a review see Lee, 2005). The mechanisms by which Tsix RNA suppresses Xist RNA are still under investigation, but simple base pairing cannot account for several of the experimental observations.

A central unanswered question about mammalian X inactivation is the mechanism by which the Xist ncRNA establishes transcriptional silencing. It is clear that Xist must coat the chromosome in order to establish inactivation, but coating the chromosome alone is not sufficient for transcriptional silencing. It seems likely that Xist recruits protein silencing factors to the chromosome, but the order and complete complement of these silencing factors remains to be determined. In addition, the mechanism by which the Tsix ncRNA functions to establish and maintain the activate X chromosome and the interplay between Xist and Tsix will need to be explored by future research.

Antisense Igf2r RNA (Air ncRNA)

Imprinting was discovered in mice that had growth defects due to inheritance of either two maternal or paternal genomes, which indicated that cells can differentiate between genes derived from the mother versus the father (Reik and Walter, 2001). Imprinted genes typically group into clusters called imprinting centres (Sleutels and Barlow, 2002). The ncRNAs that cause the repression of these clusters are transcribed from the imprinting centre (Sleutels and Barlow, 2002).

Air ncRNA is a 108 kb polyadenylated, capped, unspliced RNA polymerase II transcript (Lyle *et al.*, 2000; Sleutels *et al.*, 2002; Wutz *et al.*, 1997; Zwart *et al.*, 2001). Air ncRNA is transcribed from an imprinting centre on the paternal chromosome 17 and causes autosomal imprinting of three protein-encoding genes found at this locus (*Igf2r*, *Slc22a2*, and *Slc22a3*) (Sleutels *et al.*, 2002). The *Igf2r*, *Slc22a2*, and *Slc22a3* genes are maternally expressed, whereas Air ncRNA is paternally expressed. The maternal promoter for Air ncRNA is CpG methylated, which inactivates Air ncRNA transcription (Li *et al.*, 1993; Lyle *et al.*, 2000). When the level of Air ncRNA is reduced in mouse cells, paternal expression of the *Igf2r*, *Slc22a2*, and *Slc22a3* genes increases (Sleutels *et al.*, 2002). The Air ncRNA is transcribed from a region that partially overlaps with the *Igf2r* gene (in the antisense direction), but does not overlap with the *Slc22a2* and *Slc22a3* genes, even though Air ncRNA inhibits their transcription. Transcriptional overlap between Air ncRNA and *Igf2r* is not necessary for repression of *Slc22a2* and *Slc22a3* (Sleutels *et al.*, 2003).

The mechanism by which Air ncRNA represses transcription of these three genes is not yet clear. One possible model is that Air ncRNA remains associated with the chromatin after it is transcribed and serves to recruit silencing factors. In this way, Air ncRNA may function similarly to Xist RNA, but on a more local scale. It has yet to be determined how many other imprinting centres express ncRNAs that function in repressing transcription of imprinted genes.

Concluding remarks

Although the understanding of how mRNA transcription is controlled in mammals has advanced tremendously, the recent discovery of multiple ncRNAs that control transcription indicates that there is much more to be learned. Moreover, the fact that the limited number of ncRNA tran-

scriptional regulators discovered to date target diverse stages in the process of transcription suggests that we have only obtained a glimpse of this new class of transcriptional regulators. Given the magnitude and diversity of protein factors that function in mammalian mRNA transcription, and knowing that protein transcription factors were sought after and studied extensively for decades, it seems reasonable to imagine that many additional ncRNA transcriptional regulators will be discovered and characterized in the coming years. In hindsight, that ncRNAs would function as transcription factors in mammalian cells should have been expected. We have appreciated the roles of functional ncRNAs in cellular processes such as translation and splicing for some time; the process of transcription should not be any different. A key discovery that foretells the likelihood that many ncRNA transcriptional regulators exist in human cells is the realization that much of genome is actively transcribed into RNAs that do not encode proteins (Bertone *et al.*, 2004; Cawley *et al.*, 2004; Cheng *et al.*, 2005; Mattick, 2005; Okazaki *et al.*, 2002; Washietl *et al.*, 2005). It seems likely that transcriptional regulators will represent a significant subset of these yet uncharacterized ncRNAs.

References

Allen, T.A., Von Kaenel, S., Goodrich, J.A., and Kugel, J.F. (2004). The SINE-encoded mouse B2 RNA represses mRNA transcription in response to heat shock. Nat. Struct. Mol. Biol. *11*, 816–821.

Bertone, P., Stolc, V., Royce, T.E., Rozowsky, J.S., Urban, A.E., Zhu, X., Rinn, J.L., Tongprasit, W., Samanta, M., Weissman, S., *et al.* (2004). Global identification of human transcribed sequences with genome tiling arrays. Science *306*, 2242–2246.

Bieniasz, P.D., Grdina, T.A., Bogerd, H.P., and Cullen, B.R. (1999). Recruitment of cyclin T1/P-TEFb to an HIV type 1 long terminal repeat promoter proximal RNA target is both necessary and sufficient for full activation of transcription. Proc. Natl. Acad. Sci. USA *96*, 7791–7796.

Blume, S.W., Meng, Z., Shrestha, K., Snyder, R.C., and Emanuel, P.D. (2003). The 5′-untranslated RNA of the human dhfr minor transcript alters transcription pre-initiation complex assembly at the major (core) promoter. J. Cell. Biochem. *88*, 165–180.

Brown, C.J., and Willard, H.F. (1994). The human X-inactivation centre is not required for maintenance of X-chromosome inactivation. Nature *368*, 154–156.

Cawley, S., Bekiranov, S., Ng, H.H., Kapranov, P., Sekinger, E.A., Kampa, D., Piccolboni, A.,

Sementchenko, V., Cheng, J., Williams, A.J., *et al.* (2004). Unbiased mapping of transcription factor binding sites along human chromosomes 21 and 22 points to widespread regulation of noncoding RNAs. Cell *116*, 499–509.

Chen, R., Yang, Z., and Zhou, Q. (2004). Phosphorylated positive transcription elongation factor b (P-TEFb) is tagged for inhibition through association with 7SK snRNA. J. Biol. Chem. *279*, 4153–4160.

Cheng, J., Kapranov, P., Drenkow, J., Dike, S., Brubaker, S., Patel, S., Long, J., Stern, D., Tammana, H., Helt, G., *et al.* (2005). Transcriptional maps of 10 human chromosomes at 5-nucleotide resolution. Science *308*, 1149–1154.

Chong, J.A., Tapia-Ramirez, J., Kim, S., Toledo-Aral, J.J., Zheng, Y., Boutros, M.C., Altshuller, Y.M., Frohman, M.A., Kraner, S.D., and Mandel, G. (1995). REST: a mammalian silencer protein that restricts sodium channel gene expression to neurons. Cell *80*, 949–957.

Chow, J.C., Yen, Z., Ziesche, S.M., and Brown, C.J. (2005). Silencing of the mammalian X chromosome. Annu. Rev. Genomics Hum. Genet. *6*, 69–92.

Dingwall, C., Ernberg, I., Gait, M.J., Green, S.M., Heaphy, S., Karn, J., Lowe, A.D., Singh, M., Skinner, M.A., and Valerio, R. (1989). Human immunodeficiency virus 1 tat protein binds *trans*-activation-responsive region (TAR) RNA *in vitro*. Proc. Natl. Acad. Sci. USA *86*, 6925–6929.

Egloff, S., Van Herreweghe, E., and Kiss, T. (2006). Regulation of polymerase II transcription by 7SK snRNA: two distinct RNA elements direct P-TEFb and HEXIM1 binding. Mol. Cell. Biol. *26*, 630–642.

Espinoza, C.A., Allen, T.A., Hieb, A.R., Kugel, J.F., and Goodrich, J.A. (2004). B2 RNA binds directly to RNA polymerase II to repress transcript synthesis. Nat. Struct. Mol. Biol. *11*, 822–829.

Espinoza, C.A., Goodrich, J.A., and Kugel, J.F. (2007). Characterization of the structure, function, and mechanism of B2 RNA, an ncRNA repressor of RNA polymerase II transcription. RNA *13*, 583–596.

Feng, J., Bi, C., Clark, B.S., Mady, R., Shah, P., and Kohtz, J.D. (2006). The Evf-2 noncoding RNA is transcribed from the Dlx-5/6 ultraconserved region and functions as a Dlx-2 transcriptional coactivator. Genes Dev. *20*, 1470–1484.

Feng, S., and Holland, E.C. (1988). HIV-1 tat *trans*-activation requires the loop sequence within tar. Nature *334*, 165–167.

Fornace, A.J., Jr., and Mitchell, J.B. (1986). Induction of B2 RNA polymerase III transcription by heat shock: enrichment for heat shock induced sequences in rodent cells by hybridization subtraction. Nucl. Acids Res. *14*, 5793–5811.

Garber, M.E., Wei, P., and Jones, K.A. (1998). HIV-1 Tat interacts with cyclin T1 to direct the P-TEFb CTD kinase complex to TAR RNA. Cold Spring Harb. Symp. Quant. Biol. *63*, 371–380.

Ghanem, N., Jarinova, O., Amores, A., Long, Q., Hatch, G., Park, B.K., Rubenstein, J.L., and Ekker, M. (2003). Regulatory roles of conserved intergenic domains in vertebrate Dlx bigene clusters. Genome Res. *13*, 533–543.

Hendrickson, S.L., Wu, J.S., and Johnson, L.F. (1980). Cell cycle regulation of dihydrofolate reductase mRNA metabolism in mouse fibroblasts. Proc. Natl. Acad. Sci. USA *77*, 5140–5144.

Hogan, P.G., Chen, L., Nardone, J., and Rao, A. (2003). Transcriptional regulation by calcium, calcineurin, and NFAT. Genes Dev. *17*, 2205–2232.

Huang, Y., Myers, S.J., and Dingledine, R. (1999). Transcriptional repression by REST: recruitment of Sin3A and histone deacetylase to neuronal genes. Nat. Neurosci. *2*, 867–872.

Kaneko, S., and Manley, J.L. (2005). The mammalian RNA polymerase II C-terminal domain interacts with RNA to suppress transcription-coupled 3' end formation. Mol. Cell *20*, 91–103.

Kramerov, D.A., and Vassetzky, N.S. (2005). Short retroposons in eukaryotic genomes. Int. Rev. Cytol. *247*, 165–221.

Kuwabara, T., Hsieh, J., Nakashima, K., Taira, K., and Gage, F.H. (2004). A small modulatory dsRNA specifies the fate of adult neural stem cells. Cell *116*, 779–793.

Kwek, K.Y., Murphy, S., Furger, A., Thomas, B., O'Gorman, W., Kimura, H., Proudfoot, N.J., and Akoulitchev, A. (2002). U1 snRNA associates with TFIIH and regulates transcriptional initiation. Nat. Struct. Biol. *9*, 800–805.

Lanz, R.B., McKenna, N.J., Onate, S.A., Albrecht, U., Wong, J., Tsai, S.Y., Tsai, M.J., and O'Malley, B.W. (1999). A steroid receptor coactivator, SRA, functions as an RNA and is present in an SRC-1 complex. Cell *97*, 17–27.

Lanz, R.B., Razani, B., Goldberg, A.D., and O'Malley, B.W. (2002). Distinct RNA motifs are important for coactivation of steroid hormone receptors by steroid receptor RNA activator (SRA). Proc. Natl. Acad. Sci. USA *99*, 16081–16086.

Laspia, M.F., Rice, A.P., and Mathews, M.B. (1989). HIV-1 Tat protein increases transcriptional initiation and stabilizes elongation. Cell *59*, 283–292.

Lee, J.T. (2005). Regulation of X-chromosome counting by Tsix and Xite sequences. Science *309*, 768–771.

Lee, J.T., Davidow, L.S., and Warshawsky, D. (1999). Tsix, a gene antisense to Xist at the X-inactivation centre. Nat. Genet. *21*, 400–404.

Lee, J.T., and Lu, N. (1999). Targeted mutagenesis of Tsix leads to nonrandom X inactivation. Cell *99*, 47–57.

Li, E., Beard, C., and Jaenisch, R. (1993). Role for DNA methylation in genomic imprinting. Nature *366*, 362–365.

Li, Q., Cooper, J.J., Altwerger, G.H., Feldkamp, M.D., Shea, M.A., and Price, D.H. (2007). HEXIM1 is a promiscuous double-stranded RNA-binding protein and interacts with RNAs in addition to 7SK in cultured cells. Nucleic Acids Res. *35*, 2503–2512.

Liu, W.M., Chu, W.M., Choudary, P.V., and Schmid, C.W. (1995). Cell stress and translational inhibitors transiently increase the abundance of mammalian SINE transcripts. Nucleic Acids Res. *23*, 1758–1765.

Lonard, D.M., and O'Malley, B.W. (2005). Expanding functional diversity of the coactivators. Trends Biochem. Sci. *30*, 126–132.

Lunyak, V.V., Burgess, R., Prefontaine, G.G., Nelson, C., Sze, S.H., Chenoweth, J., Schwartz, P., Pevzner, P.A., Glass, C., Mandel, G., *et al.* (2002). Corepressor-dependent silencing of chromosomal regions encoding neuronal genes. Science 298, 1747–1752.

Lyle, R., Watanabe, D., te Vruchte, D., Lerchner, W., Smrzka, O.W., Wutz, A., Schageman, J., Hahner, L., Davies, C., and Barlow, D.P. (2000). The imprinted antisense RNA at the Igf2r locus overlaps but does not imprint Mas1. Nat. Genet. 25, 19–21.

Mancebo, H.S., Lee, G., Flygare, J., Tomassini, J., Luu, P., Zhu, Y., Peng, J., Blau, C., Hazuda, D., Price, D., *et al.* (1997). P-TEFb kinase is required for HIV Tat transcriptional activation *in vivo* and *in vitro*. Genes Dev. 11, 2633–2644.

Martianov, I., Ramadass, A., Serra Barros, A., Chow, N., and Akoulitchev, A. (2007). Repression of the human dihydrofolate reductase gene by a non-coding interfering transcript. Nature 445, 666–670.

Masters, J.N., and Attardi, G. (1985). Discrete human dihydrofolate reductase gene transcripts present in polysomal RNA map with their 5′ ends several hundred nucleotides upstream of the main mRNA start site. Mol. Cell. Biol. 5, 493–500.

Mattick, J.S. (2005). The functional genomics of noncoding RNA. Science 309, 1527–1528.

Morimoto, R.I. (1998). Regulation of the heat shock transcriptional response: cross talk between a family of heat shock factors, molecular chaperones, and negative regulators. Genes Dev. 12, 3788–3796.

Nguyen, V.T., Kiss, T., Michels, A.A., and Bensaude, O. (2001). 7SK small nuclear RNA binds to and inhibits the activity of CDK9/cyclin T complexes. Nature 414, 322–325.

O'Gorman, W., Thomas, B., Kwek, K.Y., Furger, A., and Akoulitchev, A. (2005). Analysis of U1 small nuclear RNA interaction with cyclin H. J. Biol. Chem. 280, 36920–36925.

Okazaki, Y., Furuno, M., Kasukawa, T., Adachi, J., Bono, H., Kondo, S., Nikaido, I., Osato, N., Saito, R., Suzuki, H., *et al.* (2002). Analysis of the mouse transcriptome based on functional annotation of 60,770 full-length cDNAs. Nature 420, 563–573.

Peterlin, B.M., and Price, D.H. (2006). Controlling the elongation phase of transcription with P-TEFb. Mol. Cell 23, 297–305.

Plath, K., Talbot, D., Hamer, K.M., Otte, A.P., Yang, T.P., Jaenisch, R., and Panning, B. (2004). Developmentally regulated alterations in Polycomb repressive complex 1 proteins on the inactive X chromosome. J. Cell. Biol. 167, 1025–1035.

Reik, W., and Walter, J. (2001). Genomic imprinting: parental influence on the genome. Nat. Rev. Genet. 2, 21–32.

Richter, S., Ping, Y.H., and Rana, T.M. (2002). TAR RNA loop: a scaffold for the assembly of a regulatory switch in HIV replication. Proc. Natl. Acad. Sci. USA 99, 7928–7933.

Roebuck, K.A., and Saifuddin, M. (1999). Regulation of HIV-1 transcription. Gene Expr. 8, 67–84.

Roy, S., Delling, U., Chen, C.H., Rosen, C.A., and Sonenberg, N. (1990). A bulge structure in HIV-1

TAR RNA is required for Tat binding and Tat-mediated *trans*-activation. Genes Dev. 4, 1365–1373.

Sandelin, A., Bailey, P., Bruce, S., Engstrom, P.G., Klos, J.M., Wasserman, W.W., Ericson, J., and Lenhard, B. (2004). Arrays of ultraconserved non-coding regions span the loci of key developmental genes in vertebrate genomes. BMC Genomics 5, 99.

Schoenherr, C.J., and Anderson, D.J. (1995a). Silencing is golden: negative regulation in the control of neuronal gene transcription. Curr. Opin. Neurobiol. 5, 566–571.

Schoenherr, C.J., and Anderson, D.J. (1995b). The neuron-restrictive silencer factor (NRSF): a coordinate repressor of multiple neuron-specific genes. Science 267, 1360–1363.

Selby, M.J., Bain, E.S., Luciw, P.A., and Peterlin, B.M. (1989). Structure, sequence, and position of the stem–loop in tar determine transcriptional elongation by tat through the HIV-1 long terminal repeat. Genes Dev. 3, 547–558.

Shamovsky, I., Ivannikov, M., Kandel, E.S., Gershon, D., and Nudler, E. (2006). RNA-mediated response to heat shock in mammalian cells. Nature 440, 556–560.

Shilatifard, A., Conaway, R.C., and Conaway, J.W. (2003). The RNA polymerase II elongation complex. Annu. Rev. Biochem. 72, 693–715.

Slansky, J.E., and Farnham, P.J. (1996). Transcriptional regulation of the dihydrofolate reductase gene. Bioessays 18, 55–62.

Sleutels, F., and Barlow, D.P. (2002). The origins of genomic imprinting in mammals. Adv. Genet. 46, 119–163.

Sleutels, F., Tjon, G., Ludwig, T., and Barlow, D.P. (2003). Imprinted silencing of Slc22a2 and Slc22a3 does not need transcriptional overlap between Igf2r and Air. EMBO J. 22, 3696–3704.

Sleutels, F., Zwart, R., and Barlow, D.P. (2002). The noncoding Air RNA is required for silencing autosomal imprinted genes. Nature 415, 810–813.

Teixeira, A., Tahiri-Alaoui, A., West, S., Thomas, B., Ramadass, A., Martianov, I., Dye, M., James, W., Proudfoot, N.J., and Akoulitchev, A. (2004). Autocatalytic RNA cleavage in the human beta-globin pre-mRNA promotes transcription termination. Nature 432, 526–530.

Thomas, M.C., and Chiang, C.M. (2006). The general transcription machinery and general cofactors. Crit. Rev. Biochem. Mol. Biol. 41, 105–178.

Washietl, S., Hofacker, I.L., Lukasser, M., Huttenhofer, A., and Stadler, P.F. (2005). Mapping of conserved RNA secondary structures predicts thousands of functional noncoding RNAs in the human genome. Nat. Biotechnol. 23, 1383–1390.

Weeks, K.M., Ampe, C., Schultz, S.C., Steitz, T.A., and Crothers, D.M. (1990). Fragments of the HIV-1 Tat protein specifically bind TAR RNA. Science 249, 1281–1285.

West, S., Zaret, K., and Proudfoot, N.J. (2006). Transcriptional termination sequences in the mouse serum albumin gene. RNA 12, 655–665.

Will, C.L., and Luhrmann, R. (2001). Spliceosomal UsnRNP biogenesis, structure and function. Curr. Opin. Cell Biol. *13*, 290–301.

Willingham, A.T., Orth, A.P., Batalov, S., Peters, E.C., Wen, B.G., Aza-Blanc, P., Hogenesch, J.B., and Schultz, P.G. (2005). A strategy for probing the function of noncoding RNAs finds a repressor of NFAT. Science *309*, 1570–1573.

Wutz, A., Rasmussen, T.P., and Jaenisch, R. (2002). Chromosomal silencing and localization are mediated by different domains of Xist RNA. Nat. Genet. *30*, 167–174.

Wutz, A., Smrzka, O.W., Schweifer, N., Schellander, K., Wagner, E.F., and Barlow, D.P. (1997). Imprinted expression of the Igf2r gene depends on an intronic CpG island. Nature *389*, 745–749.

Yang, Z., Zhu, Q., Luo, K., and Zhou, Q. (2001). The 7SK small nuclear RNA inhibits the CDK9/cyclin T1 kinase to control transcription. Nature *414*, 317–322.

Yik, J.H., Chen, R., Nishimura, R., Jennings, J.L., Link, A.J., and Zhou, Q. (2003). Inhibition of P-TEFb (CDK9/Cyclin T) kinase and RNA polymerase II transcription by the coordinated actions of HEXIM1 and 7SK snRNA. Mol. Cell *12*, 971–982.

Zaratiegui, M., Irvine, D.V., and Martienssen, R.A. (2007). Noncoding RNAs and gene silencing. Cell *128*, 763–776.

Zerucha, T., Stuhmer, T., Hatch, G., Park, B.K., Long, Q., Yu, G., Gambarotta, A., Schultz, J.R., Rubenstein, J.L., and Ekker, M. (2000). A highly conserved enhancer in the Dlx5/Dlx6 intergenic region is the site of cross-regulatory interactions between Dlx genes in the embryonic forebrain. J. Neurosci. *20*, 709–721.

Zhao, X., Patton, J.R., Davis, S.L., Florence, B., Ames, S.J., and Spanjaard, R.A. (2004). Regulation of nuclear receptor activity by a pseudouridine synthase through posttranscriptional modification of steroid receptor RNA activator. Mol. Cell *15*, 549–558.

Zhou, Q.P., Le, T.N., Qiu, X., Spencer, V., de Melo, J., Du, G., Plews, M., Fonseca, M., Sun, J.M., Davie, J.R., et al. (2004). Identification of a direct Dlx homeodomain target in the developing mouse forebrain and retina by optimization of chromatin immunoprecipitation. Nucl. Acids Res. *32*, 884–892.

Zhu, Y., Pe'ery, T., Peng, J., Ramanathan, Y., Marshall, N., Marshall, T., Amendt, B., Mathews, M.B., and Price, D.H. (1997). Transcription elongation factor P-TEFb is required for HIV-1 tat transactivation *in vitro*. Genes Dev. *11*, 2622–2632.

Zwart, R., Sleutels, F., Wutz, A., Schinkel, A.H., and Barlow, D.P. (2001). Bidirectional action of the Igf2r imprint control element on upstream and downstream imprinted genes. Genes Dev. *15*, 2361–2366.

Pyknons as Putative Novel and Organism-specific Regulatory Motifs

10

Isidore Rigoutsos

Abstract

In recent work, we introduced a new type of putative regulatory motif that we named 'pyknon' from the Greek adjective for dense. By definition, pyknons are variable length sequences with a statistically significant number of intact copies in the intergenic and intronic regions of the genome and additional copies in the untranslated or amino acid coding regions of known transcripts. Even though the original presentation discussed pyknons in the context of the human genome, pyknons likely represent a more general architectural component of eukaryotic genomes. The exact role of pyknons is currently unclear, but the findings so far support a regulatory responsibility. In this chapter, we review the process that led to this discovery, the results so far, and briefly present some thoughts about the possibility that pyknons hint at a previously unseen layer of cell process regulation.

Introduction

Recent years have seen a renewed research interest in the analysis of intergenic and intronic regions. This is probably a consequence of the discovery of the phenomenon of RNA interference whereby short, approximately 22 nucleotide-long RNAs (= microRNAs) that are derived from intergenic or intronic precursor sequences (= microRNA precursors) can down-regulate the expression of gene transcripts (Baulcombe, 1996; Baulcombe, 1999; Elbashir *et al.*, 2001; Fire *et al.*, 1998; Hamilton and Baulcombe, 1999; Lau *et al.*, 2001; Lee and Ambros, 2001; Napoli *et al.*, 1990; Reinhart *et al.*, 2000; Slack *et*

al., 2000; Voinnet and Baulcombe, 1997). Micro-RNAs implement this down-regulation either by directing the degradation of the messenger RNA or through translational inhibition; however, the actual mechanism(s) continue to be a matter of discussion (Bagga *et al.*, 2005; Bartel, 2004; Lytle *et al.*, 2007; Pillai *et al.*, 2005; Pillai *et al.*, 2007). In plants, this down-regulation has been shown to be effected through the amino acid coding region (CDS) of a transcript (Floyd and Bowman, 2004; Reinhart *et al.*, 2002; Rhoades *et al.*, 2002) whereas in animals all of the currently available examples involve microRNAs acting on the 3' untranslated regions (3'UTRs) of coding transcripts (Bartel, 2004; Johnston and Hobert, 2003; Moss *et al.*, 1997; Poy *et al.*, 2004; Slack *et al.*, 2000; Zhang *et al.*, 2006). Such is the importance of this discovery that it commanded the 2006 Nobel Prize in Medicine.

There are currently differing hypotheses as to the breadth of this RNA-based level of regulation in animals. Earlier work suggested that as much as 25–30% of the protein-coding transcripts of the human genome may be under the control of microRNAs and that these transcripts are involved mainly in developmental control and physiological processes (Bartel, 2004). More recently, our own analysis of several model organisms provided evidence that nearly all coding transcripts are under the control of microRNAs and other short RNAs (Miranda *et al.*, 2006; Tsirigos and Rigoutsos, unpublished data). In particular, in Miranda *et al.* (2006), we demonstrated experimentally the distinct possibility that a microRNA could be regulating a few thousand genes, a substantial departure from previous studies that had estimated this number to be approximately a few hundred (Bentwich, 2005; Lim et al., 2005; Rajewsky, 2006).

Widely differing estimates exist also for the number of microRNAs that carry out this regulation. In very early analyses, the number of microRNAs in a higher eukaryotic organism such as *H. sapiens* was projected to be ~300 (Lim et al., 2003). Not long thereafter, this number was revised upwards to a few thousand for the human genome (Bentwich et al., 2005; Berezikov et al., 2006; Berezikov et al., 2005). Our own estimates, to which we have arrived using two distinct approaches (Miranda et al., 2006; Rigoutsos et al., 2006), place the number of human microRNAs substantially higher, to several tens of thousands. Our analysis of other organisms of interest such as mouse, fruit-fly, worm, etc. generated estimates that are also much higher than previously reported numbers. It is important to point out that, unlike other published methods for determining novel microRNAs and their targets, our method does not require that the underlying sequences be evolutionarily conserved. As of Spring 2007, the public database known as miRBase (Griffiths-Jones, 2006) contained 470 predicted or validated human, 380 mouse, 78 fruit-fly and 134 worm microRNAs. These microRNAs arise from longer precursors, approximately 70 to 100 nucleotides in length, that are found in the genome's intergenic and intronic regions, i.e. from regions whose significance and role continue to remain unclear.

With respect to the analysis of a genome's non-coding regions, little happened for almost 30 years. It has been argued that Susumu Ohno's 1972 paper (Ohno, 1972), in which he coined the phrase 'junk DNA' to summarily refer to all areas of the genome that did not code for proteins, deterred scientists from mounting any significant research efforts towards the study of these regions. In fact, for many years the search for functioning motifs had been confined to the neighbourhoods immediately upstream of the 5'UTRs of genes (Brazma et al., 1998b; Ettwiller et al., 2003; Lenhard et al., 2003; Sinha and Tompa, 2002; Wasserman and Sandelin, 2004). More recently, research activity began exploring the 3'UTRs of genes, and, motifs

analogous to the *cis*-motifs of promoter regions were identified and found to be functionally relevant (Knaut et al., 2002; Lai et al., 1998; Yaffe et al., 1985).

Following the realization of the importance of RNAi as a regulatory process, the fact that the same microRNA sequences were found present in different organisms, and the discovery of the importance of 3'UTRs in the context of RNAi, led researchers to hypothesize that such motifs, if present, ought to be conserved across phylogenetically proximal genomes. Indeed, analysis of orthologous, multiply aligned 3'UTRs from several genomes revealed the presence of several new conserved elements that were also shown to be linked to RNAi (Xie et al., 2005).

Parallel studies that went beyond the immediate neighbourhoods of genes were also successful in identifying potentially functional regions that were conserved across genomes (Bejerano et al., 2004; Dubchak et al., 2000; Frazer et al., 2001; Jareborg et al., 1999; Miziara et al., 2004). However, the majority of the analyses that attempted to shed light on genomic regions that were not proximal to genes were limited by computational requirements and, consequently, focused on only a small number of organisms at a time (Boffelli et al., 2003; Dermitzakis et al., 2004; Kellis et al., 2003; Thomas et al., 2003; Wasserman et al., 2000).

In the last several years, transcription studies showed that a considerable part of the human genome is transcribed (Cheng et al., 2005; Kapranov et al., 2007). These studies had several immediate consequences. First, they provided more evidence that the so-called junk DNA is transcriptionally active. Second, they emphasized the fact that, for the most part, the function of these regions continues to elude us. Lastly, they served as a renewed 'call to arms' for scientists worldwide to make sense of these regions.

Our own interest in the study of intergenic and intronic regions arose from earlier exposure to tantalizing hypotheses about the importance and role of these regions (Mattick, 2004; Mattick, 2007; Mattick and Makunin, 2006) and also as a result of our own parallel studies of RNA interference (Miranda et al., 2006). The latter were suggesting the existence of a comparatively large number of putative microRNAs in the model animal genomes. With a length averaging approximately 70 nucleotides, the precursors for these predicted microRNAs ought to occupy a very sizeable portion of the intergenic and intronic portions of the corresponding genomes. This of course was suggesting that these regions act as depots for important regulatory molecules such as microRNAs. We thus wondered what other signals might be lurking in these regions. We set out to answer this question using motif discovery techniques of the type that we and others have been developing for more than a decade (Jensen et al., 2006; Jonassen et al., 1995; Parida et al., 2000; Parida et al., 2001; Rigoutsos and Floratos, 1998; Rigoutsos et al., 2000; Smith et al., 1990; Styczynski et al., 2004; Tompa et al., 2005).

The problem

We sought to identify motifs, i.e. putative signals, that could be located in an unsupervised manner with the help of the pattern discovery schemes. The goal underlying this effort was to determine whether any discovered motifs could be used to link, in some manner, the non-coding and coding regions of a given genome. Due to its importance, we set out to analyse the human genome first.

The key idea

Before we began our analysis, we imposed four constraints that were meant to differentiate our effort from what had been done up to that point. First, we sought motifs that went beyond the boundaries of a chromosome. Second, instead of 'hypothesizing-and-verifying' we adopted an unsupervised pattern discovery approach. The key characteristic of using a discovery scheme is that it can identify *variable-length* motifs that satisfy a minimum-length requirement. The remaining two constraints that we imposed represent a substantial departure from all previous work. In particular, we confined ourselves to working with a single genome thus obviating the need for cross-genome comparisons. Finally, we set out to address the motif discovery problem by approaching it in a backward manner. Indeed, instead of discovering motifs in (or near) genic regions and then attempting to associate them with the activity of entities embedded elsewhere in the genome, we sought to identify motifs in the intergenic/intronic regions and attempted to elucidate their possible links with known exons.

It is important to note here that the manner in which we do our computational analysis should *not* be taken to imply that we have assumed a 'direction' in the flow of information or a cause-effect relationship between non-genic and genic regions of the genome. In other words, the fact that we seek motifs in the intergenic and intronic regions of a genome does not mean that these regions are the source of regulatory activity and that the exonic regions are the recipients of such activity. As will become apparent later in the presentation, it is entirely possible that both the intergenic/intronic and the exonic regions act simultaneously as sources and recipients of regulatory activity, possibly entangled in a putative feedback loop. The currently available data do not permit us to draw general rules or to resolve the direction of information flow.

The methodology

In 1998, we published a combinatorial algorithm for the discovery of patterns in one-dimensional streams that we named *Teiresias*, after the blind seer of Greek mythology (Rigoutsos and Floratos, 1998). This algorithm solved the following problem: 'given a database D comprising multiple, variable-length, unbounded streams of events chosen from a set E of permissible events, find and report all combinations of events that satisfy a 'minimum density constraint' and a 'minimum number of appearances constraint' and are maximal in size and composition. The set E is frequently referred to as the 'alphabet.' Example event sets E include the four nucleotides {A, C, G, T}, the 20 amino acids {A, C, D, E,..., Y}, the set of 26 letters of the English alphabet, categorical values from a finite collection, e.g. {red, green, blue, white, black}, etc. The minimum support constraint is implemented using a user-defined parameter K: no patterns will be reported unless they have a minimum of K instances in the database D. The minimum density constraint is implemented with the help of two user-defined parameters, L and W (where $L \leq W$): *Teiresias* will only seek and report $<L,W>$ patterns, i.e. combinations comprising literals and wild-cards such that any L consecutive (but not necessarily contiguous) literals span at most W positions.

[KR]..F.[ILMV]..H is an example of a <2,4> pattern that can capture sequences of events in D each of which begins with either event K or R, followed by any two permissible events, followed by event F followed by any event, followed by exactly one of I, L, M or V, followed by any two events, and ending with event H. *Teiresias* was the first algorithm of its kind to guarantee that the reported patterns are length-maximal and composition-maximal. What this means from a practical standpoint is that every discovered pattern *p* that is reported to appear *k* times in the processed input is 'as long as it can be' and 'as dense as it can be': any pattern *p'* that satisfies the density constraint and is derived by adding a prefix or suffix to a discovered pattern *p* must appear in the input strictly fewer than *k* times; moreover, any pattern *p''* that is generated by dereferencing one or more wild-cards of a discovered pattern *p* must also appear in the input strictly fewer than *k* times.[1]

What sets *Teiresias* apart from other pattern discovery algorithms is its speed and the ability to guarantee the reporting of *all* patterns that are contained in D and which satisfy the density, maximality and support constraints. For the purposes of our genomic analysis, we adapted the algorithm so that it could handle the very large input that corresponds to a eukaryotic genome (each strand is treated separately) while keeping memory and intermediate disk storage requirements to a minimum.

What type of patterns

The type of expressions that a pattern discovery algorithm can discover and report can be varied and ranges from simple to more complicated ones. These expressions include 'solid', 'rigid', 'flexible' and other. An example of a solid pattern is AB-CDEF: the pattern's literals occupy consecutive positions in all of the instances of solid patterns that are present in the processed input. On the other hand, rigid patterns are more permissive in that they comprise literals as well as wild-cards. Wild-cards are typically represented by a "." and act as placeholders for exactly one event from the

permitted alphabet. AB...C.....D.E....F is an example of such a rigid pattern: across all instances of this pattern, B and C are always separated by exactly three other events, C and D are always separated by exactly five other events etc. Each group of consecutive wild-cards is also known as the 'gap.' Flexible patterns represent a variant of rigid patterns in that the gaps have widths which range between a minimum and a maximum value. A.{1,3}B.C.{4}D.{3,5}E is such an example: in this case, the instances of the pattern in the processed input contain between 1 and 3 unspecified events separating A and B, exactly 1 unspecified event separating B and C, exactly four unspecified events separating C and D etc. Even more permissive variants of these three types replace a literal by a 'bracketed' expression, e.g. [ILMV], that is meant to denote that the corresponding position of the pattern is occupied by exactly one of the events listed within the brackets, in this case exactly one of I, L, M or V. The <2,4> pattern [KR]..F.[ILMV]..H that we discussed above is such a variant.

It should be stressed here that these are notations and that there are multiple ways for 'writing' patterns. For example, using a notation that was popularized by the PROSITE database (Bairoch, 1991; Hulo *et al.*, 2006) the above rigid and flexible examples could be rewritten as A-B-x(3)-C-x(5)-D-x-E-x(4)-F and A-x(1,3)-B-x-C-x(4)-D-x(3,5)-E respectively. Additionally, it is important to note that even more permissive categories of patterns exist. However, the overwhelming majority of the algorithms that have been published in the literature have focused on the discovery of solid, rigid and flexible patterns (Jensen *et al.*, 2006; Parida *et al.*, 2001; Rigoutsos *et al.*, 2000; Styczynski *et al.*, 2004). For a more detailed discussion of these and other related topics such the *NP-hardness* of the pattern discovery problem the reader is referred elsewhere (Brazma *et al.*, 1998a; Rigoutsos *et al.*, 2000). We conclude this section by mentioning that in our analysis of the human genome that gave rise to pyknons, we worked only with solid patterns.

---

1 Note that if a pattern *p* satisfies the <*L,W*> density constraint, then any pattern *p'* that is derived from *p* by dereferencing one of more of *p*'s wild cards continues to satisfy the <*L,W*> constraint.

Minimum length and how to ensure statistical significance

Intuitively, it is easy to see that the larger the number of literals comprising a pattern the more unlikely it is that the pattern will be encountered in a given database purely by accident. Intimately linked to this observation are the notions of a pattern's sensitivity and specificity. Let's elaborate with a specific example from the amino acid sequences of leghaemoglobins. The pattern P.L.[AG]HA..[ILMV]F captures a conserved region of leghaemoglobins and the bold-faced residue is the distal histidine that provides a stabilizing hydrogen bond that prevents haem loss. Variants of this pattern include P.L.[AG]HA.K[ILMV]F.[ILMV], P.L.[AG]HA..[ILMV]F.LV.[DE].A.QL, P.L.[AG]HAEK[ILMV]F[AG][ILMV] V.DSA.QL[KR]..G as well as many others. The longer the variant the less sensitive it is in identifying distant members of the leghaemoglobin class. On the other hand, longer variants will not in general appear in random strings of amino acids, and this makes them good, albeit not sensitive, descriptors of the class at hand.

In earlier work (Rigoutsos *et al.*, 2006), we described a simulation study where we examined the distribution of the number of copies of 16-mers (fixed-length tuples) in random shufflings of the human genome. The simulation intended to determine the likelihood that some 16-mer will appear more than a given number of times in this shuffled input. Our analysis showed that the majority of 16-mers will have zero or one copy in this random input and that no 16-mer has more than ~23 copies. Note that any 16-mers that NSEG (Wootton and Federhen, 1996) deemed to be of low-complexity were not included in this analysis. As the reader may surmise from the above discussion, if we were to repeat the above analysis using fixed-length tuples longer than 16 letters, we would find the resulting distribution to be even narrower and shifted towards even fewer copies in the random database. So by requiring during the pattern discovery phase that all patterns are *at least* 16 nucleotides long and appear *at least* 40 times in the processed input, we ensure minimal overlap with the distribution that one expects from random inputs of the same size.[2] Additionally, we estimated the probability that patterns with this minimum length and number of copies occur accidentally using three different schemes and found it to be exceptionally small (Rigoutsos *et al.*, 2006).

Preparing the input

Prior to analysing the input, we pre-processed it and excluded all those nucleotide segments that are found opposite known exonic regions. The justification for this choice will become clear in view of the relationships, which we will describe later, and which allow us to link non-genic and genic regions of the genome. Removal of these 'exon-mirrors' ensures that, to the extent allowed by the current annotation, the processed input is free of segments that are the reverse complement of known 5' untranslated, amino acid coding, or 3' untranslated regions.

Results

Even with the stringent criteria that we set, we found a very large number of patterns in the human intergenic and intronic regions that satisfied them. The resulting set, P_{init}, contained more than 60 million motifs of variable length. However, for the most part, these motifs are short: 80% of the motifs are ≤ 50 nucleotides long, 15% are between 51 and 100 nucleotides, and the remaining 5% are > 100 nucleotides long. It should also be pointed out that the discovered motifs are overlapping, i.e. a given nucleotide position in the processed input can participate in two or more motifs that have different sets of instances, different support etc.

A large number of pyknons

As with our simulation, we first removed low-complexity patterns from our collection using NSEG, and then proceeded to identify additional instances of the remaining patterns inside the 5'UTRs, CDSs and 3'UTRs of known genes. This gave rise to three sets, $P_{5'UTR}$, P_{CDS} (a.k.a. P_{CODING}) and $P_{3'UTR}$. We refer to the

---

2 In terms of the *Teiresias* parameters, their values were: $L = 16$, $W = 16$ and $K = 40$.

patterns-members of these three sets as pyknons. By definition, these three sets contain nucleotide sequences that were originally discovered to have 40 or more copies in the intergenic and intronic regions of the human genome and which have additional instances inside known exons. The members of P_{init} were processed in order of decreasing value of the product 'length-of-pattern × number-of-intergenic-and-intronic-copies-of-pattern-in-P_{init}.' This heuristic method appears to work well. For a detailed discussion on how we treated collisions between pyknons and candidate pyknons, the reader is referred to the supplement of Rigoutsos et al. (2006).

Distinct sets of pyknons

The first surprising observation was that these three sets of patterns are largely disjoint. In other words, the pyknons in each of $P_{5'UTR}$, P_{CDS} and $P_{3'UTR}$ are distinct at the sequence level. Indeed, the sum of the cardinalities of the three sets is equal to 134 207 whereas the cardinality of their union is equal to 127 998. This was rather unexpected and suggested a clear distinction among the pyknon sequences found in each of these three regions. Moreover, pyknons were found to be non-redundant; in other words, at the sequence level, they are distinct and remain so even if we permit a very substantial number of letter replacements. Indeed, we attempted to cluster the union of $P_{5'UTR}$, P_{CDS} and $P_{3'UTR}$: two pyknons p_1 and p_2 would be placed in the same cluster if and only if p_1 and p_2 or p_1 and the reverse complement of p_2 differed by at most X% of their positions. For X = 10%, 20% and 30%, the 127 998 pyknons formed 89 159, 44 417 and 32 621 clusters respectively, suggesting that they are rather distinct nucleotide sequences.

Pyknons are present in nearly all genes

Even though the total number of pyknons was high, it could have well been the case that their instances were contained in only a small number of human genes. Instead, we found that one or more pyknon instances were present in 20 059 human genes or 30 675 human gene transcripts. In fact, the pyknons had a total of 226 874 instances in human 5'UTRs, CDSs and 3'UTRs.

In other words, nearly every human gene contains one or more pyknon instances in its untranslated or amino acid coding regions. Of the 30 675 gene transcripts contained in the ENSEMBL release we analyzed, more than 16 000 (~52%) contain four or more pyknon instances in their 5'UTR, CDS or 3'UTR. Approximately 2 200 gene transcripts (7.7%) contain 20 or more pyknon instances in them.

Pyknons arrange combinatorially

The observation that some genes contained multiple, non-overlapping pyknon instances in them led us to examine the kind of arrangements in which these instances were found. This resulted in yet another surprising observation: in genes that contained multiple pyknons we found what appears to be a combinatorial arrangement with sets of specific pyknons appearing in different genes in different copy numbers and order. Fig. 10.1, reproduced from (Rigoutsos et al., 2006), shows such a combinatorial example involving the 3'UTRs of transcripts from ten human genes. This figure permits several observations. First, it is clear that these 3'UTRs can be decomposed using pyknon sequences. Second, the same pyknons can be found conserved in apparently unrelated genes. And third, those of the pyknons that are conserved and present in multiple 3'UTRs do not always appear in the same order or with the same number of copies. The exceptional sequence conservation of pyknons in conjunction with their combinatorial arrangement in various genes strongly suggests their participation in a putative regulatory layer. It should be noted that similar combinatorial arrangements exist in CDSs and 5'UTRs and corresponding examples can be found in the supplement of (Rigoutsos et al., 2006).

Functional connections

The conjecture of pyknons' participation in a previously unsuspected regulatory layer was strengthened further by the fact that we found pyknons to be significantly enriched or depleted in the sequences (untranslated or amino acid coding) of genes belonging to specific biological processes. For a given GO process (Ashburner et al., 2000), we generally observed distinct levels of enrichment or depletion of pyknon

Figure 10.1 Combinatorial arrangements of pyknons in 3'UTRs – reproduced from Rigoutsos et al., 2006. (Copyright 2006, National Academy of Sciences U.S.A.). A colour version of this figure is located in the plate section at the back of the book.

instances across the three regions of interest, namely 5'UTRs, CDSs and 3'UTRs. Moreover, we encountered cases where one of the three regions, say 5'UTRs, of a given GO process was significantly enriched (respectively depleted) in pyknons, a second region was depleted (respectively enriched) and the third region showed neither enrichment nor depletion.

Lack of correlation and an extremum property of pyknons

In those cases where multiple pyknons arrange themselves in an apparently combinatorial manner, there is still the possibility that these arrangements are part of larger-size blocks that are shared between intergenic/intronic and exonic regions. For example, this could be happening with those pyknons having a single copy in the analysed exonic regions. Given the large number of pyknons and the even larger number of their non-exonic instances, we could not address this question at the level of individual pyknons but rather by treating the pyknons as a set. We reasoned that if a pyknon instance in an exonic region is part of a larger sequence fragment that somehow found its way into or was hi-jacked out of an exon (Iwashita *et al.*, 2003; Jiang *et al.*, 2004; Lev-Maor *et al.*, 2003), then a slight modification of the pyknon instances in exons should not affect the counts of their intergenic and intronic copies.

We induced such a modification by sliding a window with the same length as the pyknon by d positions to the left or to the right and generating a new pyknon sequence. If shifted to right, the latter would be missing a prefix of d letters compared to the original pyknon while having a new suffix of d letters, analogously for left shifts. We determined the count of intergenic/intronic copies for all such derived sequences and histogrammed their counts. We examined separately the impact of a shift on the intergenic and intronic counts; Fig. 10.2 shows the results of this analysis. As can be seen, even a two-nucleotide shift leads to a very substantial decrease in both the intergenic and intronic counts, thus suggesting that the exonic pyknon sequences correspond to a sort of extremum that involves multiple copies of given nucleotide segments and which maximizes the total number of copies in the non-exonic regions of the human genome.

The 'parking-spot' concept and consequences

Having established that the pyknons represent a sort of genome-wide extremum that maximizes the number of intergenic/intronic copies of the involved sequences, we sought to determine whether these sequences act as placeholders for

a region that is in fact wider than the pyknon's extent and whose sequence is not conserved. The heuristic approach that we used to order the members of P_{init} before seeking their instances in exons makes this a possibility. Fig. 10.3 illustrates this situation using an artificially created example involving 11 patterns. If CCAACG-GCGATA were found in, say, a 3'UTR then this motif would become a pyknon and claim the positions corresponding to all its 3'UTR instances while simultaneously excluding its own longer variants from further consideration. From this artificial example, one might be inclined to view CCAACGGCGATA as too stringent and instead select AGACCAACGGCGATACTAT as a more appropriate consensus for this motif. It is conceivable that many pyknons fall in this category. To determine whether this is indeed the case, we examined the distances between consecutive pyknons (top of Fig. 10.4); clearly, this can be done using only those genes containing multiple pyknon instances in them. The bottom of Fig. 10.4 shows the histogram of the distance that separates the first location of consecutive pyknons and is particularly intriguing. There is an evident bias in the relative placement of consecutive pyknons for each of the three examined regions, i.e. 5'UTRs, CDS and 3'UTRs. The three curves exhibit qualitative similarities that are even more striking when one recalls that these biases arise from the contents of $P_{5'UTR}$, P_{CDS} and $P_{3'UTR}$, which, as we have already seen, contain effectively distinct sequences.

An RNAi connection and novel classes of short RNAs

The kind of conservation that we encountered in the context of pyknons is unlike the sequence conservation that is typical of transcription factor binding sites (Tompa *et al.*, 2005). This, together with the observation that pyknons are logically distinct entities and the fact that the biases of Fig. 10.4 highlight length-ranges that are reminiscent of the RNAi context led us at the time to investigate the matter further.

First, we examined if the pyknons relate at the sequence level to known microRNAs from animals, plants or viruses by comparing them to the 2634 microRNAs contained in the June 2005 release of the miRBase collection (Griffiths-

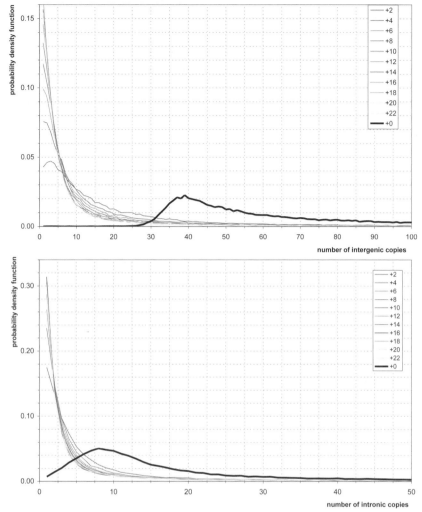

Figure 10.2 Pyknons and their natural boundaries. Probability density functions for the number of *intergenic* (top) and *intronic* (bottom) copies of the variable-length strings derived by counting the instances of 3′UTR-conserved pyknons after having shifted them to the right by *d* positions. In both cases, the black curves correspond to *d* = 0, i.e. to the pyknons in $P_{3'UTR}$. (Copyright 2006, National Academy of Sciences, U.S.A.). A colour version of this figure is located in the plate section at the back of the book.

# of copies	pattern
25	**CCAACGGCGATA**
10	AGA**CCAACGGCGATA**CTAT
9	AGA**CCAACGGCGATA**CTATA
9	CAGA**CCAACGGCGATA**CTAT
8	CAGA**CCAACGGCGATA**CTATA
8	ACAGA**CCAACGGCGATA**CTATAC
7	TACAGA**CCAACGGCGATA**CTAT
7	CAGA**CCAACGGCGATA**CTATACA
5	TACAGA**CCAACGGCGATA**CTATA
4	TACAGA**CCAACGGCGATA**CTATACA
4	CTACAGA**CCAACGGCGATA**CTATA

Figure 10.3 The parking spot concept. The figure is adapted from Supp-Fig. 6 of Rigoutsos *et al.*, 2006 (Copyright 2006, National Academy of Sciences, U.S.A.).

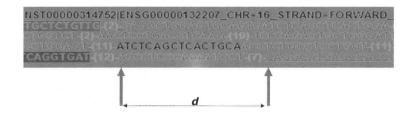

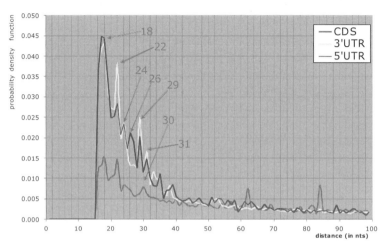

Figure 10.4 The natural boundaries of pyknons. The figure is adapted from Fig. 2 of Rigoutsos *et al.*, 2006 (copyright 2006, National Academy of Sciences, U.S.A.).

Jones, 2006). During the sequence comparison, we allowed for 30% of mismatches. Interestingly, we found that 689 pyknons clustered with 1084 known microRNAs from other animals and human viruses, providing additional support to the conjectured connection with short RNAs.

We explored this microRNA link further by generating the reverse complement of each of the 127 998 pyknons, identifying all their intergenic/intronic copies (≥ 40 copies for each pyknon, by definition) and using the Vienna package (Hofacker, 1994) to examine the secondary structure that the immediate neighbourhood of these copies would have had, had it been transcribed. In particular, we sought to determine if any of these neighbourhoods folded into the 'hairpin' secondary structure that is typical of microRNA precursors. Indeed, ~30% of the pyknons were found contained in ~382 000 neighbourhoods each of which gave rise to precursor-like hairpins when folded; moreover, the majority of these neighbourhoods had lengths ranging from 60 to 80 nucleotides, again typical of microRNA precursors.

Based on these multiple, computationally derived connections to RNA interference, we reasoned that a way to interpret the biases shown in Fig. 10.4 was to assume the existence of additional short RNAs that are linked to RNA interference and belong to molecular classes that are distinguishable by length. In fact, we argued in (Rigoutsos *et al.*, 2006) that these classes must have lengths corresponding to the most prominent of the peaks of Fig. 10.4, i.e. lengths of 18, 22, 24, 26, 29, 30 and 31 nucleotides.

Subsequent evidence for the existence of the predicted molecules and the connection with RNAi

Soon after the publication of our findings, three different groups published results from Sanger sequencing and deep-sequencing of RNA, from human, mouse and rat testes, reporting the discovery of a new class of short RNAs (Aravin *et al.*, 2006; Girard *et al.*, 2006; Lau *et al.*, 2006), dubbed piRNAs for Piwi-interacting RNAs. Even though their biogenesis is distinct from that of microRNAs, piRNAs bind to members of the Argonaute family just like microRNAs. Most importantly for our discussion, the lengths of these new short RNAs exhibited a bimodal distribution with peaks at 26 nucleotides and 29–31 nucleotides respectively, precisely as our analysis had anticipated. The functional role of piRNAs continues to remain unclear.

Organism specificity and implications

Finally, we investigated the inter-genomic properties of the human pyknons in order to determine the extent of their conservation in other genomes. We examined the conservation of human pyknon instances separately for the intergenic/intronic and the exonic regions of a genome.

Summarizing the above results, we know that all the instances of the human pyknon set cover almost 700 million out of the 6 billion positions of the human genome. To this coverage, one should add those *constrained* genomic positions that are opposite an intergenic/intronic instance of a pyknon, i.e. the ones located on the opposite strand. Doing so adds another ~200 million positions, for a grand total of ~900 million human intergenic/intronic positions that are linked to pyknon instances. However, when we repeated this calculation by searching for instances of human pyknons in the intergenic/intronic regions of other genomes, we found that the coverage that is achieved by human pyknons in other genomes is substantially smaller by comparison, and decreases as the phylogenetic distance from the human genome increases.

A similar situation arose when we searched the exons of other genomes for instances of human pyknons. For this analysis, $P_{5'UTR}$ pyknons were sought in the 5'UTRs of known genes from each of the other genomes in turn, P_{CDS} pyknons were sought in the CDSs of known genes from each of other genomes in turn, etc. We were surprised to discover that even the exonic instances of pyknons were not conserved in other genomes. This was a somewhat unexpected observation especially since it also held true for the P_{CDS} pyknons: one might have anticipated that the extensive sequence similarity at the amino acid level among orthologous genes would translate to a similar conservation at the nucleic acid level but this turned out to not be the case.

The immediate conjecture that arises from these results is that pyknons are organism-specific. We are in the process of deriving pyknon collections for a number of vertebrate genomes and our preliminary results so far suggest that a given genome's pyknons have very little overlap with the pyknons of another genome (A. Tsirigos, T. Huynh, Z. Mourelatos and I. Rigoutsos, unpublished data).

Discussion

The kind of intra-genomic sequence conservation, which the pyknons helped uncover, is unlike the motif conservation that one encounters in the context of, for example, transcription factors. This fact in conjunction with our computational findings suggested the possible involvement of small RNAs. In particular, the findings of Fig. 10.4 provided strong support for the existence of short, active RNA molecules with specific lengths and led us to put forth this hypothesis and to conjecture specific lengths for these molecules. Our hypothesis received subsequent support through the discovery of the class of piRNAs whose lengths were indeed among those that we had anticipated. Currently we are in the process of determining whether pyknons also anticipated the actual sequences of piRNAs, and preliminary analyses suggest this to be the case.

There are a few additional conjectures that one can put forth in view of the results from our human pyknon analysis. The piRNA sequences that were reported in the literature exhibited the same bimodal length distribution (peaks at lengths of 26 and 29–31 nucleotides) for all three studied animals namely human, mouse and rat indicating length conservation. On the other hand, the available piRNAs show very little sequence conservation across the three genomes. Thus, we conjecture that pyknon sets from different genomes will have little overlap and that the corresponding biases in the relative placement in exons will mimic those we observed in the human genome.

As best as we can tell, the pyknons are likely an amalgam of distinct classes of active molecules. The piRNAs would correspond to one such member-class. Currently, it is not clear how many more classes are captured by the pyknon collection and what the functional roles of these classes are. It is a fact that Fig. 10.4 has several more peaks of interest that correspond to lengths of 18, 22 and 24 nucleotides. As we discussed above, shifting the exonic instances of the human pyknons by d leads to new sequences that have drastically fewer intergenic/intronic copies. Consequently, we conjecture that these shorter lengths also indicate true, distinct classes of short RNAs possibly with biogenesis and functional roles that could very well be distinct from those of miRNAs and piRNAs.

Of particular interest is the class of 22-mers. To date, the only 22-mers that we are aware of correspond to the well-studied class of micro-RNAs. However, and with the exception of the case shown in Yekta et al. (2004), the rest of the experimentally validated examples that have been reported in the literature involve heteroduplexes where the microRNA and its target share relatively little in terms of common sequence. This is clearly orthogonal to the findings of our analysis that led to Fig. 10.4. This leads us to one more conjecture according to which one or both of the following is true: (a) there exists a class of short RNAs that are 22-nucleotide long and whose biogenesis and functionality are likely distinct from those of microRNAs; (b) the universe of a genome's microRNAs is much larger than currently believed and includes microRNAs that share extensive sequence complementarity with their intended targets.

The analysis raises the possibility that, in a given genome, the class of microRNAs is likely more abundant than currently believed. Indeed, we have generated support for this possibility in two ways. One is pyknon based, was described in Rigoutsos et al. (2006) and is summarized above. The second was described in Miranda et al. (2006) in the context of a method for finding microRNA precursors. The employed methodologies are independent yet they derived very similar estimates for the number of human microRNAs, namely ~37 000 and ~55 000 respectively. It remains to be seen whether these predictions will be validated or refuted beyond any doubt. It is intriguing nonetheless, that using two entirely different approaches we have arrived at approximately same estimate for the number of putative microRNAs.

In closing, we would like to emphasize the fact that human pyknons have very few instances in other genomes. This holds true of the human pyknons in the P_{CDS} subset as well, which in turn suggests a strong, genome-specific nature for these sequences. Their conspicuous placement in human exonic regions, their association with specific biological processes, and the rest of the accumulated evidence strongly suggest a regulatory role for pyknons. If the lack of cross-genome conservation that we observed for human pyknons persists with the pyknons of other genomes, then it follows that some aspects of cell process regulation have been implemented using substantially different sequences across organisms. This may prove to be an obstacle in the research community's efforts to understand the molecular causes of human disease as it might limit the usefulness of experimental studies that rely on the use of mice or other animals.

Acknowledgement

The author would like to thank John Mattick for many inspiring conversations that they have had over time about the importance of the so-called junk DNA.

References

Aravin, A., Gaidatzis, D., Pfeffer, S., Lagos-Quintana, M., Landgraf, P., Iovino, N., Morris, P., Brownstein, M.J., Kuramochi-Miyagawa, S., Nakano, T., et al. (2006). A novel class of small RNAs bind to MILI protein in mouse testes. Nature 442, 203–207.

Ashburner, M., Ball, C.A., Blake, J.A., Botstein, D., Butler, H., Cherry, J.M., Davis, A.P., Dolinski, K., Dwight, S.S., Eppig, J.T., et al. (2000). Gene ontology: tool for the unification of biology. The Gene Ontology Consortium. Nat. Genet. 25, 25–29.

Bagga, S., Bracht, J., Hunter, S., Massirer, K., Holtz, J., Eachus, R., and Pasquinelli, A.E. (2005). Regulation by let-7 and lin-4 miRNAs results in target mRNA degradation. Cell 122, 553–563.

Bairoch, A. (1991). PROSITE: a dictionary of sites and patterns in proteins. Nucleic Acids Res. 19, S2241–S2245.

Bartel, D.P. (2004). MicroRNAs: genomics, biogenesis, mechanism, and function. Cell 116, 281–297.

Baulcombe, D.C. (1996). RNA as a target and an initiator of post-transcriptional gene silencing in transgenic plants. Plant Mol. Biol. 32, 79–88.

Baulcombe, D.C. (1999). Gene silencing: RNA makes RNA makes no protein. Curr. Biol. 9, R599–601.

Bejerano, G., Pheasant, M., Makunin, I., Stephen, S., Kent, W.J., Mattick, J.S., and Haussler, D. (2004). Ultraconserved elements in the human genome. Science 304, 1321–1325.

Bentwich, I. (2005). Prediction and validation of microRNAs and their targets. FEBS Lett. *579*, 5904–5910.

Bentwich, I., Avniel, A., Karov, Y., Aharonov, R., Gilad, S., Barad, O., Barzilai, A., Einat, P., Einav, U., Meiri, E., *et al.* (2005). Identification of hundreds of conserved and nonconserved human microRNAs. Nat. Genet. *37*, 766–770.

Berezikov, E., Cuppen, E., and Plasterk, R.H. (2006). Approaches to microRNA discovery. Nat. Genet. *38* Suppl. 1, S2–7.

Berezikov, E., Guryev, V., van de Belt, J., Wienholds, E., Plasterk, R.H., and Cuppen, E. (2005). Phylogenetic shadowing and computational identification of human microRNA genes. Cell *120*, 21–24.

Boffelli, D., McAuliffe, J., Ovcharenko, D., Lewis, K.D., Ovcharenko, I., Pachter, L., and Rubin, E.M. (2003). Phylogenetic shadowing of primate sequences to find functional regions of the human genome. Science *299*, 1391–1394.

Brazma, A., Jonassen, I., Eidhammer, I., and Gilbert, D. (1998a). Approaches to the automatic discovery of patterns in biosequences. J. Comp. Biol. *5*, 279–305.

Brazma, A., Jonassen, I., Vilo, J., and Ukkonen, E. (1998b). Predicting gene regulatory elements in silico on a genomic scale. Genome Res. *8*, 1202–1215.

Cheng, J., Kapranov, P., Drenkow, J., Dike, S., Brubaker, S., Patel, S., Long, J., Stern, D., Tammana, H., Helt, G., *et al.* (2005). Transcriptional maps of 10 human chromosomes at 5-nucleotide resolution. Science *308*, 1149–1154.

Dermitzakis, E.T., Kirkness, E., Schwarz, S., Birney, E., Reymond, A., and Antonarakis, S.E. (2004). Comparison of human chromosome 21 conserved nongenic sequences (CNGs) with the mouse and dog genomes shows that their selective constraint is independent of their genic environment. Genome Res. *14*, 852–859.

Dubchak, I., Brudno, M., Loots, G.G., Pachter, L., Mayor, C., Rubin, E.M., and Frazer, K.A. (2000). Active conservation of noncoding sequences revealed by three-way species comparisons. Genome Res. *10*, 1304–1306.

Elbashir, S.M., Lendeckel, W., and Tuschl, T. (2001). RNA interference is mediated by 21- and 22-nucleotide RNAs. Genes Dev. *15*, 188–200.

Ettwiller, L.M., Rung, J., and Birney, E. (2003). Discovering novel *cis*-regulatory motifs using functional networks. Genome Res. *13*, 883–895.

Fire, A., Xu, S., Montgomery, M.K., Kostas, S.A., Driver, S.E., and Mello, C.C. (1998). Potent and specific genetic interference by double-stranded RNA in *Caenorhabditis elegans*. Nature *391*, 806–811.

Floyd, S.K., and Bowman, J.L. (2004). Gene regulation: ancient microRNA target sequences in plants. Nature *428*, 485–486.

Frazer, K.A., Sheehan, J.B., Stokowski, R.P., Chen, X., Hosseini, R., Cheng, J.F., Fodor, S.P., Cox, D.R., and Patil, N. (2001). Evolutionarily conserved sequences on human chromosome 21. Genome Res. *11*, 1651–1659.

Girard, A., Sachidanandam, R., Hannon, G.J., and Carmell, M.A. (2006). A germline-specific class of small RNAs binds mammalian Piwi proteins. Nature *442*, 199–202.

Griffiths-Jones, S. (2006). miRBase: the microRNA sequence database. Method. Mol. Biol. *342*, 129–138.

Hamilton, A.J., and Baulcombe, D.C. (1999). A species of small antisense RNA in posttranscriptional gene silencing in plants. Science *286*, 950–952.

Hofacker, I.L., Fontana, W., Stadler, P., Bonhoeffer, L.S., Tacker, M. & Schuster, P. (1994). Fast folding and comparison of RNA secondary structures. Monatshefte f Chemie *125*, 167–188.

Hulo, N., Bairoch, A., Bulliard, V., Cerutti, L., De Castro, E., Langendijk-Genevaux, P.S., Pagni, M., and Sigrist, C.J. (2006). The PROSITE database. Nucleic Acids Res. *34*, D227–230.

Iwashita, S., Osada, N., Itoh, T., Sezaki, M., Oshima, K., Hashimoto, E., Kitagawa-Arita, Y., Takahashi, I., Masui, T., Hashimoto, K., and Makalowski, W. (2003). A transposable element-mediated gene divergence that directly produces a novel type bovine Bcnt protein including the endonuclease domain of RTE-1. Mol. Biol. Evol *20*, 1556–1563.

Jareborg, N., Birney, E., and Durbin, R. (1999). Comparative analysis of noncoding regions of 77 orthologous mouse and human gene pairs. Genome Res. *9*, 815–824.

Jensen, K.L., Styczynski, M.P., Rigoutsos, I., and Stephanopoulos, G.N. (2006). A generic motif discovery algorithm for sequential data. Bioinformatics *22*, 21–28.

Jiang, N., Bao, Z., Zhang, X., Eddy, S.R., and Wessler, S.R. (2004). Pack-MULE transposable elements mediate gene evolution in plants. Nature *431*, 569–573.

Johnston, R.J., and Hobert, O. (2003). A microRNA controlling left/right neuronal asymmetry in *Caenorhabditis elegans*. Nature *426*, 845–849.

Jonassen, I., Collins, J.F., and Higgins, D.G. (1995). Finding flexible patterns in unaligned protein sequences. Protein Sci *4*, 1587–1595.

Kapranov, P., Cheng, J., Dike, S., Nix, D.A., Duttagupta, R., Willingham, A.T., Stadler, P.F., Hertel, J., Hackermueller, J., Hofacker, I.L., *et al.* (2007). RNA Maps Reveal New RNA Classes and a Possible Function for Pervasive Transcription. Science.

Kellis, M., Patterson, N., Endrizzi, M., Birren, B., and Lander, E.S. (2003). Sequencing and comparison of yeast species to identify genes and regulatory elements. Nature *423*, 241–254.

Knaut, H., Steinbeisser, H., Schwarz, H., and Nusslein-Volhard, C. (2002). An evolutionary conserved region in the vasa 3′UTR targets RNA translation to the germ cells in the zebrafish. Curr. Biol. *12*, 454–466.

Lai, E.C., Burks, C., and Posakony, J.W. (1998). The K box, a conserved 3′ UTR sequence motif, negatively regulates accumulation of enhancer of split complex transcripts. Development *125*, 4077–4088.

Lau, N.C., Lim, L.P., Weinstein, E.G., and Bartel, D.P. (2001). An abundant class of tiny RNAs with probable regulatory roles in *Caenorhabditis elegans*. Science *294*, 858–862.

Lau, N.C., Seto, A.G., Kim, J., Kuramochi-Miyagawa, S., Nakano, T., Bartel, D.P., and Kingston, R.E. (2006).

Characterization of the piRNA complex from rat testes. Science *313*, 363–367.

Lee, R.C., and Ambros, V. (2001). An extensive class of small RNAs in *Caenorhabditis elegans*. Science *294*, 862–864.

Lenhard, B., Sandelin, A., Mendoza, L., Engstrom, P., Jareborg, N., and Wasserman, W.W. (2003). Identification of conserved regulatory elements by comparative genome analysis. J. Biol. 2, 13.

Lev-Maor, G., Sorek, R., Shomron, N., and Ast, G. (2003). The birth of an alternatively spliced exon: 3′ splice-site selection in Alu exons. Science *300*, 1288–1291.

Lim, L.P., Glasner, M.E., Yekta, S., Burge, C.B., and Bartel, D.P. (2003). Vertebrate microRNA genes. Science *299*, 1540.

Lim, L.P., Lau, N.C., Garrett-Engele, P., Grimson, A., Schelter, J.M., Castle, J., Bartel, D.P., Linsley, P.S., and Johnson, J.M. (2005). Microarray analysis shows that some microRNAs downregulate large numbers of target mRNAs. Nature *433*, 769–773.

Lytle, J.R., Yario, T.A., and Steitz, J.A. (2007). Target mRNAs are repressed as efficiently by microRNA-binding sites in the 5′ UTR as in the 3′ UTR. Proc. Natl. Acad. Sci. USA *104*, 9667–72.

Mattick, J.S. (2004). RNA regulation: a new genetics? Nat. Rev. Genet. *5*, 316–323.

Mattick, J.S. (2007). A new paradigm for developmental biology. J. Exp Biol. *210*, 1526–1547.

Mattick, J.S., and Makunin, I.V. (2006). Non-coding RNA. Hum Mol Genet. *15 Spec No 1*, R17–29.

Miranda, K.C., Huynh, T., Tay, Y., Ang, Y.S., Tam, W.L., Thomson, A.M., Lim, B., and Rigoutsos, I. (2006). A pattern-based method for the identification of MicroRNA binding sites and their corresponding heteroduplexes. Cell *126*, 1203–1217.

Miziara, M.N., Riggs, P.K., and Amaral, M.E. (2004). Comparative analysis of noncoding sequences of orthologous bovine and human gene pairs. Genet. Mol Res. *3*, 465–473.

Moss, E.G., Lee, R.C., and Ambros, V. (1997). The cold shock domain protein LIN-28 controls developmental timing in *C. elegans* and is regulated by the lin-4 RNA. Cell *88*, 637–646.

Napoli, C., Lemieux, C., and Jorgensen, R. (1990). Introduction of a Chimeric Chalcone Synthase Gene into Petunia Results in Reversible Co-Suppression of Homologous Genes in *trans*. Plant Cell *2*, 279–289.

Ohno, S. (1972). So much 'junk' DNA in our genome. Brookhaven Symp Biol. *23*, 366–370.

Parida, L., Rigoutsos, I., Floratos, A., Platt, D.E., and Gao, Y. (2000). Pattern discovery on character sets and real valued data: linear bound on irredundant motifs and an efficient polynomial time algorithm. Proceedings 11th Annual ACM/SIAM Symposium on Discrete Algorithms (SODA '00).

Parida, L., Rigoutsos, I., and Platt, D. (2001). An output-sensitive flexible pattern discovery algorithm. Proceedings 12th Annual Symposium on Combinatorial Pattern Matching (CPM 2001).

Pillai, R.S., Bhattacharyya, S.N., Artus, C.G., Zoller, T., Cougot, N., Basyuk, E., Bertrand, E., and Filipowicz, W. (2005). Inhibition of translational initiation

by Let-7 MicroRNA in human cells. Science *309*, 1573–1576.

Pillai, R.S., Bhattacharyya, S.N., and Filipowicz, W. (2007). Repression of protein synthesis by miRNAs: how many mechanisms? Trends Cell. Biol. *17*, 118–126.

Poy, M.N., Eliasson, L., Krutzfeldt, J., Kuwajima, S., Ma, X., Macdonald, P.E., Pfeffer, S., Tuschl, T., Rajewsky, N., Rorsman, P., and Stoffel, M. (2004). A pancreatic islet-specific microRNA regulates insulin secretion. Nature *432*, 226–230.

Rajewsky, N. (2006). microRNA target predictions in animals. Nat. Genet. *38 Suppl 1*, S8-S13.

Reinhart, B.J., Slack, F.J., Basson, M., Pasquinelli, A.E., Bettinger, J.C., Rougvie, A.E., Horvitz, H.R., and Ruvkun, G. (2000). The 21-nucleotide let-7 RNA regulates developmental timing in *Caenorhabditis elegans*. Nature *403*, 901–906.

Reinhart, B.J., Weinstein, E.G., Rhoades, M.W., Bartel, B., and Bartel, D.P. (2002). MicroRNAs in plants. Genes Dev. *16*, 1616–1626.

Rhoades, M.W., Reinhart, B.J., Lim, L.P., Burge, C.B., Bartel, B., and Bartel, D.P. (2002). Prediction of plant microRNA targets. Cell *110*, 513–520.

Rigoutsos, I., and Floratos, A. (1998). Combinatorial pattern discovery in biological sequences: The TEIRESIAS algorithm. Bioinformatics *14*, 55–67.

Rigoutsos, I., Floratos, A., Parida, L., Gao, Y., and Platt, D. (2000). The emergence of pattern discovery techniques in computational biology. Metabolic Engineering 2, 159–177.

Rigoutsos, I., Huynh, T., Miranda, K., Tsirigos, A., McHardy, A., and Platt, D. (2006). Short blocks from the noncoding parts of the human genome have instances within nearly all known genes and relate to biological processes. Proc. Natl. Acad. Sci. USA *103*, 6605–6610.

Sinha, S., and Tompa, M. (2002). Discovery of novel transcription factor binding sites by statistical over-representation. Nucleic Acids Res. *30*, 5549–5560.

Slack, F.J., Basson, M., Liu, Z., Ambros, V., Horvitz, H.R., and Ruvkun, G. (2000). The lin-41 RBCC gene acts in the *C. elegans* heterochronic pathway between the let-7 regulatory RNA and the LIN-29 transcription factor. Mol. Cell *5*, 659–669.

Smith, H.O., Annau, T.M., and Chandrasegaran, S. (1990). Finding sequence motifs in groups of functionally related proteins. Proc. Natl. Acad. Sci. USA *87*, 826–830.

Styczynski, M.P., Jensen, K.L., Rigoutsos, I., and Stephanopoulos, G.N. (2004). An extension and novel solution to the (l,d)-motif challenge problem. Genome Inform Ser Workshop Genome Inform *15*, 63–71.

Thomas, J.W., Touchman, J.W., Blakesley, R.W., Bouffard, G.G., Beckstrom-Sternberg, S.M., Margulies, E.H., Blanchette, M., Siepel, A.C., Thomas, P.J., McDowell, J.C., et al. (2003). Comparative analyses of multi-species sequences from targeted genomic regions. Nature *424*, 788–793.

Tompa, M., Li, N., Bailey, T.L., Church, G.M., De Moor, B., Eskin, E., Favorov, A.V., Frith, M.C., Fu, Y., Kent, W.J., et al. (2005). Assessing computational tools for the discovery of transcription factor binding sites. Nat. Biotechnol. *23*, 137–144.

Voinnet, O., and Baulcombe, D.C. (1997). Systemic signalling in gene silencing. Nature *389*, 553.

Wasserman, W.W., Palumbo, M., Thompson, W., Fickett, J.W., and Lawrence, C.E. (2000). Human-mouse genome comparisons to locate regulatory sites. Nat. Genet. *26*, 225–228.

Wasserman, W.W., and Sandelin, A. (2004). Applied bioinformatics for the identification of regulatory elements. Nat. Rev. Genet. *5*, 276–287.

Wootton, J.C., and Federhen, S. (1996). Analysis of compositionally biased regions in sequence databases. Method. Enzymol *266*, 554–571.

Xie, X., Lu, J., Kulbokas, E.J., Golub, T.R., Mootha, V., Lindblad-Toh, K., Lander, E.S., and Kellis, M. (2005). Systematic discovery of regulatory motifs in human promoters and 3′ UTRs by comparison of several mammals. Nature *434*, 338–345.

Yaffe, D., Nudel, U., Mayer, Y., and Neuman, S. (1985). Highly conserved sequences in the 3′ untranslated region of mRNAs coding for homologous proteins in distantly related species. Nucleic Acids Res. *13*, 3723–3737.

Yekta, S., Shih, I.H., and Bartel, D.P. (2004). MicroRNA-directed cleavage of HOXB8 mRNA. Science *304*, 594–596.

Zhang, B., Pan, X., Cobb, G.P., and Anderson, T.A. (2006). Plant microRNA: a small regulatory molecule with big impact. Dev. Biol. *289*, 3–16.

RNA-Mediated Recognition of Chromosomal DNA

11

David R. Corey

Abstract

Designed molecules that recognize specific sequences within chromosomal DNA would provide useful probes for natural cellular processes, tools for laboratory experimentation and lead compounds for therapeutic development. We initially discovered that duplex DNA could be recognized by conjugates consisting of DNA oligonucleotides and cationic proteins or peptides. We subsequently observed similarly efficient recognition by neutral peptide nucleic acids (PNAs). Taking these studies as a starting point, we examined whether duplex RNAs could also mediated efficient recognition of duplex DNA. We found that RNAs can target transcription start sites and either inhibit or activate gene expression. These data, together with that from other laboratories, indicate that promoter-targeted RNAs can be powerful tools for regulating gene expression.

Introduction

Duplex RNAs are versatile agents for silencing gene expression and have become commonly used laboratory tools for research in cultured mammalian cells (Tang, 2004). The use of RNAs in vertebrate animals, however, has been much more limited and RNAi has not become a routine strategy for controlling gene expression *in vivo* (Corey, 2007).

Limited progress with animal studies has not prevented the initiation of clinical trials. Sirna Therapeutics and Acuity Pharmaceuticals are conducting trials that use siRNAs to target VEGF to treat macular degeneration. These trials take advantage of the small, isolated volume within the cornea. ALN RSV01 from Alnylam Pharmaceuticals is designed to treat RSV, a target in the lung that may be susceptible to local delivery by inhalation. These trials will be important steps towards translating the potency of RNA into clinical benefits for patients, but they will not directly address the challenges of systemic administration or whether duplex RNA can be a therapeutic option for a wide spectrum of disease.

Duplex RNA will be like any other class of drugs – achieving its full therapeutic potential will require exploring many different options for optimizing efficacy. One option is to widen the scope of RNA-mediated recognition to chromosomal DNA. The purpose of this review is to briefly recount how our laboratory initiated studies using synthetic agents to recognize chromosomal DNA. We will then describe recent work suggesting that RNA-mediated recognition of chromosomal DNA can inhibit or activate gene expression in mammalian cells.

Recognition of DNA by oligonucleotide conjugates

Our interest in improving the hybridization properties of nucleic acids began with studies of recognition by oligonucleotide-staphylococcal nuclease conjugates (Fig. 11.1A). The purpose of these studies was to develop agents capable of sequence-specifically cleaving DNA and we demonstrated that the conjugates could ef-

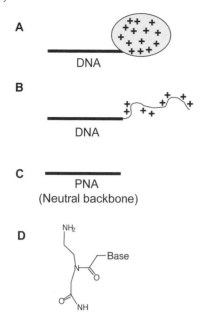

Figure 11.1 Strategies for enhanced recognition of duplex DNA. (A) Oligonucleotide-nuclease conjugate. (B) Oligonucleotide-cationic peptide conjugates. (C) Neutral oligomers. (D) Structure of peptide nucleic acid.

ficiently hydrolyse single-stranded DNA (Corey and Schultz, 1987) or plasmid DNA (Corey *et al.* 1989).

During these studies, however, we noted an unexpected phenomenon. The attached nuclease was not only promoting DNA cleavage, it was also dramatically accelerating sequence-specific recognition of duplex DNA by the attached oligonucleotide. Association rate constant (K_a) values for binding to plasmid DNA increased by up to 10,000-fold relative to K_a values for unmodified oligonucleotides (Iyer *et al.*, 1995).

Why was staphylococcal nuclease having this dramatic effect? Inspection of its protein sequence and crystal structure (Loll and Lattman, 1989) revealed that its surface was positively charged. It was reasonable to hypothesize that the positive surface would interact with the attached oligonucleotide and/or with the phosphodiester duplex at the target sequence. We subsequently observed that attachment of short cationic peptide to oligonucleotides could also enhance recognition of duplex DNA (Corey *et al.*, 1995) (Fig. 11.1B), supporting the suggestion that simple positive charge, not three dimensional protein structure, was the most important contributor to enhanced recognition.

Recognition of DNA by PNAs

The fact that recognition of duplex DNA by oligonucleotides could be accelerated by attachment of positively charged proteins or peptides led us to hypothesize that oligonucleotide mimics that lacked negative charge would also hybridize with enhanced K_a values (Fig. 11.1C) because they would not have unfavourable electrostatic interactions with the phosphodiester backbone.

To test this hypothesis we synthesized peptide nucleic acid (PNA) oligomers (Nielsen *et al.*, 1991; Kaihatsu *et al.*, 2004) complementary to chromosomal DNA. PNAs are a nucleic acid mimic consisting of normal nucleotide bases linked by an N-(2-aminoethyl) glycine based backbone and bind complementary DNA or RNA targets with elevated affinities (Fig. 11.1D). We observed that PNAs could hybridize to complementary sequences within duplex DNA with k_a values as high as those observed for analogous oligonucleotide-peptide conjugates (Smulevitch *et al.*, 1996). These data suggested that PNAs would be excellent candidates for sequence-selective recognition of chromosomal DNA.

We developed efficient methods for delivering PNAs into cultured mammalian cells (Janowski *et al.*, 2006a) and tested whether PNAs could recognize intracellular targets. We demonstrated that PNAs complementary to the RNA template of the ribonucleoprotein telomerase could block telomerase activity and cause telomeres to shorten (Herbert *et al.*, 1999). We also showed that PNAs could target mRNA and inhibit gene expression (Doyle *et al.*, 2001; Liu *et al.*, 2003). These results suggested that PNAs could readily bind to targets in the cytoplasm and allow biological phenotypes in a predictable manner.

To test whether PNAs could also recognize chromosomal DNA we synthesized antigene PNAs (agPNAs – we use this terminology antigene to distinguish agPNAs from antisense PNAs that target mRNA) complementary to the promoter for progesterone receptor (PR). PR was chosen as a target gene because it is an important regulator of normal physiology and disease (Conneely *et al.*, 2003), assaying its expression is straightforward, and its promoter has been well characterized (Kastner *et al.*, 1990; Misrahi *et al.* 1993). Most PR protein is expressed from two separate promoters, each coding for a different isoform. The promoter for the B-isoform (PR-B) is upstream from the promoter for the A-isoform (PR-A), and the PR-B isoform is larger than PR-A isoform (Fig. 11.2A).

We introduced PNAs into cultured human T47D breast cancer cells and observed that PNAs complementary to the transcription start sites for PR-B or PR-A blocked expression of PR (Janowski *et al.* 2005a). PNAs that were not complementary to the PR gene did not inhibit PR expression. These data demonstrated that PNAs could enter cells, cross into the nucleus, and bind their target sequences in spite of pre-existing base-pairing and interactions with histones and other proteins at the target site.

Inhibiting gene expression with antigene RNAs

After demonstrating inhibition of gene expression by agPNAs we noted the appearance of two papers reporting that duplex RNAs could target promoter DNA and block gene expression (Morris *et al.* 2004; Kawasaki *et al.* 2004). These findings were novel because, while there was strong evidence that RNA could target chromosomal DNA in plants (Wassenegger *et al.*, 1994) and yeast (Volpe *et al.*, 2002), there had been

little evidence that RNA could modify or silence chromosomal DNA in mammalian cells.

These findings were intriguing, but it was difficult to compare then with our results. Both papers suggested that the mechanism of silencing involved targeting RNAs to CpG islands and inducing methylation. Our PNAs were not complementary to sequences within CpG islands and were targeted to sites much closer to the transcription start site. Subsequent examination of one of the papers revealed significant flaws (Ting *et al.*, 2005) and the work was later retracted (Taira, 2006). In spite of these uncertainties, we chose to investigate promoter-targeted RNAs to further our comparative analysis of synthetic antigene agents.

We designed duplex antigene RNAs (agRNAs) analogous in sequence to the single-stranded agPNAs targeting the PR-B transcription start site (Fig. 11.2A). We did not target the PR-A transcription start site because it is downstream from the PR-B start site and overlaps with PR-B mRNA. Standard 19 base-pair duplex RNAs (Fig. 11.2B) were introduced into T47D breast cancer cells using standard transfection methods and PR protein levels were monitored by western analysis. We observed potent inhibition of gene expression by agRNAs that were complementary to the PR promoter and no inhibition by control RNAs that were not fully complementary (Fig. 11.2C) (Janowski *et al.* 2005b).

We subsequently performed a series of experiments to investigate the mechanism of RNA-mediated gene silencing.

1 5′-RACE analysis of the PR promoter. The transcription start site for PR-B had been determined by two different laboratories in two different organisms. However, to further investigate whether the agRNAs in our experiments might be acting as standard siRNAs that target mRNA. We obtained 5′-RACE data on the PR-B transcription start site in T47D cells. Twenty clones were sequenced using a modified 5′-RACE protocol designed to maximize the likelihood that only full length (rather than truncated RNA transcripts would be detected. These data identified several

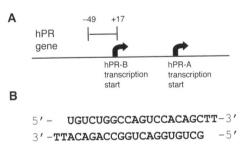

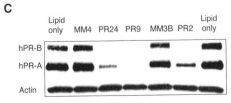

Figure 11.2 Inhibition of PR expression. (A) Structure of the PR gene promoter. The region (−49/+17) targeted by agRNAs is highlighted. (B) Typical structure of an agRNA. (C) Western analysis of PR expression upon addition to T47D cells of agRNAs PR24, PR9, and PR2 complementary to the −24/−5, −9/+10, and −2/+17 regions of the promoter respectively.

novel downstream transcription start sites, but did not identify any transcription start sites upstream from the most upstream site identified previously (Kastner et al.,1993, Misrahi et al., 1993).

2 We tested a series of duplex RNAs whose sequences were based on the most inhibitory agRNA PR9 (spanning −9 to +10) but that contained mismatched bases or were scrambled (i.e. blocks of bases interchanged). These control RNAs did not inhibit PR expression (Janowski et al. 2005b), supporting the hypothesis that inhibition was based on sequence-specific recognition of a target nucleic acid.

3 We tested twenty RNAs targeted to other sites within the PR promoter. Several of these also inhibited gene expression (Janowski et al., 2005b). The only common feature of these RNAs is that they are complementary to the PR promoter. It is unlikely that they would all be binding an unintended target and inhibiting PR expression as an off-target effect. Rather, this result suggests that inhibition requires complementarity to the PR promoter.

4 Q-PCR analysis of cells treated with agRNAs shows decreased levels of mRNA (Janowski et al., 2005b). This is consistent with a mechanism that involves recognition of promoter DNA.

5 Nuclear run-on assays revealed that agRNAs block nascent transcript synthesis but that siRNA that target mRNA do not (Janowski et al., 2006b), consistent with a mechanism that involves recognition of promoter DNA and inhibition of transcription. Other laboratories have also used nuclear run-on assays to demonstrate inhibition of transcription (Morris et al. 2004; Suzuki et al. 2005; Ting et al. 2005; Zhang et al. 2005).

6 We tested agRNAs targeting the transcription start sites of other genes, including COX-2, major vault protein (MVP), androgen receptor (AR), and huntingtin (Janowski et al., 2005b; Janowski et al., 2006b). In all cases, one or more agRNAs blocked gene expression. AR, PR, MVP, and huntingtin have TATA-less promoters, while COX-2 has a promoter that contains a TATA-box. These data suggest a general mechanism for inhibition, rather than one confined to PR.

7 We tested agRNAs complementary to sites further downstream from the transcription start site and observed inhibition of androgen receptor and huntingtin expression (Janowski et al., 2006b). These data suggest that targeting the transcription start site is not necessary for efficient inhibition of gene expression.

8 We were not able to detect histone modifications or DNA methylation at the PR promoter after treatment with agRNAs, and inhibition of AR or PR was transient (Janowski et al., 2005b; Janowski et al., 2006b). These data suggest that agRNAs do not necessarily induce permanent epigenetic changes. It remains possible that permanent changes may be induced by agRNAs targeting different sequences within AR or PR or that target other genes. Other laboratories have reported such changes (Suzuki et al., 2005; Weinberg et al., 2006)

9 Argonaute (AGO) proteins play a key role in RNA recognition during RNAi (Parker and Barford, 2006). There are four major AGO proteins in mammalian cells (AGO1–AGO4). AGO2 is known as the catalytic engine of RNAi because it acts to cleave mRNA. We observed that depleting AGO1 or AGO2 from cells leads to reversal of agRNA-mediated inhibition of gene expression (Janowski et al. 2006b). DNA Using pull down assays with anti-AGO antibodies, we observed that treating cells with agRNAs targeting the PR promoter led to association of AGO1 and AGO2 with the promoter. Rossi and colleagues observed a similar result with AGO1 (Kim et al., 2006). These findings further link mammalian post-transcriptional and transcriptional silencing pathways. Because AGO proteins are known to bind RNA (Rivas et al. 2005), these data also suggest that the mechanism involves RNA–RNA recognition. Why RNA-RNA-mediated recognition

Table 11.1 Evidence that inhibition of gene expression by agRNAs is due to association with the targeted promoter DNA

agRNAs that are not complementary to targeted promoters do not inhibit gene expression
Multiple RNAs that are complementary to target promoters inhibit gene expression. The targeted promoters are the only sequences complementary to these RNAs.
Nuclear run-on assays indicate that transcription is inhibited by agRNAs
The phenotype produced by inhibition of progesterone receptor is the same regardless of whether agRNAs or agPNAs are used, suggestion similar antigene mechanisms

might be relevant in spite of the failure to detect an mRNA species containing the target sequence will be discussed below.

Taken together, these data suggest that duplex RNA can mediate recognition of chromosomal DNA. Recognition leads to inhibition of gene expression. Several lines of evidence support the conclusion that inhibition is due to RNA-mediated recognition of the targeted promoter and is not an off-target effect caused by recognition of mRNA or toxicity due to introduction of RNA into cells (Table 11.1). We do not possess enough data to propose definitive rules for predicting ideal target sequences for agRNAs. However, our success inhibiting the expression of several genes suggests that synthesis of several RNAs that target sequences near the transcription start site is likely to afford at least one duplex capable of inhibiting gene expression.

Activating gene expression with RNA

During our experiments with agRNAs in T47D cells we observed an unexpected phenomenon. Some RNAs appeared to cause a small but reproducible increase in gene expression (1.25–1.75 fold). It was not clear whether this increase was significant because of its small magnitude. To investigate the phenomenon further, we tested agRNAs in MCF-7 breast cancer cells. MCF-7 cells express low levels of PR relative to T47D cells (Jenster et al., 1997), and we reasoned that any increase in expression might be more visible.

We introduced an RNA (PR11) into MCF-7 cells complementary to a sequence spanning the −11/+8 region surrounding the PR-B transcription start site. PR11 was chosen because it was not inhibitory in T47D cells even though it was surrounded by potent inhibitory

RNAs PR9 (−9/+10), PR10 (−10/+9), PR12 (−12/+7), and PR13 (−13/+6). We observed a substantial increase in gene expression upon addition of PR11 (Fig. 11.3) (Janowski *et al.*, 2007), suggesting that RNA could mediate gene activation.

We performed a series of experiments to afford some insights into the mechanism of gene activation (Janowski *et al.* 2007).

1 RNAs based on PR11 but containing mismatched or scrambled bases did not activate gene expression. This finding suggests that activation requires complementarity to the PR promoter.

2 We tested twenty RNAs complementary to the PR promoter and found several that activated. Database searching revealed that the only common feature of these RNAs is complementarity to the PR promoter, reinforcing the suggestion that activation requires complementarity to the PR promoter.

3 Activation was transient, with expression decreasing to basal levels after 10–14 days. This result suggest that activation is not the result of a permanent epigenetic change.

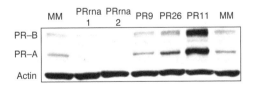

Figure 11.3 Activation of PR expression. Western analysis showing the effect of treating MCF-7 cells with mismatch–containing RNA (PRMM4), RNAs targeting PR mRNA (PRrna1 and PRrna2), and RNAs targeting the PR promoter (PR9, PR26 and PR11, targeting nucleotides −9/+10, −26/−7, and −11/+8 respectively).

Dahiya, Li, and colleagues have also observed RNA-mediated activation of gene expression, suggesting that activation may be a general phenomenon (Li *et al.*, 2006). Multiple lines of evidence suggest that the activation we observe is due to RNA-mediated recognition of the promoter and is not an off-target effect (Table 11.2). The fact that promoter-targeted RNAs can either activate or inactivate gene expression suggests that the impact of RNA-mediated recognition is extraordinarily flexible.

Models for RNA-mediated recognition of chromosomal DNA

Two models have been proposed for RNA-mediated recognition of DNA (Grewal and Moazed, 2003; Bayne and Allshire, 2005; Corey 2005; Morris, 2005). In one model RNA binds directly to duplex DNA (Fig. 11.4A). The strength of this model is that it is straightforward and the RNAs are known to be complementary to promoter sequences. The weakness is that it requires proteins capable of efficient pairing RNA with DNA.

The second model involves the duplex RNA recognizing a rare RNA transcript at the promoter rather than chromosomal DNA (Fig. 11.4 B). Recognition of chromosomal DNA would then be mediated through interactions with proteins. Several studies have appeared that suggest that much of the genome is transcribed (Dahary *et al.*, 2005; Kapranov, 2005; Lapidot and Pilpel, 2006; Willigham and Gingeras, 2006), so the presence of rare transcripts encoded by regions of the promoter that are normally not transcribed would not be surprising. In addition, involvement of AGO proteins supports the conclusion that RNA-RNA recognition is involved at one or more steps during transcriptional silencing. The weakness of this hypothesis is that it requires (i) existence of rare, previously uncharacterized transcripts, and (ii) formation of a relatively complex RNA-RNA-protein-DNA complex at the promoter.

In the latter mechanism, agRNAs enter the cells and bind RISC in the same manner as siRNAs that target mRNA. Some of the loaded RISC enters the nucleus, where it locates a rare RNA transcript analogous in sequence to promoter DNA and forms an RNA-RNA-protein complex.

The rare RNA transcript is close to the promoter during RNA synthesis. RNA and chromosomal DNA, therefore, are probably be at high effective concentrations relative to each other. These high effective concentrations would facilitate interactions between DNA-protein complexes that exist at the promoter, and the RNA–RNA–protein complex formed by the rare transcript and agRNA-loaded RISC.

For inhibition of gene expression, formation of an RNA–RNA–protein–DNA complex may form an obstruction that blocks proper assembly of transcription factors or productive engagement of RNA polymerase. For activation of gene expression, it is possible that the RNA–RNA–protein complex perturbs the balance of proteins at the inactive promoter, disrupting inactions that prevent gene expression. It is well known that small molecules can bind to transcription factors and either repress or activation gene expression, so it may reasonable to contemplate that RNA may be able to induce similar changes.

Table 11.2 Evidence that activation of gene expression by agRNAs is due to association with the targeted promoter DNA

---

agRNAs that are not complementary to targeted promoters do not activate gene expression

Multiple RNAs that are complementary to targeted promoter activate gene expression. The targeted promoters are the only sequences complementary to these RNAs

Activating agRNAs and inactive RNAs that are complementary to adjacent promoter sequences compete for the same nucleic acid target

Expression of oestrogen receptor-alpha, a key regulator of PR expression, is not altered by activating agRNAs that target PR

Activating agRNAs increase PR expression in two different cell lines, MCF7 and T47D, that regulate PR expression differently

---

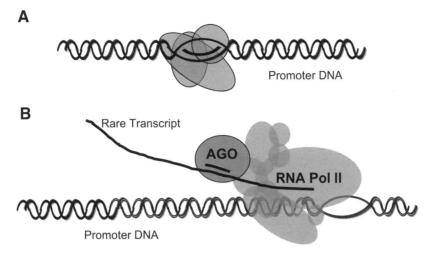

Figure 11.4 Models for RNA-mediated recognition of chromosomal DNA. (A) Direct hybridization of RNA to chromosomal DNA. (B) Recognition of a rare transcript by RNA and subsequent indirect association with chromosomal DNA.

Acknowledgements

This work was supported by the National Institutes of Health (NIGMS 60642 and 73042) and the Robert A. Welch Foundation (I-1244 to DRC).

References

Bayne, E.H., and Allshire, R.C. (2005) RNA-directed transcriptional gene silencing in mammals. Trends Genet. *21*, 370–373.

Conneely, O.M., Jericevic, B.M., and Lydon, J.P. (2003). Progesterone receptors in mammary gland development and tumorigenesis. J. Mamm. Gland Biol. Neoplasia *8*, 205–214.

Corey, D.R., and Schultz, P.G. (1987). Generation of a hybrid sequence-specific single-stranded deoxyribonuclease. Science *238*, 1401–1403.

Corey, D.R., Pei, D., and Schultz, P.G. (1989). The sequence-selective hydrolysis of duplex DNA by an oligonucleotide-directed nuclease. J. Am. Chem. Soc. *111*, 8523–8525.

Corey, D.R. (1995). 48000-fold acceleration of hybridization of chemically modified oligomers to duplex DNA. J. Am. Chem. Soc. *117*, 9373–9374.

Corey, D.R. (2005). Regulating mammalian transcription with RNA. Trends Biochem. Sci. *30*, 655–658.

Corey, D.R. (2007). RNAi learns from antisense. Nat. Chem. Biol. *3*, 8–11.

Dahary, D., Elroy-Stein, O., Sorek, R. (2005). Naturally occurring antisense: Transcriptional leakage or real overlap. Genome Res. *15*, 364–368.

Grewal, S.I.S., and Moazed, D. (2003) Heterochromatin and epigenetic control of gene expression. Science *301*, 798–802.

Iyer, M., Norton, J.C., and Corey, D.R. (1995). Accelerated hybridization of oligonucleotides to duplex DNA. J. Biol. Chem. *270*, 14712–14717.

Janowski, B.A., Kaihatsu, K., Huffman, K.E., Schwartz, J.C., Ram, R., Hardy, D., Mendelson, C.R., and Corey D.R. (2005a). Inhibiting transcription of chromosomal DNA using antigene peptide nucleic acids. Nat. Chem. Biol. *1*, 210–215.

Janowski, B.A., Huffman, K.E., Schwartz, J.C., Ram, R., Hardy, D., Shames, D, Minna, J.D., and Corey D.R. (2005b) Inhibition of gene expression at transcription start sites using antigene RNAs (agRNAs). Nat. Chem. Biol. *1*, 216–222.

Janowski, B.A., Hu, J., Corey, and D.R. (2006a) Antigene inhibition by peptide nucleic acids and duplex RNAs. Nat. Protocols *1*, 436–443.

Janowski, B.A., Huffman, K.E., Schwartz, J.C., Ram, R., Nordsell, R., Shames, D., Minna, J.D., and Corey, D.R. (2006b) Involvement of AGO1 and AGO2 in mammalian transcriptional silencing. Nat. Struct. Mol. Biol. *13*, 787–792.

Janowski, B.A., Younger, S.T., Hardy, D.B., Ram, R., Huffman, K.E., and Corey, D.R. (2007). Activating gene expression in mammalian cells with promoter-targeted duplex RNAs. Nat. Chem. Biol. in press.

Jenster, G., Spencer, T.E., Burcin, M.M., Tsai, S.Y., Tsai, M.J., and O'Malley, B.W., (1997). Steroid induction of transcription: A two step model. Proc. Natl. Acad. Sci. USA *94*, 7879–84.

Kaihatsu, K., Janowski, B.A., and Corey D.R. (2004). Recognition of Duplex DNA by Oligonucleotides and peptide nucleic acids. Chem. Biol. *11*, 749–758.

Kapranov, P., Drenkow, J., Cheng, J., Long, J., Helt, G., Dike, S., and Gingeras T.R. (2005). Examples of the complex architecture of the human transcriptome revealed by RACE and high-density tiling arrays. *Genome Res.* *15*, 987–997.

Kastner, P., Krust, A., Turcotte, B., Stropp, U., Tora, L., Gronemeyer, H., and Chambon, P. (1990). Two distinct estrogen-regulated promoters generate transcripts encoding the two functionally different human progesterone receptor isoforms A and B. EMBO J. *9*, 1603–14.

Kawasaki, H., and Taira, K. (2004). Induction of DNA methylation and gene silencing by short interfering RNAs in human cells. Nature *431*, 211–7.

Kim, D.H. Villeneuve, L.M., Morris, K.V., and Rossi, J.J. (2006) Argonaute-1 directs siRNA-mediated transcriptional gene silencing in human cells. Nat. Struc. Mol. Biol. *13*, 792–797.

Lapidot M., and Pilpel Y. (2006). Genome-wide natural antisense transcription: coupling its regulation to its different regulatory mechanisms. EMBO Rep. 7, 1216–22.

Li, L.C., Okino, S.T., Zhao, H., Pookot, D., Place, R.F., Urakami, S., Enokida, H., and Dahiya, R. (2006). Small dsRNAs induce transcriptional activation in human cells. Proc. Natl. Acad. Sci. USA *103*, 17337–17342.

Loll P.J., and Lattman E.E. (1989). The crystal structure of the ternary complex of staphylococcal nuclease, Ca^{2+}, and the inhibitor pdTp, refined at 1.65 A. Proteins 5, 183–201.

Misrahi, M., Venencie, P–Y., Saugier–Veber, P., Sar, S., Dessen, P., and Milgrom, E. (1993). Structure of the human progesterone receptor gene. Biochim. Biophys. Acta *1216*, 289–92.

Morris K.V., Chan S.W., Jacobsen S.E., and Looney D.J. (2004). Small interfering RNA-induced transcriptional silencing in human cells. Science *305*, 1289–92.

Morris, K.V. (2005). siRNA-mediated transcriptional gene silencing: the potential mechanism and a possible role in the histone code. Cell. Mol. Life Sci. *62*, 3057–66.

Nielsen, P.G., Egholm, M., Berg, R.H., and Buchardt, O. (1991). Sequence-selective recognition of DNA by strand displacement with a thymine substituted polyamide. Science *254*, 1497–1500.

Parker, J.S., and Barford, D. (2006). Argonaute: a scaffold for the function of short regulatory RNAs. Trends Biochem. Sci. *31*, 622–630

Rivas, F.V., Tolia, N.H., Song, J., Aragon, J.P., Liu, J., Hannon, G.J., and Joshua-Tor, L. (2005). Purified Argonaute2 and an siRNA form recombinant human RISC. Nat. Struc. Mol. Biol. *12*, 340–349.

Smulevitch, S.V., Simmons, C.G., Norton, J.C., Wise, T.W., and Corey, D.R. (1996). Enhanced strand invasion by oligonucleotides through manipulation of backbone charge. Nature Biotech. *14*, 1700–1705.

Suzuki K., Shijuuku, T., Fukamachi, T., Zaunders, J., Guillemin, G., Cooper, D., and Kelleher, A. (2005). Prolonged transcriptional silencing and CpG methylation induced by siRNAs targeted to the HIV-1 promoter region. J. RNAi Gene Silencing *1*, 66–78.

Taira K. (2006) Nature *441*, 1176.

Tang, G. (2004). siRNA and miRNA: an insight into RISCs. Trends Biochem. Sci. *30*, 106–114.

Ting, A.H., Schuebel, K.E., Herman, J.G., and Baylin, S.B. (2005). Short double-stranded RNA induces transcriptional gene silencing in human cells in the absence of DNA methylation. Nat. Genetics *37*, 906–910.

Volpe, T.A., Kidner, C., Hall, I.M., Teng, G., Grewal, S.I.S. Martienssen, R.A. (2002). Regulation of heterochromatic silencing and histone H3 lysine-9 methylation by RNAi. Science *297*, 1833–1837.

Wassenegger, M., Heimes, S., Riedel, L., and Sange, H.L. (1994). RNA-directed de novo methylation of genomic sequences in plants. Cell *76*, 567–576.

Weinberg, M.S., Villeneuve, L.M., Ehsani, A., Amarzguioui, M., Aagaard, L., Chen, Z.X., Riggs, A.D., Rossi, J.J., and Morris, K.V. (2006). The antisense strand of small interfering RNAs directs histone methylation and transcriptional gene silencing in human cells. RNA *12*, 256–262.

Willigham, A.T., and Gingeras, T.R. (2006). TUF love for 'junk' DNA. Cell *125*, 1215.

Zhang M-X. Ou, H., Shen, Y.H., Wang, J., Wang, J., Coselli, J., and Wang, X.L. (2005). Regulation of endothelial nitric oxide synthase by small RNA. Proc. Natl. Acad. Sci USA *102*, 16967–16972.

RNA-mediated Transcriptional Gene Silencing: Mechanism and Implications in Writing the Histone Code

12

Kevin V. Morris

Abstract

Epigenetics is the study of meiotically and mitotically heritable changes in gene expression which are not coded for in the DNA (Egger *et al.*, 2004; Jablonka and Lamb, 2003). Exactly how these epigenetic modifications are directed to the particular gene and the local chromatin has remained enigmatic. Three distinct mechanisms appear to be intricately related and implicated in initiating and/or sustaining epigenetic modifications; DNA methylation, RNA-associated silencing, and histone modifications (Egger *et al.*, 2004). Recently we and others have shown in human cells that RNA can specifically direct epigenetic modifications to targeted loci (the promoter regions) and modulate silencing. This regulatory effect is through RNA-associated silencing, can be transcriptional in nature, and is operable through an RNA interference based mechanism (RNAi) that is specifically mediated by the antisense strand of small-interfering RNAs (siRNAs) interacting with a low-copy promoter associated RNA. RNA-mediated transcriptional gene silencing, directed DNA methylation as well as the putative mechanism involved in the context of human cells will be discussed. Undoubtedly, the ramifications from these recent observations represent a paradigm shift in which a hidden layer of complexity is involved in gene regulation and is operative via the action RNA epigenetically regulating DNA.

Introduction

The recent discovery that small interfering RNAs (siRNAs) can potently and specifically knock-down gene expression represents a new paradigm in gene regulation. Initially, siRNAs were shown to knockdown the expression of a particular gene by targeting the mRNA in a post-transcriptional manner. While there are a plethora of reports applying siRNA-mediated post-transcriptional silencing (PTGS) experimentally as well as therapeutically there are apparent limitations such as the duration of the effect and a saturation of the RNA induced silencing complex (RISC). Recently, data has emerged from our lab as well as others which indicates an alternative pathway is operative in human cells where siRNAs have been shown, similar to plants, *Drosophila*, *C. elegans* and *S. pombe*, to mediate transcriptional gene silencing (TGS). Transcriptional gene silencing in human cells is operative by the antisense strand of the siRNA targeting chromatin remodelling complexes to the specific promoter region(s) via recognition of a low-copy RNA that spans the targeted promoter. The siRNA or antisense RNA targeting of genes results in epigenetic modifications at the targeted gene that ultimately lead to a re-writing of the local histone code, silent state chromatin marks, and heterochromatization of the targeted gene. The observation that siRNA directed TGS is operative via epigenetic modifications suggests that similar to plants, and *S. pombe*, human genes may also be able to be silenced more permanently or for longer periods following a single treatment. These data support the notion that a previously unrecognized layer of gene regulation is operative in the cell which involves RNA-mediated epigenetic regulation of DNA (genes).

RNAi-mediated PTGS

To date the majority of work on RNA and the regulation of gene expression is concerned mostly with active suppression of transcription in the cytoplasm. The majority of investigators utilize siRNAs targeted to a particular genes transcript and as such suppress mRNA expression in a post-transcriptional manner. In particular RNA interference (RNAi) specifically post-transcriptional gene silencing mechanisms is operative predominantly in the cytoplasm and while it is far easier to measure mRNA activity in the cytoplasm there has been an almost complete neglect with regards to what role the RNAi plays within the context of the nucleus.

RNA interference (RNAi) is the process in which double-stranded small interfering RNAs (siRNAs) modulate gene expression. RNAi was initially termed co-suppression and was first described in plants (reviewed (Tijsterman et al., 2002). The siRNAs are generated by the action of the ribonuclease (Rnase) III-type enzyme Dicer (Bernstein et al., 2001) on double-stranded RNAs. Dicer is in complex with the HIV-1 TAR RNA-binding protein (TRBP), Argonaute 2 (Ago-2), and the dsRNA-binding protein PACT (Lee et al., 2006) (Fig. 12.1). The double-stranded RNAs are processed by Dicer into small ~21–27 bp RNAs (siRNAs) which can then pair with the complementary target mRNA where cleavage of the mRNA is instigated by the action of Argonaute 2 (Ago-2) (Liu et al., 2004). The Argonaute proteins, specifically Ago-2, is the major component of the RNA induced silencing complex (RISC) and contains three highly conserved domains:

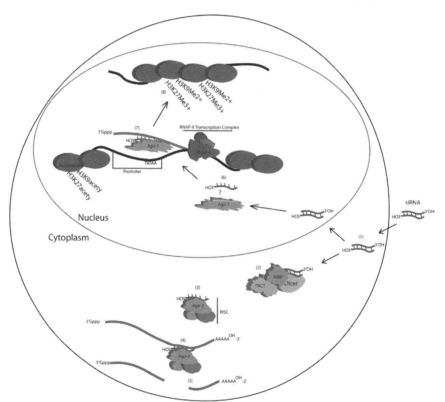

Figure 12.1 Post-transcriptional vs. transcriptional gene silencing. Synthetic siRNAs transfected directly into the cell (1) can modulate post-transcriptional gene silencing (PTGS) via Dicer and RISC as well as direct epigenetic modifications to the corresponding genomic loci (the DNA that codes for the targeted mRNA). This nuclear component appears to involve Ago-1 (Kim et al., 2006). To direct PTGS the transfected siRNA is processed by Dicer/TRBP/PACT (2) and then enters into the Ago-2 containing RISC complex (3) where RISC then somehow associates with the target mRNA (4) resulting in the target mRNA being sliced by the action of Ago-2 (5). Alternatively, synthetic siRNAs can localize to the nucleus in an Ago-1 dependent manner (6) and specifically target the homologous genomic sequence to induce chromatin modifications such as H3K9me2 or H3K27me3 (7) which are known to result in the conversion of euchromatin to heterochromatin (8) and subsequent transcriptional gene silencing.

the amino-terminal PAZ domain, the core conserved domain, and the carboxyl-terminal PIWI domain (Song *et al.*, 2004). The heart of RISC includes Ago-2 and the guide strand of the siRNA (the antisense strand relative to the targeted mRNA) (Tolia and Joshua-Tor, 2007). The slicer/Ago-2 interacts specifically with the 3′ end of one of the siRNA strands via the Ago-2 conserved PAZ domain (Song *et al.*, 2004). To date RNAi-mediated PTGS has been shown to involve siRNA targeting of mRNA and in human cells is operable in both the cytoplasm and nucleus (Langlois *et al.*, 2005; Robb *et al.*, 2005). These same mechanisms involved in siRNA-mediated PTGS are also involved in an alternative suppressive mechanism referred to as translational repression. Translation repression directed by siRNAs is typically utilized by microRNAs (miRNAs) to regulate endogenous gene expression via binding to the 3′ end of a targeted transcript and facilitating the sequestration of the transcript away from the protein translational machinery into P-bodies (Rossi, 2005). Less is currently known about the mechanism of siRNA-mediated translational repression but is an area of active investigation.

RNAi-mediated TGS

Double-stranded RNAs have also been shown to function within the nucleus to produce transcriptional gene silencing (TGS) of regions complementary to the siRNAs in *Arabidopsis*, *S. pombe*, *Drosophila* and recently in mammalian cells (reviewed in Matzke and Birchler, 2005). The initial observation of siRNA-mediated TGS was performed by Matzke and colleagues (Matzke, 1989). Interestingly the observation was essentially a null-observation as the goal was to express a transgene in doubly transformed tobacco plants. What ensued surprisingly was a suppressed phenotype of the transgene. Closer examination indicated that observed suppression of the transgene was the result of directed DNA methylation at the transgene loci (Matzke, 1989). Upon further examination it became apparent that the observed TGS and transgene methylation in plants was mediated by dsRNAs, as was substantiated in viriod-infected plants (Wassenegger, 1994), and shown to be the result of RNA-dependent DNA methyla-

tion (RdDM). The action of RdDM requires a dsRNA which is subsequently processed to yield short RNAs (Mette, 2000; Wassenegger, 1994). Serendipitously, short double-stranded RNAs were generated in the doubly transformed tobacco plant which happened to include sequences that were identical to genomic promoter utilized to express the transgene expression. The result of these small double-stranded RNAs was ultimately TGS via methylation of the homologous promoter and the subsequent reduction in transgene expression (Matzke, 1989). In general, transcriptional gene silencing in plants is carried out by a larger size class of siRNAs, 24–26 nt in length (Hamilton *et al.*, 2002; Zilberman *et al.*, 2003).

More recently, members of the Argonaute protein family in *Arabidopsis* have been shown to play an essential role in RdDM of promoter DNA and transposon silencing (Lippman *et al.*, 2003). Specifically, Ago4 is known to direct siRNA-mediated silencing, and Ago4 mutants display reactivation of silent *SUP* alleles, along with a corresponding decrease in both CpNpG DNA and H3K9 methylation (Zilberman *et al.*, 2003). Consequently in plants siRNAs that include sequences with homology to genomic promoter regions are capable of directing the methylation of the homologous promoter and subsequent transcriptional gene silencing. Transcriptional gene silencing mediated by siRNAs has also been shown in several model organisms with many conserved similarities as well as some variation on the theme.

TGS in *S. pombe*

TGS is not solely endogenous in plants and in fact appears conserved throughout many various organisms including humans. The most well-studied however is the fission yeast *S. pombe* which employs TGS via dicer generated siRNAs to specifically silence heterochromatic regions which exhibit bi-directional transcription. Interestingly, *S. pombe* lacks the epigenetic mechanism of DNA methylation but instead relies on Argonaute 1 (Ago1) to direct histone methylation and heterochromatin formation (Lippman *et al.*, 2003). Thus, S. pombe regulates gene expression of regions which exhibit bi-directional transcription via epigenetic modifications specific to the chromatin of those particular regions. The most common epigenetic histone methy-mark found in the transcriptional silencing of bidirectoinal transcripts in *S. pombe* can be found at Histone 3 Lysine-9 methylation (H3K9) (Volpe, 2002). Mutants in *dcr1* (Dicer homologue) and the only known argonaute (Ago-1) were shown to be reduced in centromeric repeat H3K9 methylation, which is necessary for centromere function (Volpe, 2002) suggesting a link between RNAi and the directed targeting of specific histone modifications to the corresponding genomic sequences. The targeted genomic sequences were in fact those regions containing homology to siRNAs which were generated by the action of *dcr1* on the bi-directional transcripts. Thus, *dcr1* and Ago-1 are required in *S. pombe* to generate the siRNA directed histone modifications which are essentially the first step in epigenetic silencing and are followed secondarily by the recruitment or interaction with Swi6 (and possible a larger complex) resulting in regulation of the hetero-chromatic state (Volpe, 2002).

More detailed evidence soon emerged and showed that *dcr1* processed double-stranded RNAs, corresponding to the centormeric repeats (bi-directional transcripts) in *S. pombe*, interacts with Ago-1, Chip1 (chromodomain protein), and Tas1 (previously uncharacterized) to form the RITS complex (Verdel *et al.*, 2004) (See Chapter 4). The presence of these siRNAs to be loaded into the RITS complex was shown to require Rdp1, Hrr1 (*helicase required* for *RNA-mediated* heterohromatin assembly 1) and Cid12 (a 38 kDa protein involved in mRNA polya-denylation) (Motamedi *et al.*, 2004). Next this siRNA loaded RITS complex can associate with the chromatin binding factors Swi6 and Clr4 (Suv39H6 human homolog) and be directed to the desired chromatin via the siRNAs and as such directly silence the bidirectonal expressed genomic regions (Verdel *et al.*, 2004) in an RNA polymerase II dependant fashion (Kato *et al.*, 2005). The siRNA targeting of RITS to the dsRNA producing genomic region is mechanis-tically active in silencing by the action of Ago-1-mediated slicing of the centromeric expressed RNAs (Irvine *et al.*, 2006) as well as silencing through the recruitment of histone methylation of the corresponding centromeric region.

TGS in *C. elegans*, *Drosophila* and *Neurospora*

RNAi-mediated TGS appears to be conserved throughout many biological systems. Clearly, the majority of work and detailed characterization of the underlying mechanism involved in TGS has been carried out in *S. pombe* and plants. However, other model organisms have demonstrated siRNA-mediated TGS with slight variations in the underlying theme. In the fungi *Neurospora crassa* the silencing of homologous sequences has been termed quelling (Romano and Macino, 1992). Quelling tends to be used interchangeably in fungi to refer to both PTGS and TGS-based mechanisms. Interestingly, in Neurospora both the PTGS and TGS pathways both utilize histone 3–lysine 9 di-methylation but yet do appear to be distinct from one another (Chicas *et al.*, 2005). Whereas in fruit fly, *Drosophila*, the two pathways of RNAi, PTGS and TGS respectively, appear connected via *piwi* proteins (Pal-Bhadra, 2002). The piwi family of proteins are a relatively recent discovered class of proteins which have several homologues (piwi/sign/ elF2C/rde1/argonaute) and are conserved from plants to animals (Fagard *et al.*, 2000; Grishok *et al.*, 2001; Tabara *et al.*, 1999). Interestingly, in the nematode C. elegans the PAZ-PIWI like protein Rde-1 plays an essential role in RNA-mediated silencing in the soma of C. *elegans*. Whereas other RNA-mediated silencing mechanisms that are operative in the germline do not appear to require Rde-1 (Dernburg *et al.*, 2000; Ketting and Plasterk, 2000). These data indicated a dis-tinct mechanistic bifurcation when dealing with germline vs. somatic cell silencing.

In C. *elegans* RNA-mediated transcriptional silencing of somatic transgenes has been shown to be the result of ADAR-encoding genes, adr-1 and adr-2 and is dependent on Rde-1(Grishok *et al.*, 2005). The observed silencing and require-ments for both adr-1 and 2 as well as Rde-1 was the result of characterizing siRNAs targeted to pre-mRNAs. The silencing corresponded with a decrease in both RNA polymerase II and acetylated histones at the targeted genomic re-gion (Grishok *et al.*, 2005). Interestingly, follow-ing an RNAi screen in C. *elegans*, genes encoding RNA-binding, Polycomb, and chromodomain proteins as well as histone methyltransferases were detected (Grishok *et al.*, 2005). These data were recapitulated by others in *C. elegans* in an interesting set of experiments that demonstrated long-term transcriptional gene silencing by RNAi. In essence one dose of siRNAs was capable of inducing gene silencing that was inherited indefinitely in the absence of the original siRNA trigger (Vastenhouw *et al.*, 2006). This observed long-term inheritance of siRNA-mediated TGS appeared to require had-4 (a class II histone deacetylase), K03D10.3 (a histone acetyltransferase of the MYST family), isw-1 (a homologue of the Yeast chromatin-remodelling ATPase ISW1), and mrg-1 (a chromodomain protein) (Vastenhouw *et al.*, 2006). Taken together these data strongly suggest that RNA-mediated TGS in *C. elegans* contains a convergence of pathways that include epigenetic modifying factors as well as RNA. Overall, when comparisons are made between plants, yeast, worms and flies, one cannot help but notice that indeed RNA is more intricately involved in the regulation of gene expression than has previously been envisaged, and this strongly suggests, provided there is some evolutionary conservation in nature, that other organisms might also utilize RNA directed mechanisms to regulate gene expression epigenetically, namely humans.

TGS in human cells

There are a distinct set of commonalities that can be discerned from the carefully executed experiments performed in plants, *S. pombe*, *C. elegans* and *Drosophila*. The quintessential observation is that RNA, in particular siRNAs, can; (1) direct transcriptional gene silencing in a specific and directed manner, (2) the directed silencing correlates with epigenetic modifications such as histone or DNA methylation at the chromatin of the particular siRNA targeted gene loci, and (3) PIWI-related proteins appear to be involved and required for siRNA directed TGS. As such, and based on the relative conservation in biology, i.e. the theory of evolution and natural selection, one cannot help but also expect to observe similarities or conserved commonalities between Plants, *C. elegans*, *S. pombe*, *Drosophila* and humans.

While much data has been generated to date with regards to siRNA directed TGS in human cells these observations of siRNA-mediated TGS in human cells has lagged behind the work done in other model organisms such as *Arabidopsis* (plants) and *S. pombe* (yeast) (Morris *et al.*, 2004). However, recent studies by our group as well as others have begun to reveal; (1) that siRNA-mediated TGS in mammalian cells does occur and (2) appears to be the result of the siRNA directed H3K9, H3K27 methylation at the corresponding siRNA targeted promoter (Buhler *et al.*, 2005; Castanotto *et al.*, 2005; Janowski, 2005; Morris *et al.*, 2004; Suzuki, 2005). While some DNA methylation has also been observed at the siRNA targeted promoters (Janowski, 2005; Kim *et al.*, 2007; Morris *et al.*, 2004; Murali, 2007; Park *et al.*, 2004; Svoboda *et al.*, 2004; Ting *et al.*, 2005) the role in which DNA methylation plays in the observed silencing is debatable (Park *et al.*, 2004; Svoboda *et al.*, 2004; Ting *et al.*, 2005) and may be contingent on the gene/promoter loci targeted. Indeed, the ability for siRNAs to direct targeted DNA methylation could result in a much greater duration of suppression, as DNA methylation tends to correlate more robustly with long-term suppressed genes than does histone methylation. However, to date little is known regarding the duration of persistence of siRNA directed TGS of RNA polymerase II promoters (RNAPII) in human cells.

The majority of siRNA-directed TGS experiments in human cells have been carried out so far with synthetic siRNAs targeted to RNAPII promoters (Castanotto *et al.*, 2005; Janowski *et al.*, 2006; Janowski, 2005; Kim *et al.*, 2006; Morris *et al.*, 2004; Suzuki, 2005; Ting *et al.*, 2005; Weinberg, 2005; Zhang, 2005) or constitutive expressing shRNA expressing stable lentiviral transduced cell lines (Castanotto *et al.*, 2005; Kim *et al.*, 2007; Murali, 2007). Nonetheless, that observation that siRNA directed TGS is operative via epigenetic modifications argues favourably for a longer-term effect relative to siRNA directed PTGS provided the siRNAs are efficiently delivered to the nucleus. It should however remain clear that experimental evidence supporting this claim is still lacking.

While the duration of siRNA directed effects remains to be determined what is evident is that siRNAs directed to RNAPII promoter regions can mediate transcriptional silencing in human cells and that a repressive histone methyl-mark is observed at the corresponding siRNA targeted promoter (Buhler et al., 2005; Han, 2007; Morris et al., 2004; Ting et al., 2005; Weinberg, 2005). Generally, to target an RNAPII promoter 3 different siRNAs are selected based on GC content (greater GC content is selected) and of these three selected siRNAs usually one will be sufficient for modulating transcriptional silencing of the particular siRNA targeted promoter. While it is still a little early to determine the proper sequence constraints for those siRNAs that are efficacious at modulating TGS some commonalities have begun to emerge. When all the previously published siRNA sequences shown to either direct TGS or the activation of gene expression (see Chapter 14) or to be unable to modulate gene expression are taken together a pattern begins to emerge. Interestingly, there appears to be a conservation of purines from nucleotide 6–15 in the targeted promoter for those siRNAs that are effective relative to those that are not capable of modulating gene expression (Fig. 12.2A). Moreover, the majority of siRNAs that are efficacious at modulating gene expression are found within ~200–300 bp upstream of the previously reported TATAA or transcriptional start site of the particular gene of interest (K.V. Morris, direct observation). These observations however should be taken with a grain of salt as they are based on no greater than 10 different siRNA or antigene (see Chapter 13) targeted genes.

While it had remained unclear how the siRNAs were effective at targeting and localizing to the particular homologous promoters recent data has begun to emerge. Through studies with a promoter targeted biotin linked siRNA specific for the EF52 siRNA target site, which was the first example of siRNA-mediated TGS in human cells (Morris et al., 2004), it became apparent that only the antisense strand of the siRNA was required to mediate histone methylation and silencing of the targeted RNAPII promoter (Weinberg, 2005) (Fig. 12.2B). Furthermore, through experiments with alpha-amanatin (an RNAPII inhibitor) it was shown that RNAPII is required for siRNA-mediated TGS (Weinberg, 2005). The use of the biotin linked EF52 siRNA allowed for various pulldown experiments to be performed as well and as such it was discovered that the DNA methyltransferase 3A (DNMT3A) co-immunoprecipitates (co-IP) with biotin linked siRNAs specifically at the H3K27me3 targeted promoter (Weinberg, 2005). Not shown in this body of work was the fact that Flag-tagged HP1 alpha, beta, and gamma as well as DNMT1 were unable to co-immunoprecipitate with the biotin linked siRNA EF52 (Fig. 12.3A–C). While Flag-tagged DNMT3A and to a lesser extent DNMT3A were functional in binding in vitro the biotin tagged siRNA EF52 (Fig. 12.3C). These studies provided the first conclusive evidence that chromatin remodelling factors, specifically DNMT3A and the resulting H3K27me3 silent state methyl-mark are localized at the siRNA targeted promoter along with the antisense strand of the siRNA (Weinberg, 2005). Other groups have shown previously that DNMT3A can bind siRNAs in vitro (Jeffery and Nakielny, 2004) as well as that DNMT3A co-immunoprecipitates with HDAC1 and Suv39H1 (Datta et al., 2003; Fuks et al., 2001; Fuks et al., 2003) (the H3K9 methyltransferase; see Chapter 2). Consequently, a link between promoter targeted RNAs and the direct localization of epigenetic gene regulatory factors at the targeted promoter was demonstrated in this work (Weinberg, 2005).

While the majority of data indicated that the antisense strand of the siRNA was directly implicated in the recruitment of silent state epigenetic marks at the targeted promoter a link to the known RNAi machinery remained lacking. However, recently Ago-1 and possibly also Ago-2 have been shown to be involved in siRNA-mediated TGS in human cells (Janowski et al., 2006; Kim et al., 2006). Ago-1 was shown to co-IP directly with RNAPII and an enrichment of EZH2 and TR BPwere also observed along with Ago-1 at the particular siRNA targeted promoters (Kim et al., 2006). Moreover, the suppression of Ago-1, by RNAi, functionally inhibited siRNA-mediated TGS in human cells (Kim et al., 2006). These data clearly indicated that in human cells, similar to observations in other organisms such as S. pombe and Arabidopsis, Argonaute proteins and histone methylation are required for transcriptional silencing, linking the RNAi and the chromatin silencing machinery.

(A)

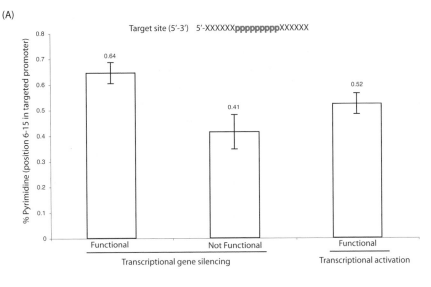

Target site (5'-3')   5'-XXXXXX**pppppppppp**XXXXXX

(B)

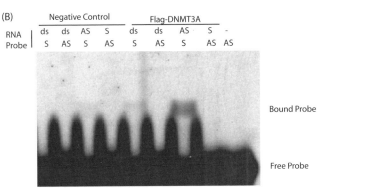

Figure 12.2 Characteristics of siRNAs involved in TGS of human cells. The siRNA targeted promoters from various published (Castanotto *et al.*, 2005; Janowski *et al.*, 2006; Janowski, 2005; Janowski *et al.*, 2007; Kim *et al.*, 2006; Li *et al.*, 2006; Morris *et al.*, 2004; Suzuki, 2005; Wang *et al.*, 2007; Weinberg, 2005) as well as unpublished works (K.V. Morris personal observation) were compared and contrasted for patterns involved in discerning the appropriate siRNAs to modulate TGS in human cells. (A) A total of 18 suppressive, 3 activating, and 12 non-functional siRNA target sites were contrasted. Notably a significant pattern emerged in the target site from bp 6–15 (5'–3' direction) where purine enrichment correlated with a functional siRNA target region in the promoter. Standard error of the means are shown. (B) Detection of the antisense strand in siRNA-mediated TGS of the EF1a promoter. A total of 4×10^6 293 HEK cells were transfected with Flag-Tag DNMT3A (15 μg) or untransfected (Negative control) and 24hrs later transfected with either double-stranded siRNA (ds), Sense stranded (S) or antisense stranded (AS) EF52 RNAs (10nM, MPG (Morris *et al.*, 2004; Morris *et al.*, 1997) and cell lysates collected, Flag-tagged DNMT3A isolated and probed for strand association. Only the antisense strand appeared to co-precipitate with Flag-tagged DNMT3A. Data is a gift compliments of Ali Ehsani (The Beckman Research Institute of the City of Hope, Duarte, CA, USA).

Model of TGS in human cells

Taken together a model for the mechanism of how siRNA directed TGS is directed and initiated in human cells has begun to emerge and appears to exhibit many similarities as well as some distinct differences with the previously established models for TGS in *S. pombe* and plants. Similar to *S. pombe* siRNA-mediated TGS in human cells involves the siRNAs, particularly the antisense strand, RNAPII, and

histone methylation (Janowski *et al.*, 2006; Kim *et al.*, 2006; Morris, 2005; Morris, 2006). However, in human cells DNMT3A has also been shown to co-immunoprecipitate along with the antisense strand of the promoter specific siRNA at the siRNA targeted promoter (Weinberg, 2005). Interestingly, DNMT3A has also been shown to bind siRNAs *in vitro* (Jeffery and Na-kielny, 2004). Similar to observations in *S. pombe* (Kato *et al.*, 2005), TGS in human cells requires

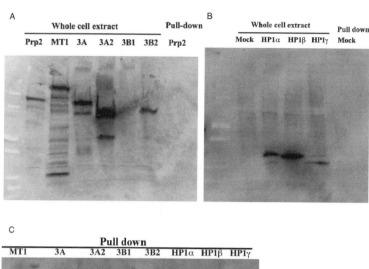

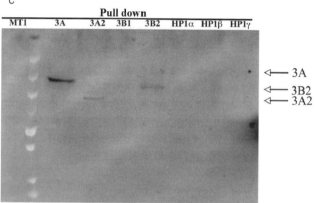

Figure 12.3 Screening DNMTs and HP1s for binding to siRNA EF52+biotin. Various Flag-tagged DNMTs (1, 3A1, 3A2, 3B1, 3B2) and the control Prp2 (A) or HP1s (alpha, beta, and gamma) (B) were transfected (15μg, Lipofectamine 2000™) in ~4×10⁶ 293 HEK cells and the relative expression determined by western blot analysis with an anti-Flag antibody. Next, the various Flag-Tagged DNMTs and HP1s from the whole cell extracts were incubated with a total of 500nM siRNA biotin labelled siRNA EF52 (Weinberg, 2005). A biotin pulldown was then performed and western blots run (C) with anti-Flag antibody on aviden pulled down extracts to determine which Flag-tagged proteins associate *in vitro* with the EF52 siRNA (Weinberg, 2005).

RNAPII (Weinberg, 2005). The requirement for RNAPII in siRNA directed transcriptional silencing of human cells and the fact that only the antisense strand of the siRNA was required to initiate the silencing provided some interesting insights. The requirement of only the antisense strand suggested that the target is also single stranded, either DNA or RNA? That fact that RNAPII is required for siRNA directed TGS provides some insights as to where in the lifecycle of the cell single stranded nucleic acid targets can be found, i.e. during transcription. As a result of these data two models began to emerge. The first model, the RNA/RNA model, proposes that there is an RNAPII expressed non-coding transcript homolgous to the targeted gene which somehow remains associated with the local chromatin corresponding to the targeted gene

(Fig. 12.4A). This non-coding RNA may remain affiliated with the nucleosome(s), and as such would permit the RNAi machinary to direct chromatin modifications of the targeted genomic region ultimately leading to TGS. This putative non-coding RNA could be envisioned to act as a local 'address' to allow chromatin and RNAi modification complexes, guided by the antinsense small RNAs, access to the targeted gene (Fig. 12.4A). Such an RNA/RNA model is supported by the observation that heterochromatin formation in mouse cells involves HP1 proteins and treatment with RNase causes a dispersion of HP1 proteins from pericentromeric foci (Muchardt et al., 2002).

An alternative model, the RNA/DNA model (Fig. 12.4B), can also be proposed and envisioned to operate in an RNAPII dependent

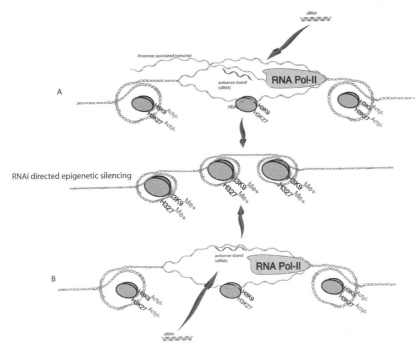

Figure 12.4 The RNA/DNA vs. RNA/RNA models of TGS in human cells. Two models for siRNA-mediated TGS were initially proposed; the RNA/DNA and RNA/RNA-mediated mode of silencing. (A) The RNA/RNA was proposed to operate via the siRNAs antisense strand interactions with a low copy promoter associated transcript. This promoter RNA would be expected to essentially span the chromatin of the targeted promoter region. (B) Alternatively, siRNA-mediated TGS might function through an RNA/DNA based mechanism. The RNA/DNA model would be hypothesized to function by the antisense strand of the siRNA gaining access to the targeted DNA via the effects of RNAPII unwinding the targeted genomic region which would then allow the targeted localization of a chromatin silencing complex to the siRNA targeted promoter ultimately leading to gene silencing.

manner. In the RNA/DNA model RNAPII localized at the siRNA-targeted promoter essentially unwinds the targeted DNA and as such allows the siRNA and RNAi machinary access to the targeted promoter. Supporting this model is the observation that RNAPII has been shown to associate with complete unfolding of 1.85 out of 3 nucleosomes upstream of the transcription start site for the PH05 promoter (Boeger *et al.*, 2005). Interestingly, the siRNA EF52 target site, shown to initiate TGS of the EF1a gene in human cells is ~1 nucleosome upstream of the TATAA transcriptional start site (Morris *et al.*, 2004). Moreover, RNAPII is associated with nearly 60 subunits and a mass in excess of 3 million Daltons (Boeger *et al.*, 2005). Overall, these reports suggested, at least in human cells, that a albeit speculative hypothesis for the role of RNAPII in siRNA-mediated TGS, might be to unwind the targeted promoter region to allow the promoter directed siRNAs access to their respective target.

While both the RNA/RNA (Fig. 12.4A) and RNA/DNA (Fig. 12.4B) models proposed seemed plausible, and in fact both models may be operative depending on where the small RNA targets the promoter, one model has been shown to be operative in promoter targeting when the siRNAs are upstream (~100 bp or more) of the TATAA. Specifically, evidence from siRNA-mediated TGS of the EF1a promoter/gene loci (Morris *et al.*, 2004), CCR5 (Kim *et al.*, 2006), RASSF1a (Castanotto *et al.*, 2005), and HIV-1 promoters (Weinberg, 2005) has begun to demonstrate a paradigm for the underlying mechanism by which the antisense strand of the siRNA can direct TGS and specifically recruit the corresponding silent histone methyl-mark to the targeted promoter. Experiments carried out on EF1a, CCR5, p16, NF-kB, and Cyclin D1 RNAPII promoters in various human cell lines and primary cells has yielded evidence that these RNAPII promoters express a corresponding low-copy RNA transcript which overlaps the respective targeted promoters (Han, 2007) (Fig. 12.5). This low-copy promoter associated RNA (pRNA) is processed in a manner similar to mRNAs, spliced and poly adenylated, but appears to contain a much extended 5′ UTR (Han, 2007). The current paradigm is that the 5′UTR (which essentially overlaps the RNAPII promoter) is recognized by the antisense strand of promoter-targeted siRNA which either recruits or contains *a priori* the appropriate chromatin remodelling factors in a complex (Fig. 12.5). As such the antisense RNA or siRNA targeted promoter undergoes chromatin remodelling and epigenetic silencing resulting in the loss of transcription (Fig. 12.5). The entire complex bound to the promoter associated RNA could be envisioned as depicted (Fig. 12.6).

The temporal aspects of the emerging mechanism might function by initially RNAPII reading through the targeted promoter, transcribing the low-copy pRNA, which can then become bound by the antisense strand of siRNA (Fig. 12.5). The pRNA and siRNA complex might then associate with the local chromatin architecture through a yet-to-be defined chromatin remodelling complex possibly containing DNMT3A. Indeed, DNMT3A has been shown to bind siRNAs (Jeffery and Nakielny, 2004), as well as co-immunoprecipitate with the antisense strand of EF1a specific EF52 siRNA and H3K27me3 at the targeted EF1-alpha promoter (Weinberg, 2005). This bound complex could then permit the docking of a chromatin remodelling complex or itself function as a chromatin remodelling

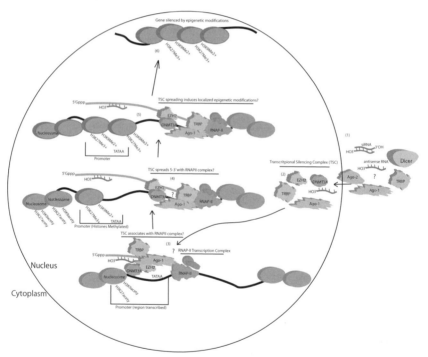

Figure 12.5 Model for RNA-mediated TGS in human cells. The *pRNA/5′UTR* model of siRNA-mediated TGS is based on data which suggests that RNAPII promoters contain a low-copy pRNA which is characterized by an extended 5′ UTR. This 5′UTR of the pRNA is recognized by the antisense strand of the siRNA during RNAPII-mediated transcription of the siRNA targeted promoter. The effector molecule for TGS in human cells is the antisense strand which recognizes the 5′ UTR. Thus, siRNAs must somehow be unwound for the antisense strand to localize to the 5′UTR of the pRNA homologous to the particular promoter target (1). The unwinding of the siRNA might be via Ago-1 (1)? Next, the antisense strand of the siRNA might associate with a transcriptional silencing complex (2) and then localize to the targeted promoter region (3). The transcriptional silencing complex (TSC) might contain EZH2, DNMT3A, TRBPand/or Ago-1 as all four molecules have been detected specifically at transcriptionally silenced promoter regions (Kim *et al.*, 2006; Weinberg, 2005). The TSC could be expected to modulate chromatin remodelling of the nucleosomes which are reconstituting the recently transcribed promoter (4) as well as spread distal in concert with the RNAPII transcriptional silencing complex epigenetically modifying the local chromatin as the complexes spread distal 5′–3′ (5). The net result would be the RNA targeted gene being silenced epigenetically (6).

Figure 12.6 Mechanistic insights into RNA-directed transcriptional gene silencing in human cells. Small antisense RNAs can be generated to specifically recognize gene promoter regions ultimately resulting in transcriptional silencing of the targeted genes expression. How the small RNAs recognize the promoter has remained an enigma. Recent direct evidence shows that a low-copy promoter associated RNA (grey) is present at the targeted gene/promoter loci and becomes bound by the small antisense RNAs (Han, 2007). The small RNAs (red) are thought to recruit or be in complex with a transcriptional silencing complex which can then direct specific epigenetic modifications to the local chromatin leading at the targeted loci leading ultimately to transcriptional silencing of the antisense RNA targeted gene. A colour version of this figure is located in the plate section at the back of the book.

complex. The chromatin remodelling complex can then initiate the writing of a silent state histone code at the targeted promoter as well as spreading of H3K9me2 and H3K27me3 distal of the siRNA targeted site in a 5′–3′ manner along with transcription (Kim *et al.*, 2006; Weinberg, 2005) (Fig. 12.5). Such a scenario would accommodate previous observations in which siRNAs have been shown to interact with various components of chromatin remodelling complexes (Janowski *et al.*, 2006; Kim *et al.*, 2006; Weinberg, 2005). Of significant interest is the role DNMT3A might play in siRNA-mediated TGS and/or antisense RNA directed chromatin modifications. DNMT3A has not only been shown to interact with siRNAs (Jeffery and Nakielny, 2004; Weinberg, 2005) but also to co-immunoprecipitate with the H3K27 methyltransferase EZH2 (Vire *et al.*, 2005), HDAC-1, and Suv39H1 (Fuks *et al.*, 2001). Moreover, EZH2 and Ago-1 have both recently been observed at siRNA targeted promoters and suppression of Ago-1 inhibits siRNA-mediated TGS (Janowski *et al.*, 2006; Kim *et al.*, 2006),

suggesting a link between RNA and chromatin remodelling components.

To date there is some precedence for the proposed model of how siRNAs modulate TGS in human cells and specifically the role of pRNAs in the fundamental mechanism (Fig. 12.6) (Han, 2007). A species of RNA, similar to pRNAs, has been observed spanning intergenic regions and appear to be coupled to chromatin remodelling complexes in controlled expression of repetitive RNA genes in human cells (Mayer *et al.*, 2006). Furthermore, in the human dihydrofolate reductase gene a non-coding upstream initiated RNA has been shown to act as a promoter interfering transcript (Martianov *et al.*, 2007) and in *S. pombe* a similar RNA has been shown to be involved in argonaute-mediated slicing and transcriptional silencing (Irvine *et al.*, 2006). Interestingly, higher order chromatin structures have been shown previously to contain an uncharacterized RNA component that may function as a scaffolding in chromatin remodelling (Maison *et al.*, 2002). These data, along with observations that antisense transcription appears to be ubiquitous in human cells (Katayama et al., 2005), are suggestive of an endogenous mechanism by which antisense RNAs can function by interactions with the pRNAs to direct specific epigenetic modifications in human genetic diseases (Cho et al., 2005; Tufarelli et al., 2003), gene regulation (Buhler et al., 2005; Zhang, 2005), and possibly be utilized in maintaining viral latency (Wang et al., 2005). Regarding a yet-to-be determined endogenous antisense RNA regulatory network in human cells it does seem rather interesting that experimentally the antisense strand of siRNAs appears to be more stable and preferentially compartmentalized within the cell (Berezhna et al., 2006), possibly implicating a role for antisense RNAs in regulating pseudogenes (Svensson et al., 2006) or retroelements in human cells (Yang and Kazazian, 2006) putatively through interactions with pRNAs spanning these genomic regions.

Endogenous small RNAs and targeted gene regulation?

While it is becoming apparent in human cells a mechanism is in place by which siRNAs or small 21 nucleotide long antisense RNAs can transcriptionally modulate gene expression (Han, 2007; Morris et al., 2004) it is less clear what the endogenous signal utilizing this pathway might be, i.e. siRNA, small antisense RNAs, micro-RNAs, non-coding RNAs, etc. Deep sequencing done in *C. elegans* has revealed a novel class 21 nucleotide long RNAs (21 U-RNAs) which contain a uridine 5'-monophophate and 3'-terminal ribose (Ruby, 2006). These 21 U-RNAs correlated with 5,700 genomic loci and were dispersed between protein-coding regions (Ruby, 2006). Overall, these data suggest that non-coding regions of the genome, while not necessarily coding for a protein, appear to express RNAs that are possibly involved in the regulation of gene expression. As many of these small 21 U-RNAs matched non-coding regions it is possible that they are involved in transcriptional modulation of *C. elegans* gene expression (Ruby, 2006). Far less is currently known in human cells and the majority of inferences have been surmised from computational based approaches (see Chapters 10–11). However, recent observations in human cells have involved the discovery and characterization of a vast array of small (21-to 26-nt), non-coding RNAs (Mattick, 2007) which strongly support the concept that there is an RNA component, possibly involved in gene regulation, i.e. siRNA-mediated TGS, within the basic fabric of the human cells (Kapranov et al., 2007; Katayama et al., 2005). Overall, these data indicate that there may be a previously unrecognized 'hidden layer of complexity' which is operative in the genome and specifically the regulation of gene expression in many organisms and strongly support a paradigm by which RNA is actively involved in the regulation of DNA. One cannot help but contemplate the concept that DNA and genomes are simply repositories of information actively being managed by RNA. Oddly enough these RNAs are encoded from the DNA/genes which they are actively managing, a sort of chicken and egg scenario.

Conclusion

The observation in human cells that siRNAs, particularly the antisense strand alone, can direct TGS and that this event involves DNMT3a, histone methylation, and RNAPII strongly suggests that siRNAs can be used to specifically direct epigenetic modifications in human cells. Indeed the modification of histone tails, such as methylation, results in a 'histone code'. The 'histone code 'hypothesis argues that the local histone environment (specifically in the nucleosomes) can have an effect on the expression profile of the corresponding local gene (Jenuwein and Allis, 2001). These 'marked' histone tails are then capable of dictating the recruitment of various specialized chromatin remodelling factors (Strahl et al., 1999; Turner, 2000) (see Chapter 2). Interestingly, the fundamental underlying mechanism responsible for governing the histone code is not yet well understood. One cannot help but make the connection that the mechanism regulating the histone code might be mediated by small RNAs and in particular small antisense RNAs. Possibly a mechanism that is weaved into the basic fabric of the cell and has to date been overlooked. While it is becoming apparent that RNAi goes beyond the confines of the cytoplasm the observations with siRNA-mediated TGS are evocative of an antisense related phenomenon that is deeply seeded in the fabric of the cell. Indeed one day it may be possible to harness RNA to direct permanent epigenetic modifications resulting in superlative control of the human genome.

Acknowledgements

I would like to thank Paula J. Olecki for graphical assistance and John J. Rossi for his valued conversations on siRNA-mediated TGS. This work is supported by NIH HLB R01 HL83473 to KVM.

References

Berezhna, S.Y., Supekova, L., Supek, F., Schultz, P.G., and Deniz, A.A. (2006). siRNA in human cells selectively localizes to target RNA sites. Proc. Natl. Acad. Sci. USA 103, 7682–7687.

Bernstein, E., Caudy, A.A., Hammond, S.M., and Hannon, G.J. (2001). Role for a bidentate ribonuclease in the initiation step of RNA interference. Nature 409, 363–366.

Boeger, H., Bushnell, D.A., Davis, R., Griesenbeck, J., Lorch, Y., Strattan, J.S., Westover, K.D., and Kornberg, R.D. (2005). Structural basis of eukaryotic gene transcription. FEBS Lett. 579, 899–903.

Buhler, M., Mohn, F., Stalder, L., and Muhlemann, O. (2005). Transcriptional silencing of nonsense codon-containing immunoglobulin minigenes. Mol. Cell 18, 307–317.

Castanotto, D., Tommasi, S., Li, M., Li, H., Yanow, S., Pfeifer, G.P., and Rossi, J.J. (2005). Short hairpin RNA-directed cytosine (CpG) methylation of the RASSF1A gene promoter in HeLa cells. Mol. Ther. 12, 179–183.

Chicas, A., Forrest, E.C., Sepich, S., Cogoni, C., and Macino, G. (2005). Small interfering RNAs that trigger posttranscriptional gene silencing are not required for the histone H3 Lys9 methylation necessary for transgenic tandem repeat stabilization in Neurospora crassa. Mol. Cell. Biol. 25, 3793–3801.

Cho, D.H., Thienes, C.P., Mahoney, S.E., Analau, E., Filippova, G.N., and Tapscott, S.J. (2005). Antisense transcription and heterochromatin at the DM1 CTG repeats are constrained by CTCF. Mol. Cell 20, 483–489.

Datta, J., Ghoshal, K., Sharma, S.M., Tajima, S., and Jacob, S.T. (2003). Biochemical fractionation reveals association of DNA methyltransferase (Dnmt) 3b with Dnmt1 and that of Dnmt 3a with a histone H3 methyltransferase and Hdac1. J. Cell. Biochem. 88, 855–864.

Dernburg, A.F., Zalevsky, J., Colaiacovo, M.P., and Villeneuve, A.M. (2000). Transgene-mediated cosuppression in the C. elegans germ line. Genes Dev. 14, 1578–1583.

Egger, G., Liang, G., Aparicio, A., and Jones, P.A. (2004). Epigenetics in human disease and prospects for epigenetic therapy. Nature 429, 457–463.

Fagard, M., Boutet, S., Morel, J.B., Bellini, C., and Vaucheret, H. (2000). AGO1, QDE-2, and RDE-1 are related proteins required for post-transcriptional gene silencing in plants, quelling in fungi, and RNA interference in animals. Proc. Natl. Acad. Sci. USA 97, 11650–11654.

Fuks, F., Burgers, W.A., Godin, N., Kasai, M., and Kouzarides, T. (2001). Dnmt3a binds deacetylases and is recruited by a sequence-specific repressor to silence transcription. EMBO J. 20, 2536–2544.

Fuks, F., Hurd, P.J., Deplus, R., and Kouzarides, T. (2003). The DNA methyltransferases associate with HP1 and the SUV39H1 histone methyltransferase. Nucleic Acids Res. 31, 2305–2312.

Grishok, A., Pasquinelli, A.E., Conte, D., Li, N., Parrish, S., Ha, I., Baillie, D.L., Fire, A., Ruvkun, G., and Mello, C.C. (2001). Genes and mechanisms related to RNA interference regulate expression of the small temporal RNAs that control C. elegans developmental timing. Cell 106, 23–34.

Grishok, A., Sinskey, J.L., and Sharp, P.A. (2005). Transcriptional silencing of a transgene by RNAi in the soma of C. elegans. Genes Dev. 109, 683–696

Hamilton, A., Voinnet, O., Chappell, L., and Baulcombe, D. (2002). Two classes of short interfering RNA in RNA silencing. EMBO J. 21, 4671–4679.

Han, J., D. Kim, and K.V. Morris (2007). Promoter-associated RNA is required for RNA-directed transcriptional gene silencing in human cells. Proc. Natl. Acad. Sci. USA 104, 12422–12427.

Irvine, D.V., Zaratiegui, M., Tolia, N.H., Goto, D.B., Chitwood, D.H., Vaughn, M.W., Joshua-Tor, L., and Martienssen, R.A. (2006). Argonaute slicing is required for heterochromatic silencing and spreading. Science 313, 1134–1137.

Jablonka, E., and Lamb, M.J. (2003). Epigenetic heredity in evolution. Tsitologiia 45, 1057–1072.

Janowski, B.A., Huffman, K.E., Schwartz, J.C., Ram, R., Nordsell, R., Shames, D.S., Minna, J.D., and Corey, D.R. (2006). Involvement of AGO1 and AGO2 in mammalian transcriptional silencing. Nat. Struct. Mol. Biol. 9, 787–792.

Janowski, B.A., Huffman, K.E., Schwartz, J.C., Ram, R., D. Hardy, D.S. Shames, J.D. Minna, D.R. Corey. (2005). Inhibiting gene expression at transcription start sites in chromosomal DNA with antigene RNAs. Nature Chemical Biology 1, 210–215.

Janowski, B.A., Younger, S.T., Hardy, D.B., Ram, R., Huffman, K.E., and Corey, D.R. (2007). Activating gene expression in mammalian cells with promoter-targeted duplex RNAs. Nat. Chem. Biol.

Jeffery, L., and Nakielny, S. (2004). Components of the DNA methylation system of chromatin control are RNA-binding proteins. J. Biol. Chem. 279, 49479–49487.

Jenuwein, T., and Allis, C.D. (2001). Translating the histone code. Science 293, 1074–1080.

Kapranov, P., Cheng, J., Dike, S., Nix, D.A., Duttagupta, R., Willingham, A.T., Stadler, P.F., Hertel, J., Hackermueller, J., Hofacker, I.L., et al. (2007). RNA maps reveal new RNA classes and a possible function for pervasive transcription. Science 316, 1484–1488.

Katayama, S., Tomaru, Y., Kasukawa, T., Waki, K., Nakanishi, M., Nakamura, M., Nishida, H., Yap, C.C., Suzuki, M., Kawai, J., et al. (2005). Antisense transcription in the mammalian transcriptome. Science *309*, 1564–1566.

Kato, H., Goto, D.B., Martienssen, R.A., Urano, T., Furukawa, K., and Murakami, Y. (2005). RNA polymerase II is required for RNAi-dependent heterochromatin assembly. Science *309*, 467–469.

Ketting, R.F., and Plasterk, R.H. (2000). A genetic link between co-suppression and RNA interference in C. elegans. Nature *404*, 296–298.

Kim, D.H., Villeneuve, L.M., Morris, K.V., and Rossi, J.J. (2006). Argonaute-1 directs siRNA-mediated transcriptional gene silencing in human cells. Nat. Struct. Mol. Biol..

Kim, J.W., Zhang, Y.H., Zern, M.A., Rossi, J.J., and Wu, J. (2007). Short hairpin RNA causes the methylation of transforming growth factor-beta receptor II promoter and silencing of the target gene in rat hepatic stellate cells. Biochem. Biophys. Res. Commun. *359*, 292–297.

Langlois, M.A., Boniface, C., Wang, G., Alluin, J., Salvaterra, P.M., Puymirat, J., Rossi, J.J., and Lee, N.S. (2005). Cytoplasmic and nuclear retained DMPK mRNAs are targets for RNA interference in myotonic dystrophy cells. J. Biol. Chem. *280*, 16949–16954.

Lee, Y., Hur, I., Park, S.Y., Kim, Y.K., Suh, M.R., and Kim, V.N. (2006). The role of PACT in the RNA silencing pathway. EMBO J. *25*, 522–532.

Li, L.C., Okino, S.T., Zhao, H., Pookot, D., Place, R.F., Urakami, S., Enokida, H., and Dahiya, R. (2006). Small dsRNAs induce transcriptional activation in human cells. Proc. Natl. Acad. Sci. USA *103*, 17337–17342.

Lippman, Z., May, B., Yordan, C., Singer, T., and Martienssen, R. (2003). Distinct Mechanisms Determine Transposon Inheritance and Methylation via Small Interfering RNA and Histone Modification. PLoS Biol. *1*, E67.

Liu, J., Carmell, M.A., Rivas, F.V., Marsden, C.G., Thomson, J.M., Song, J.J., Hammond, S.M., Joshua-Tor, L., and Hannon, G.J. (2004). Argonaute2 Is the Catalytic Engine of Mammalian RNAi. Science *305*, 1437–1441.

Maison, C., Bailly, D., Peters, A.H., Quivy, J.P., Roche, D., Taddei, A., Lachner, M., Jenuwein, T., and Almouzni, G. (2002). Higher-order structure in pericentric heterochromatin involves a distinct pattern of histone modification and an RNA component. Nat. Genet *19*, 19.

Martianov, I., Ramadass, A., Serra Barros, A., Chow, N., and Akoulitchev, A. (2007). Repression of the human dihydrofolate reductase gene by a non-coding interfering transcript. Nature *7128*, 526–530.

Mattick, J.S. (2007). A new paradigm for developmental biology. J. Exp Biol. *210*, 1526–1547.

Matzke, M.A., and Birchler, J.A. (2005). RNAi-mediated pathways in the nucleus. Nat. Rev. Genet *6*, 24–35.

Matzke, M.A., M. Primig, J. Trnovsky, and A.J.M. Matzke. (1989). Reversible methylation and inactivation of marker genes in sequentially transformed tobacco plants. The EMBO Journal *8*, 643–649.

Mayer, C., Schmitz, K.M., Li, J., Grummt, I., and Santoro, R. (2006). Intergenic Transcripts Regulate the Epigenetic State of rRNA Genes. Mol Cell *22*, 351–361.

Mette, M.F., W. Aufsatz, J. Van der Winden, A.J.M. Matzke, and M.A. Matzke. (2000). Transcriptional silencing and promoter methylation triggered by double-stranded RNA. The EMBO Journal *19*, 5194–5201.

Morris, K.V. (2005). siRNA-mediated transcriptional gene silencing: the potential mechanism and a possible role in the histone code. Cell. Mol. Life Sci. *62*, 3057–3066.

Morris, K.V. (2006). Therapeutic potential of siRNA-mediated transcriptional gene silencing. Biotechniques Suppl., 7–13.

Morris, K.V., Chan, S.W., Jacobsen, S.E., and Looney, D.J. (2004). Small interfering RNA-induced transcriptional gene silencing in human cells. Science *305*, 1289–1292.

Morris, M.C., Vidal, P., Chaloin, L., Heitz, F., and Divita, G. (1997). A new peptide vector for efficient delivery of oligonucleotides into mammalian cells. Nucleic Acids Res. *25*, 2730–2736.

Motamedi, M.R., Verdel, A., Colmenares, S.U., Gerber, S.A., Gygi, S.P., and Moazed, D. (2004). Two RNAi complexes, RITS and RDRC, physically interact and localize to noncoding centromeric RNAs. Cell *119*, 789–802.

Muchardt, C., Guilleme, M., Seeler, J.S., Trouche, D., Dejean, A., and Yaniv, M. (2002). Coordinated methyl and RNA binding is required for heterochromatin localization of mammalian HP1alpha. EMBO Rep. *3*, 975–981.

Murali, S., K. Pulukuri, and J.S. Rao. (2007). Small Interfering RNA-directed reversal of Urokinase plasminogen activator demethylation inhibits prostate tumor growth and metastasis. Cancer Res. *67*, 6637–6646.

Pal-Bhadra, M., U. Bhadra, J.A. Birchler. (2002). RNAi related mechanisms affect both transcriptional and posttranscriptional transgene silencing in *drosophila*. Molecular Cell *9*, 315–327.

Park, C.W., Chen, Z., Kren, B.T., and Steer, C.J. (2004). Double-stranded siRNA targeted to the huntingtin gene does not induce DNA methylation. Biochem. Biophys. Res. Commun. *323*, 275–280.

Robb, G.B., Brown, K.M., Khurana, J., and Rana, T.M. (2005). Specific and potent RNAi in the nucleus of human cells. Nat. Struct. Mol. Biol. *12*, 133–137.

Romano, N., and Macino, G. (1992). Quelling: transient inactivation of gene expression in Neurospora crassa by transformation with homologous sequences. Mol Microbiol *6*, 3343–3353.

Rossi, J.J. (2005). RNAi and the P-body connection. Nat. Cell. Biol. *7*, 643–644.

Ruby, J.G., C. Jan, C. Player, M.J. Axtell, W. Lee, C. Nusbaum, H. Ge, and D.P. Bartel. (2006). Large-scale sequencing reveals 21U-RNAs and additional MICRORNAS and endogenous siRNAs in C. elegans. Cell *127*, 1193–1207.

Song, J.J., Smith, S.K., Hannon, G.J., and Joshua-Tor, L. (2004). Crystal structure of Argonaute and its implications for RISC slicer activity. Science *305*, 1434–1437.

Strahl, B.D., Ohba, R., Cook, R.G., and Allis, C.D. (1999). Methylation of histone H3 at lysine 4 is highly conserved and correlates with transcriptionally active nuclei in tetrahymena [In Process Citation]. Proc. Natl. Acad. Sci. USA 96, 14967–14972.

Suzuki, K., T. Shijuuku, T. Fukamachi, J. Zaunders, G. Guillemin, D. Cooper, and A. Kelleher. (2005). Prolonged transcriptional silencing and CpG methylation induced by siRNAs targeted to the HIV-1 promoter region. J. RNAi Gene Silencing 1, 66–78.

Svensson, O., Arvestad, L., and Lagergren, J. (2006). Genome-wide survey for biologically functional pseudogenes. PLoS Comput Biol. 2, e46.

Svoboda, P., Stein, P., Filipowicz, W., and Schultz, R.M. (2004). Lack of homologous sequence-specific DNA methylation in response to stable dsRNA expression in mouse oocytes. Nucleic Acids Res. 32, 3601–3606.

Tabara, H., Sarkissian, M., Kelly, W.G., Fleenor, J., Grishok, A., Timmons, L., Fire, A., and Mello, C.C. (1999). The rde-1 gene, RNA interference, and transposon silencing in C. elegans. Cell 99, 123–132.

Tijsterman, M., Ketting, R.F., and Plasterk, R.H. (2002). The genetics of RNA silencing. Annu Rev. Genet 36, 489–519.

Ting, A.H., Schuebel, K.E., Herman, J.G., and Baylin, S.B. (2005). Short double-stranded RNA induces transcriptional gene silencing in human cancer cells in the absence of DNA methylation. Nat. Genet 37, 906–910.

Tolia, N.H., and Joshua-Tor, L. (2007). Slicer and the argonautes. Nat. Chem. Biol. 3, 36–43.

Tufarelli, C., Stanley, J.A., Garrick, D., Sharpe, J.A., Ayyub, H., Wood, W.G., and Higgs, D.R. (2003). Transcription of antisense RNA leading to gene silencing and methylation as a novel cause of human genetic disease. Nat. Genet 34, 157–165.

Turner, B.M. (2000). Histone acetylation and an epigenetic code. Bioessays 22, 836–845.

Vastenhouw, N.L., Brunschwig, K., Okihara, K.L., Muller, F., Tijsterman, M., and Plasterk, R.H. (2006). Gene expression: long-term gene silencing by RNAi. Nature 442, 882.

Verdel, A., Jia, S., Gerber, S., Sugiyama, T., Gygi, S., Grewal, S.I., and Moazed, D. (2004). RNAi-mediated targeting of heterochromatin by the RITS complex. Science 303, 672–676.

Vire, E., Brenner, C., Deplus, R., Blanchon, L., Fraga, M., Didelot, C., Morey, L., Van Eynde, A., Bernard, D., Vanderwinden, J.M., et al. (2005). The Polycomb group protein EZH2 directly controls DNA methylation. Nature.

Volpe, T.A., C. Kidner, I.M. Hall, G. Teng, S.I.S. Grewal, R.A. Martienssen. (2002). Regulation of Heterochromatic Silencing and Histone H3 Lysine-9 Methylation by RNAi. Science 297, 1833–1837.

Wang, Q.Y., Zhou, C., Johnson, K.E., Colgrove, R.C., Coen, D.M., and Knipe, D.M. (2005). Herpesviral latency-associated transcript gene promotes assembly of heterochromatin on viral lytic-gene promoters in latent infection. Proc. Natl. Acad. Sci. USA 102, 16055–16059.

Wang, X., Feng, Y., Pan, L., Wang, Y., Xu, X., Lu, J., and Huang, B. (2007). The proximal GC-rich region of p16(INK4a) gene promoter plays a role in its transcriptional regulation. Mol Cell Biochem.

Wassenegger, M., M.W. Graham, M.D. Wang. (1994). RNA-directed de novo methylation of genomic sequences in plants. Cell 76, 567–576.

Weinberg, M.S., L.M. Villeneuve, A. Ehsani, M. Amarzguioui, L. Aagaard, Z. Chen, A.D. Riggs, J.J. Rossi, and K.V. Morris. (2005). The antisense strand of small interfering RNAs directs histone methylation and transcriptional gene silencing in human cells. RNA 12.

Yang, N., and Kazazian, H.H., Jr. (2006). L1 retrotransposition is suppressed by endogenously encoded small interfering RNAs in human cultured cells. Nat. Struct. Mol. Biol..

Zhang, M., H. Ou, Y.H. Shen, J. Wang, J. Wang, J. Coselli, X.L. Wang (2005). Regulation of endothelial nitric oxide synthase by small RNA. Proc. Natl. Acad. Sci. USA 102, 16967–16972.

Zilberman, D., Cao, X., and Jacobsen, S.E. (2003). ARGONAUTE4 control of locus-specific siRNA accumulation and DNA and histone methylation. Science 299, 716–719.

Small RNA-mediated gene activation 13

Long-Cheng Li

Abstract

Small double-stranded RNA (dsRNA), such as small interfering RNA (siRNA) and microRNA (miRNA), have been found to be the trigger of an evolutionary conserved mechanism known as RNA interference (RNAi). RNAi invariably leads to gene silencing via remodelling chromatin to thereby suppress transcription, degrading complementary mRNA, or blocking protein translation. Recent discoveries now suggest that dsRNAs may also act as small activating RNA (saRNA). By targeting sequences in gene promoters, saRNAs readily induce target gene expression in a phenomenon referred to as dsRNA-induced transcriptional activation (RNAa). This discovery expands our knowledge on both the functionality and complexity small RNAs have in regulating gene expression. In this chapter, we will examine the evidence accumulated thus far on RNA molecules that positively regulate gene expression and function.

Introduction

RNA is a nucleic acid polymer synthesized by the cell in order to relay genetic information written within the DNA. RNAs are split into two distinct classes including messenger RNAs (mRNAs), which are translated into proteins, and non-coding RNAs (ncRNAs), which do not code for proteins. For years it was thought that ncRNAs served only as accessory components in protein complexes that functioned in such fundamental processes as RNA splicing (e.g. spliceosomal RNAs), RNA editing, and protein synthesis (e.g. tRNAs and rRNAs). However, over time it became apparently clear that RNAs possessed other surprising functions that were typically only ascribed to DNA or protein molecules. For example, RNA is the genetic core in some viruses, similar to DNA in higher organisms. Other ncRNA molecules can function as catalysts (ribozyme) (Steitz and Moore, 2003), regulatory elements that monitor cellular changes in metabolites and chemicals (riboswitch) (Mandal and Breaker, 2004), or molecular sensors that responds to internal changes in cellular temperature (RNA thermometers) (Narberhaus et al., 2006). Recently, the role of small ncRNAs has generated great interest. Following the initial discovery of RNA interference (RNAi) (Elbashir et al., 2001; Fire et al., 1998), it has been shown that short ~21-nt double-stranded RNAs (dsRNAs), such as small interfering RNAs (siRNAs) and microRNAs (miRNAs), regulate gene expression at multiple levels including chromatin architecture, DNA modification, transcription, mRNA decay, and translation. Each form of RNA-dependent regulation invariably leads to the silencing or elimination of target sequences and is collectively called RNAi or RNA silencing (Bagga et al., 2005; Morris et al., 2004; Olsen and Ambros, 1999; Volpe et al., 2002). The most well-known and characterized mechanism of RNAi is siRNA-mediated mRNA degradation; commonly employed to generate transient knock-outs in cell culture.

RNAi is conserved in most organisms and believed to have evolved as a form of innate immunity against viruses and other invading pathogens. However, such examples are rare in higher eukaryotes. It is possible that evolution has adapted this defence mechanism as a general means of regulating gene expression. Recent discoveries now indicate that small dsRNAs can also activate target gene expression (Janowski *et al.*, 2007; Li *et al.*, 2006). This, of course, raises the question as to whether or not this phenomenon occurs naturally and to what regards it is related to RNAi. In this chapter, we will examine evidence accumulated thus far that RNA positively regulates genomic structure, function, and gene expression.

Small non-coding RNAs (ncRNAs)

A large portion of the eukaryotic genome is transcribed as non-coding RNAs. It is estimated that 98% of the human genomic transcripts represent RNA that does not encode protein (Mattick, 2005). ncRNAs can derived from independently transcribed RNAs and introns of protein coding transcripts. Based on their function, ncRNAs can be grouped into two major classes: housekeeping ncRNAs and regulatory ncRNAs (reviewed in Prasanth and Spector, 2007; Szymanski *et al.*, 2003). The former group of ncRNAs is generally constitutively expressed and is essential for cell vitality such as tRNAs, rRNAs, snRNAs, snoRNAs, and RNase P RNAs. Regulatory ncRNAs range in size from about 20 nucleotide (nt) to over 10,000 nt and are expressed at certain stages of development or as a response to external stimuli. Those with a size of 20- to 30-nt are called small ncRNAs that exist in several classes including miRNAs (Lee *et al.*, 1993), siRNAs (Fire *et al.*, 1998), repeat-associated siRNAs (rasiRNAs) (Lau *et al.*, 2006) and Piwi-interacting RNAs (piRNAs) (Lau *et al.*, 2006). Large regulatory ncRNAs can regulate the expression of other genes in diverse processes such as dosage compensation, genomic imprinting, and hormonal response. Many of these regulations occur at the transcriptional level via diverse mechanisms leading to either transcriptional silencing or activation (reviewed in Goodrich and Kugel, 2006). Over the past several decades hypotheses, supported by scarce evidence, have been raised that ncRNAs are able to positively affect gene transcription.

Historical views of RNA activation

In 1969, Britten and Davidson (Britten and Davidson, 1969) proposed a theory of gene regulation in higher eukaryotic cells in which a class of RNA termed activator RNA, transcribed from redundant genomic sequences and restricted to the nucleus, activated a battery of protein-coding genes by forming sequence-specific complexes with their respective regulatory sequences. They further defined that the RNA-mediated regulation was accomplished by sequence-specific binding of an activator RNA to DNA and by specific activation of otherwise repressed sites, rather than by repression of otherwise active sites (Britten and Davidson, 1969). In 1985, another similar hypothesis was put forward by Sekeris that a processed fragment of heterogeneous nuclear RNA (HnRNA), acts as 'signal' RNA and activated ribosomal gene transcription by base-pair interaction with the non-transcribed region of the ribosomal gene (Sekeris, 1985). In both hypotheses, the mediator of gene regulation is small ncRNAs that act by base-pairing with their targets. Such property of the mediator is highly reminiscent of what is now known as miRNA, a naturally occurring small ncRNA molecule. In addition, both hypotheses indicated that the effect of the regulation is gene activation.

RNA, dosage compensation and transcriptionally active chromatin marks

Many organisms have heteromorphic sex chromosomes with females having two X chromosomes while males only one. To ensure both sexes express the same amount of X-linked gene products, different dosage compensation mechanisms are implemented in different organisms. In mammals, genes on one of the female X chromosomes are inactivated by DNA methylation. In hermaphrodite worms, genes on both female X chromosomes are downregulated by half. Whereas in flies, transcription from the single male X chromosome is doubled.

This transcription upregulation is mediated by the dosage compensation complex (DCC), which contains two non-coding RNAs known as roX1 and roX2, along with other core proteins such as the male-specific lethal (MSL) proteins, MLE (maleless, an RNA helicase), MOF (males absent on the first), etc. The role of roX RNAs is to direct DCC to the X chromosome where DCC further spreads into the flanking chromatin and causes hyperacetylation at histone 4 lysine 16, which is carried out by MOF (reviewed in Bernstein and Allis, 2005; Gilfillan et al., 2004). It is unclear whether the roX RNAs act like small dsRNAs by self-folding into stem–loop structures which might further be processed into small dsRNA, similar to how miRNAs are processed. Although roX RNAs are not conserved in mammals, interestingly most DCC proteins are well conserved (Smith et al., 2005) and responsible for the majority of histone H4 acetylation at lysine 16, a positive mark of transcription (Smith et al., 2005; Taipale et al., 2005). Of particular interest is the MOF protein that contains a histone acetyltransferase (HAT) domain and a chromodomain. The latter has been found to function as RNA interaction modules (Neal et al., 2000). Their interaction with noncoding RNA may target chromatin modifying activity to specific chromosomal sites (Akhtar et al., 2000).

Another natural ncRNA molecule that positively regulates gene expression is the steroid receptor RNA activator (SRA). SRA, about ~700-nucleotide in size, functions as a transcriptional co-activator for steroid-hormone receptors such as progesterone receptor (PR), oestrogen receptor (ER), glucocorticoid receptor (GR), and androgen receptor (AR) (Lanz et al., 1999). It acts as part of a ribonucleoprotein complex that interacts with the transcriptional machinery. Interestingly, a set of stem–loop structures within SRA are required for its coactivation function (Lanz et al., 2002). It is possible that the stem–loop structure can give rise to dsRNAs via RNase III processing (Danin-Kreiselman et al., 2003) and the resulted dsRNAs are the actual guide molecule.

RNA and DNA methylation

DNA methylation is an important epigenetic mechanism of gene regulation. Similar to RNAi, DNA methylation has been regarded primarily as a defence mechanism against genome parasites (Yoder et al., 1997). DNA methylation has evolved to assume other functions such as genomic imprinting, X chromosome inactivation, tissue-specific gene expression (Chan et al., 2000), and is even associated with active gene transcription (Chan et al., 2000; Hellman and Chess, 2007; Labhart, 1994; Stein et al., 1983). In plants, it has been known that RNA can induce de novo DNA methylation, a phenomenon referred to as RNA-directed DNA methylation (RdDM), in which dsRNAs generated from viral RNA or transgenes carrying a perfect inverted repeat (IR) direct cytosine methylation of DNA regions with which they share sequence identity (reviewed in Wassenegger, 2000). In human, DNA hypermethylation-related gene silencing and hypomethylation-related gene activation have been implicated in cancer and other diseases (reviewed in Li et al., 2005). However, a fundamental question remains open of how DNA methylation is controlled. In other words, why is the promoter of some genes methylated while others not and why is a particular gene silenced by hypermethylation in one type of cancer while not in others? Attempts to identify sequence elements in promoter sequence that may guide DNA methylation have been largely unsuccessful. Since the discovery of RNAi and miRNA, there has been new hope that the puzzle may eventfully be solved by linking small RNA to DNA methylation. Small RNA, with its short sequence signature and capacity of inducing gene silencing, is a likely molecular signal that may dictate DNA methylation changes. It is, however, a controversial issue whether RdDM exists in mammals. There are reports that dsRNAs targeting gene promoters reduce gene expression by inducing partial promoter methylation in human cells (Castanotto et al., 2005; Morris et al., 2004), while others could not confirm such observation (Janowski et al., 2005; Ting et al., 2005). It is unclear whether this discrepancy is due to the cell-type or gene specificity of such changes.

RNA has also been associated with DNA demethylation. An early observation found that in an in vitro system RNA is required for DNA demethylation because existence of RNase in the system inhibits DNA demethylation (Weiss et al., 1996). Imamura et al (Imamura et al., 2004) identified an antisense transcript that may induce DNA demethylation. This antisense transcript is originated from the CpG island of the rat sphingosine kinase-1 (Sphk1) locus. The CpG island region contains a tissue-dependent differentially methylated region (T-DMR). Methylation of the T-DMR is related to the regulation of Sphk1 expression in normal tissues. Forced expression of this antisense transcript that overlaps the T-DMR dramatically decreases CpG methylation in the T-DMR. Although dsRNA has been found to activate gene transcription (Janowski et al., 2007; Li et al., 2006), no evidence supports that RNA-induced DNA demethylation is involved in this process (Li et al., 2006). Nevertheless, based on the similarity between DNA methylation- and small RNA-mediated gene regulation mechanisms: both were originated as a host defence mechanism, both have evolved to serve as a general means of gene regulation, their relationship warrants further study.

dsRNA-induced transcriptional gene activation

It has been known that RNAi functions in the nucleus leading to the downregulation of gene transcription (Morris et al., 2004; Ting et al., 2005) or the degradation of nuclear RNAs (Robb et al., 2005). Two very recently studies have surprisingly demonstrated that gene transcription can be turned on by targeting gene promoter sequences using small dsRNAs, also known as small activating RNAs (saRNAs) (Janowski et al., 2007; Li et al., 2006). In the study reported by our group (Li et al., 2006) the original goal was to understand how promoter specific DNA methylation is controlled in mammalian cells especially in cancer cells and whether small RNAs dictate sequence-specific DNA methylation. We designed two dsRNAs targeting the promoter of the human E-cadherin gene which is frequently silenced by DNA methylation of a CpG island surrounding its transcription start site in many types of cancer cells. The dsRNA design was carried out by following the rational siRNA design rules for mRNA sequence knockdown (Reynolds et al., 2004). These rules exclude target sites with high GC content, favour targets with a less thermodynamically stable 3′ end than

its 5′ end. As a result, our two initial dsRNAs were picked from a region that is over 200 bp upstream the transcription start site and outside the CpG island of the E-cadherin gene promoter. When these dsRNAs were introduced into human prostate cancer PC-3 cells which express low levels of E-cadherin with no apparent CpG island methylation, we unexpectedly found a remarkable increase in the expression of E-cadherin mRNA and protein compared to cells treated with a control dsRNA (Li et al., 2006). Further experiments with dsRNAs targeting other gene promoters including the p21$^{WAF1/Cip1}$ and the vascular endothelial growth factor (VEGF) gene also showed gene activation (Li et al., 2006). Similar observations with the progesterone receptor (PR) and the major vault protein (MVP) gene in human cells were recently reported by another group (Janowski et al., 2007), suggesting that small RNA induced transcriptional activation (RNAa) is a general phenomenon.

RNAa is potent in restoring gene expression and function

In cells with low levels of target gene expression, RNAa is potent and usually causes a 5- to 10-fold induction in gene expression on both mRNA and protein levels (Fig. 13.1). Activation of endogenous genes through RNAa helps restore their nature functions leading to phenotypic changes in transfected cells similar to the effects of ectopic gene expression using an expression vector. For example, RNAa of p21 and E-cadherin genes causes cell growth inhibition indicating the restoration of their gene function (Fig. 13.1C); PR activation by saRNA results in expression changes of PR-regulated genes (Janowski et al., 2007). Even in cells which express high levels of target genes, RNAa can further enhance their expression (Janowski et al., 2007; Li et al., 2006).

Time course of RNAa

It is well known that RNAi can be induced within hours in cells transfected with siRNAs and the effects of RNAi usually last for a period of 5–7 days (Dykxhoorn et al., 2003). In contrast, RNAa takes a very different time course. Upon transfecting saRNA into cells, it takes 48–72

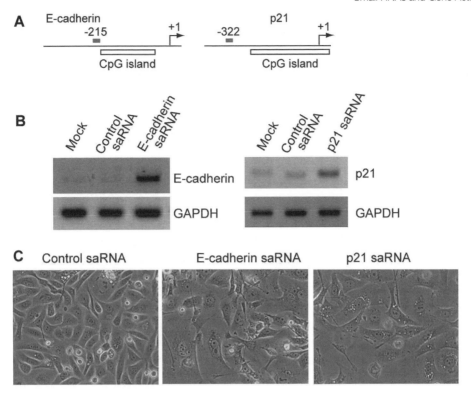

Figure 13.1 saRNAs induce sequence-specific gene expression and inhibit prostate cancer PC-3 cell growth. (A) A saRNA was designed targeting the E-cadherin and p21 promoters respectively. The target location relative to the transcription start site (+1) and CpG islands surrounding the promoters are indicated. (B) saRNAs targeting the human E-cadherin and p21 genes and a control saRNA were transfected into human PC-3 prostate cancer cells at a concentration of 50 nM. RNA was isolated from the transfected cells 72 hrs following transfection and 1 μg of RNA was reverse transcribed into cDNA using oligo dT primers. The resulting cDNA was amplified using gene specific primers and primers for the housekeeping gene GAPDH. (C) Cell images (100X) were taken 72 hrs following a saRNA transfection. saRNAs for E-cadherin and p21 caused arrested cell growth and distinct morphologic changes compared to the control saRNA transfected PC-3 cells.

h for RNAa to appear. Why does RNAa need longer time to manifest? One possible explanation is that genomic DNA targets are only accessible during cell division when nuclear membrane disappears (Morris, 2005). This is supported by the proximity between the RNAa time lag and the doubling time of the cells in which RNAa is observed.

Another interesting observation is the long-lasting effects of RNAa. It has been shown that saRNA induced gene expression lasts for 10–15 days following a single transfection of saRNA (Janowski *et al.*, 2007; Li *et al.*, 2006). This prolonged effect has been attributed to saRNA induced epigenetic changes such as demethylation of histone 3 at lysine 9 (Li *et al.*, 2006), which are considered inheritable across cell divisions (Egger *et al.*, 2004). It is likely that the actual

gene induction might extend well beyond the observed period or could be permanent, considering that cells with an activated gene as in the case of E-cadherin are selected against and eventually untransfected cells become dominant. Further experiments are needed using stable and inducible shRNA transfection to determine the nature of long-lasting RNAa effects. Nevertheless, this feature of RNAa is attracting from a clinical standpoint.

saRNA sequence specificity for RNAa

The 5′-end or the 'seed' sequence of the guide strand of a siRNA and miRNA are critical for target recognition. Similarly, mutations to the 5′-end of two saRNAs targeting the E-cadherin and p21 promoters, respectively, completely disrupted activity, while mutations to the 3′-end had a minimal effect on RNAa. Furthermore, mutations to the middle of the p21 saRNA duplex still retained low levels of activity. These findings reveal that mismatches between the target sequence and the saRNAs are capable of inducing gene expression as long as they avoid the 'seed' sequence. This suggests that RNAa shares a similar mechanism with RNAi in terms of strand recognition and preference.

RNAa's dependence on Argonaute 2 protein

Argonautes (Agos) are a large family of protein containing the PIWI and the PAZ domain and are involved in RNAi at both the transcriptional level and the posttranscriptional level. Humans have a total of four closely related Ago proteins (Ago1–4) that function in small RNA-mediated gene regulation. In posttranscriptional RNAi, Ago2 protein together with siRNA forms the RISC complex which cleaves target mRNA by virtue of Ago2's catalytic activity (Liu et al., 2004). Both Ago1 and Ago2 have been implicated in transcriptional RNAi (Janowski et al., 2006; Kim et al., 2006) although the detailed mechanisms remain elusive. The requirement for different Ago proteins in RNAa has also been tested by knocking down individual Ago members using RNAi. RNAa is totally abolished when Ago2 is knocked down. Knocking down other Ago members (Ago1, 3 and 4) has no significantly effect RNAa (Li et al., 2006). It is surprising that RNAa also requires Ago2, the only Ago member that possesses cleavage activity in human cells (Liu et al., 2004; Meister et al., 2004), which is apparently not relevant for DNA targeting. Two recent studies provide a possible explanation for this paradox. For RNAi, it was initially thought that siRNA is unwound by a helicase before assembling into the RISC complex; however, the studies by Rand et al. (2005) and Matranga et al. (2005) found that it is the Ago2 protein that cleaves and discards the passenger strand of a dsRNA the guide strand which then becomes poised for target binding. Similarly, Ago2 may be required to process saRNA into an active form as an initial step. However, it is unclear whether Ago2 is directly involved in the subsequent transcription activation at saRNA's target promoter because an attempt to locate the Ago2 protein at saRNA target sites using ChIP assays was not successful (Janowski et al., 2007). Other Ago proteins may also be involved in RNAa although they are not as critical as Ago2 (Li et al., 2006).

Epigenetic changes associated with RNAa

Some critical questions regarding RNAa are how saRNA targeting of promoter activates gene expression and what is the nature of the effector complex if there is any. To answer these questions, chromatin immunoprecipitation (ChIP) assays have been used to map the chromatin changes associated with saRNA's target regions in gene promoters such as acetylation and methylation and different types of chromatin modifications have been identified with a particular target promoter. For example, loss of histone 3 di- and tri-methylation at lysine 9 (H3K9) is associated with RNAa of E-cadherin in PC-3 prostate cancer cells, whereas no significant changes in methylation of H3K4 (Li et al., 2006) and global acetylation at H3 and H4 (Li, unpublished observations) have been found. In another study, reduced acetylation at H3K9 and H3K4, and increased di- and tri-methylation at H4K4 are associated with RNAa of the PR gene in MCF-7 breast cancer cells. The diversity in chromatin modifications may reflect gene specific epigenetic changes associated with RNAa (Janowski et al., 2007). Even so the question remains open as to whether these changes are the cause of gene activation or secondary to transcriptional activation.

Target location effects

dsRNA target location has a great impact on which direction transcription is regulated (i.e., upregulation or downregulation) and the magnitude of the regulation. Sifting of a target site for a few bases may change a saRNA into a siRNA. It is unclear what renders different locations with varying dsRNA targeting activity. One common observation is that dsRNAs targeting CpG islands or GC-rich areas tend to function as siRNA (Morris et al., 2004; Ting et al., 2005) and these regions should be avoided as saRNA targets (Li et al., 2006). One handy explanation for this observation is that dsRNAs that target CpG island regions may downregulate gene expression by inducing CpG methylation, However, as discussed above whether dsRNAs induce DNA methylation in mammalian cells has not yet been confirmed. An alternative explanation is that different target locations render corresponding dsRNAs with varying thermodynamic properties which affect their fate in the dsRNA processing machinery because the thermodynamic end stability of a dsRNA determines which strand is used by Ago protein as the guide strand (Ma *et al.*, 2005).

miRNAs and positive gene regulation

miRNAs are a large class of small regulatory RNAs that are transcribed by RNA polymerase II as long primary miRNAs (pri-miRNAs) in size ranging from hundreds to thousands of nucleotides from the introns of protein coding mRNAs or introns and exons of non-coding RNAs (reviewed in Du and Zamore, 2005). The pri-miRNAs are processed by an RNase III endonuclease Drosha and DGCR8 into about 70-nt stem–loop structures called precursor miRNA (pre-miRNAs), which are then exported into the cytoplasm by Exportin-5. In the cytoplasm, pre-miRNAs are further processed into ~22-nt miRNA/miRNA* double-stranded duplexes by a second RNase III endonuclease called Dicer-1. miRNAs have been found mainly as the mediator of posttranscriptional gene silencing through one of two mechanisms: inhibiting mRNA translation at the initiation and elongation steps, or by affecting mRNA levels by inducing deadenylation, decapping and decaying of target mRNAs within the P-bodies (Liu *et al.*, 2005).

Target mRNA cleavage is not required for miRNA-mediated gene silencing. Mature miRNAs facilitate their activity on a wide range of transcripts by partial complementary to target transcripts. Mismatches between target transcripts and miRNAs are quite common and tolerable for activity. As mentioned above, mismatches between targeted promoter sequences and saRNAs still retain saRNA's ability to induce gene expression. This is remarkably similar to miRNA target recognition and suggests that miRNAs with homology to promoter sequences may also function to activate gene expression. In addition, ample evidence has shown that miRNAs are involved in diverse biological processes such as organism development, morphogenesis, stem cell division, and disease aetiology such as diabetics, and cancer (reviewed in Ambros, 2004; Calin and Croce, 2006), however only a few miRNA targets have been experimentally verified as miRNA-silenced genes. Given the functional complexity of miRNA-mediated regulation, it is difficult to imagine that the regulation is only one-directional – silencing. Similar to exogenously introduced saRNAs, it is highly likely that miRNAs also positively regulate gene activity.

The initial evidence that miRNAs positively regulate target sequences came from a study by Jopling *et al.* (2005). In this study, a miRNA (miR-122) specifically expressed in human liver was found to enhance hepatitis C viral gene transcription by targeting its 5′ non-coding gene (Jopling *et al.*, 2005). Sequestration of miR-122 in liver cells resulted in marked loss of autonomously replicating hepatitis C viral RNAs. It is however nuclear how this miRNA enhances gene transcription by targeting 5′ instead of 3′ non-coding sequences.

Many published studies have tried to survey and identify potential miRNA target genes using high throughput cDNA microarray technology. In these studies, apart from genes that are thought to be downregulated by microRNAs, many other genes show a divergent expression change following manipulation of miRNAs or the miRNA machinery. This group of genes is usually regarded as secondary effects and largely ignored (Giraldez *et al.*, 2006; Krutzfeldt *et al.*, 2005; Lim *et al.*, 2005; Rehwinkel *et al.*, 2006; Schmitter *et al.*, 2006). However, genes with a divergent expression profile sometimes outnumber genes that are downregulated by miRNAs (Krutzfeldt *et al.*, 2005; Schmitter *et al.*, 2006) and many of these miRNA-upregulated genes are significantly overrepresented in distinct gene ontology categories (Krutzfeldt *et al.*, 2005). These observations strongly suggest that the divergence in miRNA-regulated gene expression cannot simply be explained as secondary effects and may instead represent the flip side of miRNA's role – positively regulation of gene expression.

Mechanisms of action and proposed models for RNAa

Based on published observations on RNAa (Janowski *et al.*, 2007; Li *et al.*, 2006) and other related mechanisms, we propose a model for RNAa in mammalian cells (Fig. 13.2). In this model, the small RNA trigger is either exogenously introduced dsRNAs or naturally occurring small RNAs such as miRNAs. Once in the cytoplasm, these dsRNAs are loaded onto the Ago proteins, preferably Ago2, and are then processed by Ago2 into an active single-stranded guide RNA by cleaving and subsequently discarding the other strand of the RNA duplex. Because a promoter-targeting saRNA has no mRNA target in the cytoplasm, the RNA guide strand then directs the Ago protein to its DNA target during cell division when the nuclear envelope disappears. This notion is supported by the time lag required for RNAa to appear

following a saRNA transfection, although active transportation of the RNA-Ago complex is also possible. Once bound to DNA, the Ago protein serves as a chromatin remodelling platform and recruits histone modifying activities such as HAT, histone demethylases, leading to transcriptionally active chromatin. The binding of RNA to DNA may also help DNA helix opening which is required for transcription initiation. Alternatively, the RNA-Ago complex is tethered through base-pairing to nascent transcripts transcribed from the target region and further recruits active chromatin modifying activities leading to enhanced transcription.

Conclusions and future directions

An increasing body of evidence suggests that RNAa is a general mechanism of gene regulation mediated by small RNA molecules and may con-

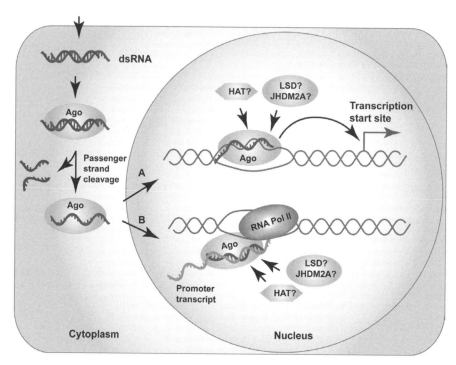

Figure 13.2 Proposed models for RNAa. Exogenously introduced or naturally occurring dsRNAs are loaded onto an Ago protein where the passenger strand is cleaved and discarded. The resulted guide-strand-Ago2 complex then gains access to chromosomal DNA during cell mitosis when the nuclear envelope disappears or through active transportation. The Ago-bound saRNA may then bind to its DNA target (A) or cryptic promoter transcripts derived from the dsRNA target region. In (A), the saRNA binds to one of the DNA strand where Ago acts as a reaction platform for recruiting histone modifying activities, such as histone demethylase [lysine specific demethylase (LSD), JHDM2A] and histone acetyltransferases (HATs), resulting in transcriptionally active chromatin. In (B), the guide-strand-Ago complex binds to nascent promoter transcripts and further recruits histone modifying enzymes as in (A) resulting in a transcriptionally active promoter.

stitutes an integrate network of gene regulation with small RNA induced gene silencing mechanisms. Before RNAa can be fully harnessed for therapy and be used as a tool for gene manipulation, there is urgent need for answers to questions such as: What is the actual target molecule of a promoter dsRNA, DNA or nascent RNA? What is the nature of the effector complex which implements the dsRNA triggered sequence-specific effects and why does promoter targeting of different regions by dsRNAs have varying or even opposite effects on gene transcription? An understanding of these questions will help design effective RNAa targets on many different genes and leads to the solving of more questions of biological and clinical significance such as: Is RNAa a naturally occurring mechanism whereby flexible gene regulation at transcriptional level is possible? If yes, what are the trigger RNA molecules? Does dysregulation of such mechanism play roles in diseases such as cancer?

References

Akhtar, A., Zink, D., and Becker, P.B. (2000). Chromodomains are protein-RNA interaction modules. Nature 407, 405–409.

Ambros, V. (2004). The functions of animal microRNAs. Nature 431, 350–355.

Bagga, S., Bracht, J., Hunter, S., Massirer, K., Holtz, J., Eachus, R., and Pasquinelli, A.E. (2005). Regulation by let-7 and lin-4 miRNAs results in target mRNA degradation. Cell 122, 553–563.

Bernstein, E., and Allis, C.D. (2005). RNA meets chromatin. Genes Dev. 19, 1635–1655.

Britten, R.J., and Davidson, E.H. (1969). Gene regulation for higher cells: a theory. Science 165, 349–357.

Calin, G.A., and Croce, C.M. (2006). MicroRNA signatures in human cancers. Nat. Rev. Cancer 6, 857–866.

Castanotto, D., Tommasi, S., Li, M., Li, H., Yanow, S., Pfeifer, G.P., and Rossi, J.J. (2005). Short hairpin RNA-directed cytosine (CpG) methylation of the RASSF1A gene promoter in HeLa cells. Mol. Ther. 12, 179–183.

Chan, M.F., Liang, G., and Jones, P.A. (2000). Relationship between transcription and DNA methylation. Current Topics Microbiol. Immunol. 249, 75–86.

Danin-Kreiselman, M., Lee, C.Y., and Chanfreau, G. (2003). RNAse III-mediated degradation of unspliced Pre-mRNAs and lariat introns. Mol. Cell 11, 1279–1289.

Du, T., and Zamore, P.D. (2005). microPrimer: the biogenesis and function of microRNA. Development 132, 4645–4652.

Dykxhoorn, D.M., Novina, C.D., and Sharp, P.A. (2003). Killing the messenger: short RNAs that silence gene expression. Nat. Rev. Mol. Cell. Biol. 4, 457–467.

Egger, G., Liang, G., Aparicio, A., and Jones, P.A. (2004). Epigenetics in human disease and prospects for epigenetic therapy. Nature 429, 457–463.

Elbashir, S.M., Harborth, J., Lendeckel, W., Yalcin, A., Weber, K., and Tuschl, T. (2001). Duplexes of 21-nucleotide RNAs mediate RNA interference in cultured mammalian cells. Nature 411, 494–498.

Fire, A., Xu, S., Montgomery, M.K., Kostas, S.A., Driver, S.E., and Mello, C.C. (1998). Potent and specific genetic interference by double-stranded RNA in Caenorhabditis elegans. Nature 391, 806–811.

Gilfillan, G.D., Dahlsveen, I.K., and Becker, P.B. (2004). Lifting a chromosome: dosage compensation in Drosophila melanogaster. FEBS Lett. 567, 8–14.

Giraldez, A.J., Mishima, Y., Rihel, J., Grocock, R.J., Van Dongen, S., Inoue, K., Enright, A.J., and Schier, A.F. (2006). Zebrafish MiR-430 promotes deadenylation and clearance of maternal mRNAs. Science 312, 75–79.

Goodrich, J.A., and Kugel, J.F. (2006). Non-coding-RNA regulators of RNA polymerase II transcription. Nat. Rev. Mol. Cell. Biol. 7, 612–616.

Hellman, A., and Chess, A. (2007). Gene body-specific methylation on the active X chromosome. Science 315, 1141–1143.

Imamura, T., Yamamoto, S., Ohgane, J., Hattori, N., Tanaka, S., and Shiota, K. (2004). Non-coding RNA directed DNA demethylation of Sphk1 CpG island. Biochem. Biophys. Res. Commun. 322, 593–600.

Janowski, B.A., Huffman, K.E., Schwartz, J.C., Ram, R., Hardy, D., Shames, D.S., Minna, J.D., and Corey, D.R. (2005). Inhibiting gene expression at transcription start sites in chromosomal DNA with antigene RNAs. Nat. Chem. Biol. 1, 216–222.

Janowski, B.A., Younger, S.T., Hardy, D.B., Ram, R., Huffman, K.E., and Corey, D.R. (2007). Activating gene expression in mammalian cells with promoter-targeted duplex RNAs. 3, 166–173.

Jopling, C.L., Yi, M., Lancaster, A.M., Lemon, S.M., and Sarnow, P. (2005). Modulation of hepatitis C virus RNA abundance by a liver-specific MicroRNA. Science 309, 1577–1581.

Kim, D.H., Villeneuve, L.M., Morris, K.V., and Rossi, J.J. (2006). Argonaute-1 directs siRNA-mediated transcriptional gene silencing in human cells. Nat. Struct. Mol. Biol. 13, 793–797.

Krutzfeldt, J., Rajewsky, N., Braich, R., Rajeev, K.G., Tuschl, T., Manoharan, M., and Stoffel, M. (2005). Silencing of microRNAs in vivo with 'antagomirs'. Nature 438, 685–689.

Labhart, P. (1994). Negative and positive effects of CpG-methylation on Xenopus ribosomal gene transcription in vitro. FEBS Lett. 356, 302–306.

Lanz, R.B., McKenna, N.J., Onate, S.A., Albrecht, U., Wong, J., Tsai, S.Y., Tsai, M.J., and O'Malley, B.W. (1999). A steroid receptor coactivator, SRA, functions as an RNA and is present in an SRC-1 complex. Cell 97, 17–27.

Lanz, R.B., Razani, B., Goldberg, A.D., and O'Malley, B.W. (2002). Distinct RNA motifs are important for coactivation of steroid hormone receptors by steroid receptor RNA activator (SRA). Proc. Natl. Acad. Sci. USA 99, 16081–16086.

Lau, N.C., Seto, A.G., Kim, J., Kuramochi-Miyagawa, S., Nakano, T., Bartel, D.P., and Kingston, R.E. (2006). Characterization of the piRNA complex from rat testes. Science 313, 363–367.

Lee, R.C., Feinbaum, R.L., and Ambros, V. (1993). The C. elegans heterochronic gene lin-4 encodes small RNAs with antisense complementarity to lin-14. Cell 75, 843–854.

Li, L.C., Carroll, P.R., and Dahiya, R. (2005). Epigenetic changes in prostate cancer: implication for diagnosis and treatment. J. Natl. Cancer Inst. 97, 103–115.

Li, L.C., Okino, S.T., Zhao, H., Pookot, D., Place, R.F., Urakami, S., Enokida, H., and Dahiya, R. (2006). Small dsRNAs induce transcriptional activation in human cells. Proc. Natl. Acad. Sci. USA 103, 17337–17342.

Lim, L.P., Lau, N.C., Garrett-Engele, P., Grimson, A., Schelter, J.M., Castle, J., Bartel, D.P., Linsley, P.S., and Johnson, J.M. (2005). Microarray analysis shows that some microRNAs downregulate large numbers of target mRNAs. Nature 433, 769–773.

Liu, J., Carmell, M.A., Rivas, F.V., Marsden, C.G., Thomson, J.M., Song, J.J., Hammond, S.M., Joshua-Tor, L., and Hannon, G.J. (2004). Argonaute2 is the catalytic engine of mammalian RNAi. Science 305, 1437–1441.

Liu, J., Rivas, F.V., Wohlschlegel, J., Yates, J.R., 3rd, Parker, R., and Hannon, G.J. (2005). A role for the P-body component GW182 in microRNA function. Nat. Cell. Biol. 7, 1261–1266.

Ma, J.B., Yuan, Y.R., Meister, G., Pei, Y., Tuschl, T., and Patel, D.J. (2005). Structural basis for 5'-end-specific recognition of guide RNA by the A. fulgidus Piwi protein. Nature 434, 666–670.

Mandal, M., and Breaker, R.R. (2004). Gene regulation by riboswitches. Nat. Rev. Mol. Cell. Biol. 5, 451–463.

Matranga, C., Tomari, Y., Shin, C., Bartel, D.P., and Zamore, P.D. (2005). Passenger-strand cleavage facilitates assembly of siRNA into Ago2-containing RNAi enzyme complexes. Cell 123, 607–620.

Mattick, J.S. (2005). The functional genomics of noncoding RNA. Science 309, 1527–1528.

Meister, G., Landthaler, M., Patkaniowska, A., Dorsett, Y., Teng, G., and Tuschl, T. (2004). Human Argonaute2 mediates RNA cleavage targeted by miRNAs and siRNAs. Mol. Cell 15, 185–197.

Morris, K.V. (2005). siRNA-mediated transcriptional gene silencing: the potential mechanism and a possible role in the histone code. Cellular and molecular life sciences 62, 3057–3066.

Morris, K.V., Chan, S.W., Jacobsen, S.E., and Looney, D.J. (2004). Small interfering RNA-induced transcriptional gene silencing in human cells. Science 305, 1289–1292.

Narberhaus, F., Waldminghaus, T., and Chowdhury, S. (2006). RNA thermometers. FEMS Microbiol. Rev. 30, 3–16.

Neal, K.C., Pannuti, A., Smith, E.R., and Lucchesi, J.C. (2000). A new human member of the MYST family of histone acetyl transferases with high sequence similarity to Drosophila MOF. Biochim. Biophys. Acta 1490, 170–174.

Olsen, P.H., and Ambros, V. (1999). The lin-4 regulatory RNA controls developmental timing in Caenorhabditis elegans by blocking LIN-14 protein synthesis after the initiation of translation. Dev. Biol. 216, 671–680.

Prasanth, K.V., and Spector, D.L. (2007). Eukaryotic regulatory RNAs: an answer to the 'genome complexity' conundrum. Genes Dev. 21, 11–42.

Rand, T.A., Petersen, S., Du, F., and Wang, X. (2005). Argonaute2 cleaves the anti-guide strand of siRNA during RISC activation. Cell 123, 621–629.

Rehwinkel, J., Natalin, P., Stark, A., Brennecke, J., Cohen, S.M., and Izaurralde, E. (2006). Genome-wide analysis of mRNAs regulated by Drosha and Argonaute proteins in Drosophila melanogaster. Mol. Cell. Biol. 26, 2965–2975.

Reynolds, A., Leake, D., Boese, Q., Scaringe, S., Marshall, W.S., and Khvorova, A. (2004). Rational siRNA design for RNA interference. Nat. Biotechnol. 22, 326–330.

Robb, G.B., Brown, K.M., Khurana, J., and Rana, T.M. (2005). Specific and potent RNAi in the nucleus of human cells. Nat. Struct. Mol. Biol. 12, 133–137.

Schmitter, D., Filkowski, J., Sewer, A., Pillai, R.S., Oakeley, E.J., Zavolan, M., Svoboda, P., and Filipowicz, W. (2006). Effects of Dicer and Argonaute downregulation on mRNA levels in human HEK293 cells. Nucleic Acids Res. 34, 4801–4815.

Sekeris, C.E. (1985). The role of HnRNA in the control of ribosomal gene transcription. J. Theor. Biol. 114, 601–604.

Smith, E.R., Cayrou, C., Huang, R., Lane, W.S., Cote, J., and Lucchesi, J.C. (2005). A human protein complex homologous to the Drosophila MSL complex is responsible for the majority of histone H4 acetylation at lysine 16. Mol. Cell. Biol. 25, 9175–9188.

Stein, R., Sciaky-Gallili, N., Razin, A., and Cedar, H. (1983). Pattern of methylation of two genes coding for housekeeping functions. Proc. Natl. Acad. Sci. USA 80, 2422–2426.

Steitz, T.A., and Moore, P.B. (2003). RNA, the first macromolecular catalyst: the ribosome is a ribozyme. Trends Biochem. Sci. 28, 411–418.

Szymanski, M., Barciszewska, M.Z., Zywicki, M., and Barciszewski, J. (2003). Noncoding RNA transcripts. J. Appl. Genet. 44, 1–19.

Taipale, M., Rea, S., Richter, K., Vilar, A., Lichter, P., Imhof, A., and Akhtar, A. (2005). hMOF histone acetyltransferase is required for histone H4 lysine 16 acetylation in mammalian cells. Mol. Cell. Biol. 25, 6798–6810.

Ting, A.H., Schuebel, K.E., Herman, J.G., and Baylin, S.B. (2005). Short double-stranded RNA induces transcriptional gene silencing in human cancer cells in the absence of DNA methylation. Nat. Genet. 37, 906–910.

Volpe, T.A., Kidner, C., Hall, I.M., Teng, G., Grewal, S.I., and Martienssen, R.A. (2002). Regulation of heterochromatic silencing and histone H3 lysine-9 methylation by RNAi. Science *297*, 1833–1837.

Wassenegger, M. (2000). RNA-directed DNA methylation. Plant Mol. Biol. *43*, 203–220.

Weiss, A., Keshet, I., Razin, A., and Cedar, H. (1996). DNA demethylation *in vitro*: involvement of RNA. Cell *86*, 709–718.

Yoder, J.A., Walsh, C.P., and Bestor, T.H. (1997). Cytosine methylation and the ecology of intragenomic parasites. Trends Genet. *13*, 335–340.

Therapeutic Potential of RNA-mediated Control of Gene Expression: Options and Designs

14

L Scherer and JJ Rossi

Abstract

We review factors to consider when choosing an mRNA knockdown approach in human cells using therapeutic expressed RNAs. We emphasize methods that use RNA triggers of endogenous cellular pathways to target and degrade mRNAs–RNAu, RNase P external guide sequences, tRNAse Z^L small guide sequences, and RNAi– rather than those RNAs with intrinsic activity, such as ribozymes and aptamers. The range of available methods may be particularly important in combinatorial approaches to inhibit viruses prone to mutational escape, such as HIV-1 and HCV.

Scope

The scope of this review will be limited to therapeutic expressed RNAs, as opposed to exogenously delivered synthetic molecules. The focus will be on factors to consider when choosing a method to knockdown an mRNA in human cells, although the concepts are generally relevant to other mammalian cells as well. We will not consider delivery methods, except as it pertains to introducing the genes for RNAi triggers into target cells. In addition, we place emphasis on those methods that take advantage of endogenous cellular pathways. Other RNA-based inhibitory molecules with intrinsic activity, such as ribozymes, are considered elsewhere in this volume, and aptamers have been the subject of recent review (Que-Gewirth and Sullenger, 2007). In particular, we will consider the use of various knockdown methods from the perspective of designing gene therapeutic strategies for challenging therapeutic targets such as HIV-1, where the emerging consensus suggests the necessity of a combinatorial, or multiplexing approach. Sections begin with a brief introduction to the basic underlying biology of the method where necessary, followed by a discussion of applications from the current literature and lastly, a general overview of practical aspects and design considerations.

General considerations

As a conceptual framework for subsequent material, we begin with a general discussion of the factors that should be considered for applications. This will be referred to in later sections, where specific methods are discussed. Many of these issues are interrelated or may involve tradeoffs in application, and should be evaluated accordingly. This overview is not intended to be comprehensive, but to provide the reader with examples that illustrate specific points.

Factors influencing the choice of target mRNA and sequence

Selection of a target sequence within an mRNA is a major consideration. The necessity to discriminate between closely related mRNAs can limit the pool of available sequences specific to the desired transcript. For instance, selecting one splice isoform among several requires limiting the choice of a sequence to a unique exon. Targeting a single member of a multiple gene family must be restricted to unique, or non-homologous regions. The only region available to specifically degrade the transcript of an oncogene resulting from a translation, such as bcr/abl, is the sequence that spans the fusion joint; otherwise, the parental endogenous genes may be inhibited as well. Sequence choices in these situations may be severely constrained and will also influence the choice of method.

The underlying biology of the target mRNA in its cellular biochemical and physiological context is an important consideration in choosing a knockdown method. Factors include the target mRNA expression level and desired degree of knockdown. To illustrate, in one scenario the goal may be to return the level of an overexpressed endogenous transcript to normal levels. This situation would not necessarily require a highly efficient knockdown method. In contrast, the complete elimination of an oncogenic or viral transcript would be ideal, and potentially require a different approach. In addition, an early, low level, viral regulatory transcript might be a preferable target to a late, relatively abundant structural gene. If the virus is also highly prone to mutational escape, it may be necessary to target sequences that are highly conserved across viral clades. Other considerations are the temporal, developmental, and cell type expression profile and subcellular localization of the target RNA, which may necessitate the use of particular promoters, tethered sequences in *cis* that direct localization to a particular cellular compartment or transcript, or delivery vectors.

Knockdown method

Specificity
The specificity or discrimination of a method is a major concern as off target effects are highly undesirable in therapeutic applications. For methods that rely on complementarity to recognize and cleave the target RNA, specificity is a function of a complex interplay of a number of factors, including: (1) the total sequence space that is accessible by the method (influenced by location requirements in a transcript and accessibility); (2) the sequence length that can be targeted by a single effector RNA; (3) the effect of base pair mismatches between target and effector; and (4) the overall efficiency of the method. To put this issue into perspective, the human genome is approximately 3×10^9 base pairs; an estimated

1.2% (International Human Genome Sequencing Consortium 2004) or 3.6×10^7 base pairs code for protein. However, the amount of the non-repetitive genome that is transcribed into functional, but non-coding RNA is significantly higher in both nuclear and cytoplasmic compartments (Willingham and Gingeras, 2006). The total sequence complexity of RNA at any given time will also reflect the cell type and cell state. The optimal length of complementarity between substrate and effector required for full activity differs between methods and sets the maximum specificity in a given target population. In general, the shorter the target sequence, the greater the possibility that a given sequence will be present in multiple mRNAs, increasing the potential for off-target effects. The frequency of a 10 nucleotide sequence (excluding factors such as species-specific sequence and base composition biases) is 1 in 4^{10} or $1/1.0 \times 10^6$; likewise, 12 bases, $1/1.7 \times 10^7$; 15 bases, $1/1.1 \times 10^9$; and 19 bases, $1/2.7 \times 10^{11}$. A knockdown method that tolerates mismatches, particularly multiple mismatches, effectively increases the number of potential sequences targeted, and reduces specificity. If, in addition, the method is very efficient, any deleterious off-target effects are likely to be exacerbated. This becomes a thorny problem for inhibition of HIV and HCV, where balancing the need for efficient knockdown and prevention of mutational escape on the one hand and avoidance of off-target effects on the other can be challenging.

Knockdown efficiency
For the sake of discussion, we divide RNA effectors into two categories: autonomous effectors that have an intrinsic activity, such as aptamers and ribozymes; and RNA triggers that utilize or recruit an endogenous cellular pathway, such as RNA interference (RNAi). Intrinsic RNA effectors typically reach their target by diffusion and efficiency is proportional to expression level; this is particularly true for aptamers, which unlike ribozymes, lack innate catalytic potential. Efficiency may be increased by employing techniques that improve effector:mRNA co-localization. These include using promoters or tethered sequences that direct expression to a specific sub-cellular compartment or target. RNA effec-

tors such as siRNAs that utilize an endogenous pathway tend to be inherently more efficient. It is crucial, however, that both the target and effector be expressed in the cellular compartment containing the pathway components.

Toxicity

Off-target effects can lead to a number of undesirable consequences including cellular toxicity. A screen of the genome to eliminate targets that may potentially cross-react may reduce the problem, although this can be complicated if the knockdown method tolerates mismatches. Also, RNA effectors present in excess can swamp the very cellular pathway they utilize for their effect, displacing endogenous RNAs requiring access to the same pathway to perform their normal physiological function. Some RNAs can induce cellular immune responses. Performance of appropriate control experiments can identify many of these problems in the early development phase of an application. In addition, it may be wise to confirm the biological effects of mRNA reduction by targeting a separate sequence within the same target. Some toxicity issues can be circumvented by the use of inducible or tissue-specific promoters. For clinical applications, more stringent screens for off-target effects may be necessary, including microarray analysis and ultimately, testing using *in vivo* model systems.

Delivery

While not a focus of this review, the anticipated mode of cellular delivery has significant implications for the choice of RNA knockdown methodology and expression cassette. Delivery vectors differ in their size capacity for inserted DNA, ease of production and introduction into the cell type of interest. Issues of effector design that can impact delivery will be mentioned as they arise.

Ribozymes and aptamers: RNA therapeutics with intrinsic inhibitory properties

Ribozymes

Ribozymes are RNAs that act as enzymes, even in the complete absence of proteins, and contain complementary sequences that determine their target specificity. Ribozyme applications are covered in depth elsewhere in this volume, but for the purposes of comparison we note that the hairpin and hammerhead ribozyme catalytic motifs are most often employed against specific cellular and viral targets. Efficient target cleavage can be difficult to achieve *in vivo*, but may be improved by targeting accessible sequences or the use of: (1) promoters that mediate high levels of expression in the appropriate subcellular compartment; (2) appended sequences that direct the ribozyme to the same cellular or subcellular compartment as the target mRNA (Jambhekar and Derisi, 2007); (3) appended sequences that interact with a specific target. Because their activity is intrinsic, ribozymes can target transcripts in both the nucleus and cytoplasm. While the number of potential target sites is somewhat limited, the number of complementary nucleotides, and hence the specificity, is quite flexible (Bagheri and Kashani-Sabet, 2004; Bartolome *et al.*, 2004; Khan and Lal, 2003; Komatsu, 2004; Muller *et al.*, 2006; Peracchi, 2004; Rossi, 1999).

Aptamers

Aptamers are oligonucleotides designed to bind proteins with high specificity and affinity based on tertiary structure, thereby blocking protein function. Aptamers can be based on natural RNA binding motifs, such as the HIV-1 TAR element (Sullenger *et al.*, 1990), or derived *de novo*; in either case, maximal affinity is usually achieved by iterative *in vitro* selection using SELEX, systematic evolution of ligands by exponential enrichment. Aptamers have a broad array of applications in the elucidation of protein function. The majority of gene therapy research has centred on the use of synthetic aptamers for short-term protein inhibition (for recent reviews, see Famulok and Mayer, 2005; Held *et al.*, 2006; James, 2007; Que-Gewirth and Sullenger, 2007); however, there are an increasing number of studies reporting the results of expressed RNA aptamers, or intramers, in mammalian cells.

Early applications of this technology to inhibit HIV used pol III promoters to express HIV Rev-binding aptamers (RBEs). Transient co-transfection assays of U6-RBE expression plasmids and pNL4–3 infectious proviral DNA in 293 cells mediated 80–90% inhibition of HIV replication (Good *et al.*, 1997). Subsequently, an HIV-1 TAR decoy construct was developed in which the TAR sequence was embedded in the apical loop of a U16 snoRNA for nucleolar localization and expressed from a U6 promoter. In infectious HIV-1 challenge assays, CEM cells stably transduced with the U6-U16-TAR decoy in a retroviral backbone completely blocked HIV replication, while the TAR decoy alone only partially inhibited the virus (Michienzi et al., 2002). Another approach expressed an anti-HIV reverse transcriptase (RT) pseudoknot aptamer from a modified tRNA^Meti promoter, with deletions at both the 5' and 3' ends that remove the sequences that form the acceptor stem in the mature tRNA and abolish processing (Chaloin et al., 2002). In this system, the aptamer is transcribed at the 3' end of the modified tRNA^Meti chimeric transcript which is transported to the cytoplasm, the location of the target protein. Jurkat cells stably transduced with the tRNA^Meti–RT aptamer incorporated in a Moloney murine leukaemia virus MoMLV-based retroviral vector completely inhibited HIV replication after low dose of HIV infection, and showed lower titres and delayed onset of HIV titres. Another study compared the ability of aptamers against HIV-1 reverse transcriptase (RT) and shRNAs directed against *vif* or *tat/rev* mRNAs to inhibit viral replication (Joshi et al., 2005). All inhibitors were stably expressed in CEM cells from a U6 promoter in a retroviral backbone. While nearly equally efficacious at lower multiplicities of infection (MOI), the aptamers, unlike the shRNAs, maintained a degree of inhibition at higher MOIs. This result may be explained in part by the observation that viral particles produced from aptamer-expressing cells packaged the aptamer, and had a reduced replication capacity in the subsequent round of infection.

Aptamers are the only expressed RNA-based therapeutics that directly target protein, rather than RNA. They are also less prone to target mutational escape, since changes that affect the ability of the target protein to bind to the aptamer can adversely affect the endogenous protein function as well and are less likely to confer a selective advantage. Assets of aptamers include their very high specificity and avidity and the ability to reverse their inhibitory properties using anti-sense 'antidotes'. Difficulties can

arise however, as a result of flanking sequences that can alter the tertiary structure and thus, decrease or eliminate activity (Chaloin et al., 2002; Ishizaki et al., 1996) (Martell et al., 2002). When the inclusion of flanking sequences is unavoidable, SELEX can be used to optimize aptamer binding in the context of the extra sequence (Martell et al., 2002), but that must be done individually with every aptamer construct. Other alternatives are to express the aptamer (1) from an external pol III promoter such as H1 (Mi et al., 2006) or U6 (Li et al., 2005) or (2) as part of a tRNA chimera which is processed by cellular processing machinery to release the aptamer (Scherer et al., 2007) or (3) as part of a chimera with *cis*-cleaving ribozymes that release the aptamer with minimal flanking sequences (Joshi et al., 2005). Expression, preferably at high levels, in the correct cellular compartment is essential although localizing sequences can be added to the aptamer (Kolb et al., 2006; Michienzi et al., 2002), provided they do not interfere with protein binding.

RNAu: utilizing the spliceosomal machinery

Eukaryotic polyadenylation is a complex and highly regulated process dependent on *cis*-acting signals in the mRNA and at least 12 proteins (Fig. 14.1B) (reviewed in Gilmartin, 2005; Weiner, 2005). Briefly, cleavage and polyadenylation specificity factor (CPSF) consists of five proteins, including CPSF-160 which recognizes the poly(A)signal AAUAAA (or a closely related sequence) and CPSF-73, the probable endonuclease (Mandel et al., 2006) responsible for cleaving the mRNA ~20 nucleotides downstream of the poly(A) signal, typically after a CA dinucleotide. Heterotrimeric cleavage stimulatory factor (CStF) recognizes a U- or GU-rich sequence called the downstream element (DSE), positioned within 30 nucleotides downstream of poly(A) cleavage site. CPSF and CStF binding is cooperative. Other proteins include poly(A) polymerase (PAP), which adds the poly(A) tail after cleavage of the transcript, and additional multimeric cleavage factors CF1 and CFII. Other recognition/recruitment sequences in the 3' UTR, such as UGUA sequences upstream of the poly(A) signal and microRNA binding

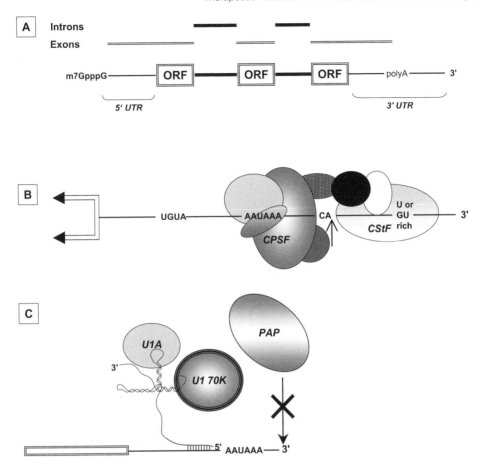

Figure 14.1 (A) Diagrammatic representation of primary mRNA transcript, showing, from 5′ to 3′: cap structure (m7GpppG); 5′ untranslated region, 5′ UTR, (thin line); open reading frame (ORF) protein coding regions (boxes); introns (heavy black line); and 3′-UTR, with polyA site indicated. Double lines above the primary transcript indicate regions corresponding to 5′-terminal, internal, and 3′-terminal exons according to the exon definition model of splicing; introns are shown as a heavy black line. (B) Expanded diagrammatic view of the 3′ UTR. The polyadenylation signal AAUAAA is recognized by cleavage and polyadenylation specificity factor (CPSF). Cleavage occurs approximately 20 nucleotides downstream of the polyA signal (arrow) typically after a CA dinucleotide. The U– or GU–rich downstream element is recognized by cleavage stimulatory factor, CStF. Not to scale. (C) Modified U1 snRNA (RNAu) inhibition. A heterologous sequence complementary to a target in the 3′-terminal exon permits cleavage downstream of the polyA signal, but bound U1 70K inhibits subsequent polyA addition by polyA polymerase (PAP). U1A, U1 snRNA binding protein (not all U1 snRNA binding proteins are shown).

sites, can affect the efficiency of polyadenylation, choice of poly(A) cleavage site and degradation in mRNA post-transcriptional regulation (Gilmartin, 2005; Hu et al., 2005; Zhao et al., 1999).

The primary cellular function of the U1 small nuclear ribonucleoprotein (snRNP) is recognition of 5′ splice signals. Mammalian U1 snRNA (164 nucleotides) provides a scaffold for associated proteins and recognizes 5′ splice junctions of mRNAs by homology to 10 nucleotides at its 5′ terminus (Will and Lührmann, 2006).

However, U1A protein and U1 snRNP participate in several additional modes of regulation at the level of mRNA 3′ end processing. In the first example, U1A splicing factor autoregulates the level of its synthesis by inhibiting polyadenylation of its own mRNA in a negative feedback loop (Boelens et al., 1993; Gunderson et al., 1994). Two U1A molecules bind to the U1A pre-mRNA and interact directly with poly(A) polymerase (PAP) to inhibit PAP activity. This results in a shorter poly(A) tail, and reduced stability of the U1A mRNA.

Another example involves the mechanism of regulating expression of papillomavirus early and late proteins (Furth *et al.*, 1994; Zheng and Baker, 2006). Both classes of mRNAs are transcribed throughout the viral life cycle; however, late gene mRNAs are unstable or not translated during the early phase of infection. This inhibition is mediated by a single consensus 5′ splice site (5′ss) sequence in bovine papillomavirus (BPV-1) or four overlapping sequences with partial homology to the consensus 5′ss in human papillomavirus HPV-16 present in the 3′ UTR upstream of the poly(A) signal of the late genes. However, these sequences do not function as splice donors. Rather, they bind U1 snRNPs and block PAP polyadenylation activity, in a manner reminiscent of U1A autoregulation, but with two differences. First, in U1A autoregulation, two molecules of U1A bind PAP; in the papilloma virus late gene polyadenylation, two sites on a single U1 70K polypeptide interact with PAP. Second, U1A protein interacts directly with its own mRNA; in papillomavirus the U1 snRNA serves as the targeting mechanism for chaperoning U1 70K, part of the U1 snRNP, to the 3′ UTR of the viral transcript (Gunderson *et al.*, 1998).

These observations led to the development of modified U1 snRNAs (mU1 snRNA) where the first 10 nucleotides that recognize the 5′ splice junction are replaced with sequences complementary to other mRNA sequences in reporter (Beckley *et al.*, 2001; Furth *et al.*, 1994; Liu *et al.*, 2002), endogenous (Fortes *et al.*, 2003; Liu *et al.*, 2004) or viral (Sajic *et al.*, 2007) genes. This type of inhibition is commonly referred to as RNAu (Fig. 14.1C) and has been used to correct splicing of thalassaemic β-globin pre-mRNA in tissue culture (Gorman *et al.*, 2000). The target site must be located in the terminal exon as defined by the exon definition model of splicing; the region between the 3′ splice acceptor of the last exon and the poly(A) site of an mRNA (Fig. 14.1A) (Niwa *et al.*, 1992). The modified U1 RNA complex anneals to the target mRNA bringing bound U1 70K protein into the vicinity of PAP action and inhibiting polyA addition. Inhibition is dependent on the AAUAAA polyadenylation mechanism as it is not observed when the poly(A) sequence and cleavage site are replaced by histone 3′ end processing signals (Fortes *et al.*, 2003). The targeted mRNAs are incompletely polyadenylated (Liu *et al.*, 2004) and become unstable, by a mechanism similar to papillomavirus late gene inhibition (Sajic *et al.*, 2007).

Ab initio design of mU1 sequences is not currently possible, due to the limited number of investigations employing RNAu. In a recent study, only 5 of 15 mU1 targeted against HIV-1 had activity (Sajic *et al.*, 2007); nonetheless, some general parameters are known. While 10 nucleotide target sequences, equivalent to the 5′ss target sequence, are typical, other sequence lengths have received limited comparative testing. Extending the anti-sense length from 10 to 16 nucleotides only slightly diminishes activity; however, fewer than 8 nucleotides and greater than 25 are ineffective. Target degradation was observed in 1 case using 60–65 nucleotides of complementarity, although an antisense, rather than a mU1-specific mechanism, was responsible (Liu *et al.*, 1997). Target site accessibility is also important (Fortes *et al.*, 2003). The heterologous sequences at the mU1 5′ end may effect transcription level and start site in some cases, although the wild-type U1A RNA polymerase II (pol II) promoter is very active, driving approximately 3×10^4 copies of U1 snRNA/cell (Dahlberg and Lund, 1987). Longer leaders may also perturb U1 snRNA structure and co-factor binding. A detailed protocol for constructing mU1 snRNAs has been published (Liu *et al.*, 2005).

A target site of 10 nucleotides does not have a high baseline of specificity, although that is somewhat ameliorated by the position dependence for the 3′ exon (although that constraint can also limit application); however, increasing the target length to 16 nucleotides does not severely compromise activity. Slightly longer sequences may retain enough activity to be effective in some applications. RNAu can discriminate single mismatches, although only a few positions and mismatch base pairs have been tested (Liu *et al.*, 2002). Intermediate levels of knockdown are possible, depending on the chosen target sequence in the mRNA (Fortes *et al.*, 2003; Sajic *et al.*, 2007). In this context, it will also be interesting to determine whether recently described

U1-like snRNAs (Kyriakopoulou *et al.*, 2006) can be adapted for RNAu. Another advantage of RNAu is that stable expression of an exogenous mU1-snRNA is unlikely to overwhelm the endogenous pathway, since the capacity of the cellular processing system is large.

Several additional factors should be considered before applying mU1 snRNAs. Retroviral vectors are a prevalent delivery method for stable expression of exogenous genes, particularly those based on lentiviruses since they can efficiently transduce non-dividing cells. Packaging of the retroviral vectors requires production of full-length genomic viral RNA from the pol II promoter in the 5′ LTR through the polyA signal in the 3′ LTR; therefore, the pol II transcription termination signal in the mU1 expression cassettes must be removed so that LTR transcript is not prematurely terminated. The

affect of any extra viral sequences at the 3′ end of the mU1 transcript on the inhibitory activity may have to be assessed. The mU1 expression cassette is relatively large, since the promoter and mU1 sequence are approximately 315 and 165 nucleotides, respectively, which may present difficulties with vectors having size constraints, particularly when employed as part of a combinatorial knockdown strategy.

Utilizing cellular pre-tRNA processing pathways for knockdown of RNA: External Guide Sequences (EGS) and small guide RNAs (sgRNAs)

Endogenous eukaryotic nuclear pre-tRNAs are transcribed by RNA polymerase III (pol III), and undergo a multi-step maturation process (Fig. 14.2) (for overviews, see Hopper and Phi-

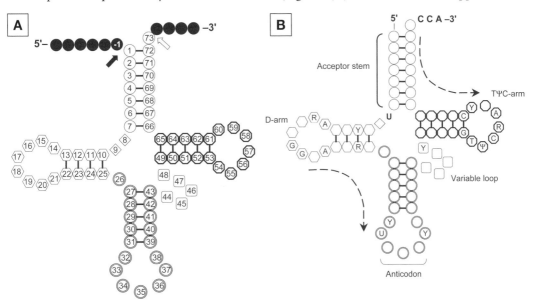

Figure 14.2 Simplified schematic of A) tRNA precursor (pre-tRNA) with standard nucleotide numbering. (Some isotypes contain additional nucleotides in the D and variable loops). In eukaryotes, the pre-tRNA is transcribed by RNA polymerase III with 5′ leader and 3′ trailer sequences (filled circles) which vary in length. RNAse P cleaves the pre-tRNA between nucleotides −1 and +1 (solid block arrow). tRNAse Z^L removes the 3′ trailer, usually after the first unpaired base, known as the discriminator base (stippled in B), on the 3′ side of the acceptor stem (open block arrow). Some pre-tRNA isoforms contain a small intron between the residues 37 and 38. The CCA nucleotidyl-transferase adds the CCA trinucleotide to the processed 3′ end to make the mature tRNA shown in B. Subsequently, an amino acid is added to the CCA by the cognate aminoacyltransferase. In addition, the tRNA undergoes a number of post-translational modifications (not shown, except for thymidine, T and pseudouridine, ψ in TψC arm, shown as UUC in subsequent figures). (B) Mature tRNA, with invariant and semi-variant bases important for tRNA tertiary structure indicated. The tRNA secondary structure domains are denoted from 5′ to 3′ in this Figure and subsequent Figures where applicable by: single circles, acceptor stem; hexagons, dihydrouridine arm (D-arm); double circles, anticodon arm; squares, variable loop; octagons, TψC arm. The acceptor and TψC stem form one continuous helix, the D-stem and anticodon stems the other helix (curved, dashed arrows), that fold at 90° into the characteristic L-shaped tertiary tRNA structure. Stippled circle, discriminator base. Y, pyrimidine. R, purine.

zicky, 2003; Nakanishi and Nureki, 2005; Wolin and Matera, 1999). The 5′ leader is removed, possibly co-transcriptionally, (Reiner et al., 2006) to produce the mature tRNA 5′ end by the ribonucleoprotein complex RNase P, which in humans consists of at least 10 proteins and the single H1 RNA component (reviewed in Evans et al., 2006; Jarrous, 2002; Kirsebom, 2002; Xiao et al., 2002). The 3′ trailer is trimmed by a 3′ tRNase endonuclease activity, tRNase Z^L in human cells, typically immediately after the discriminator base (reviewed in Levinger et al., 2004a; Morl and Marchfelder, 2001; Vogel et al., 2005). The CCA trinucleotide is added to the processed 3′ end by the template-independent CCA nucleotidyl-transferase (Weiner, 2004) prior to aminoacylation by the cognate aminoacyl-tRNA synthetase (Sprinzl, 2006; Xiong and Steitz, 2006). The CCA of mature tRNA is an anti-determinant for eukaryotic tRNase Z^L (Mohan et al., 1999; Nashimoto, 1997; Zareen et al., 2006) preventing non-productive and energetically expensive cycles of CCA addition and removal. Some pre-tRNAs have small introns that must be excised. In addition, tRNAs undergo many post-transcriptional modifications prior to aminoacylation (Engelke and Hopper, 2006; Geslain and Ribas de Pouplana, 2004; Helm, 2006).

Co-opting the pre-tRNA processing mechanism for RNA knockdown is an attractive prospect. Since there are over 400 human nuclear tRNAs genes producing 2% of the total cellular RNA, the endogenous pre-tRNA processing pathways are likely to be robust and less prone to strain by the relatively small amount of pre-tRNA-like molecules produced by stable integrants of exogenous expression cassettes. Strategies have been developed to redirect the activities of the endonucleolytic pre-tRNA processing enzymes, RNase P and tRNase Z^L toward other RNA targets by RNA external guide sequences (EGS) or small guide RNAs (sgRNAs) respectively, described below.

RNase P-based ribozymes (M1GS) and EGS

Two RNase P-based methodologies for specific RNA knockdown in eukaryotic cells have been developed: RNase P M1GS (see below) and

RNase P EGS. We discuss M1GS only briefly since it is an E. coli-derived targeted ribozyme approach, more similar to other trans ribozyme approaches discussed elsewhere in this volume. Unlike the much larger eukaryotic versions, E. coli RNase P consists of one protein, C5, and one RNA subunit, M1 RNA. M1 RNA retains partial, but substantial catalytic activity in vitro under high salt conditions; however, the C5 protein is required under physiological conditions in vitro and in vivo (Guerrier-Takada et al., 1983; Reich et al., 1988). In contrast, the human analogue of M1 RNA, H1 RNA (see above) was only recently shown to be weakly catalytically active in vitro in the absence of the protein components of the human RNase P complex (Kikovska et al., 2007). In vitro studies using model substrates to determine the essential structural characteristics for RNase P substrate recognition and cleavage revealed that the pre-tRNA D- and anticodon- (AC-) stem/loops are unnecessary for M1 cleavage in vitro. A substrate consisting of a stem–loop structure that mimics the acceptor stem/TψC stem helix is sufficient (McClain et al., 1987), and even the loop is not required (Forster and Altman, 1990). Thus, an EGS that forms a minimal complex consisting of a 5′ leader, a ~12 bp helix, and a 3′ trailer of RCCA (Fig. 14.3E) can recruit E. coli RNase P to direct cleavage of the substrate immediately 5′ to first base pair of the complex (reviewed in Guerrier-Takada and Altman, 2000; Raj and Liu, 2003). In E. coli, the CCA trinucleotide is encoded as part of the pre-tRNA, rather than being added post-transcriptionally as in most eukaryotes; the 'R' corresponds in position to the eukaryotic discriminator base. The 3′ RCCA sequence binds to a specific loop in M1 RNA (Harris et al., 1997; Kirsebom, 2002; Oh et al., 1998; Oh and Pace, 1994) and the interaction is required for E. coli RNase P cleavage in vivo (Wegscheid and Hartmann, 2006). Human RNase P cannot recognize this type of 'stem EGS' (Fig. 14.3E). However, the M1 RNA retains catalytic activity in eukaryotic cells, and when 13–18 nucleotides of continuous complementarity to the target mRNA is fused to the 3′ end of the MI RNA, can direct cleavage of the target, provided an unpaired ACCA terminates the complementary sequence. The chimera is

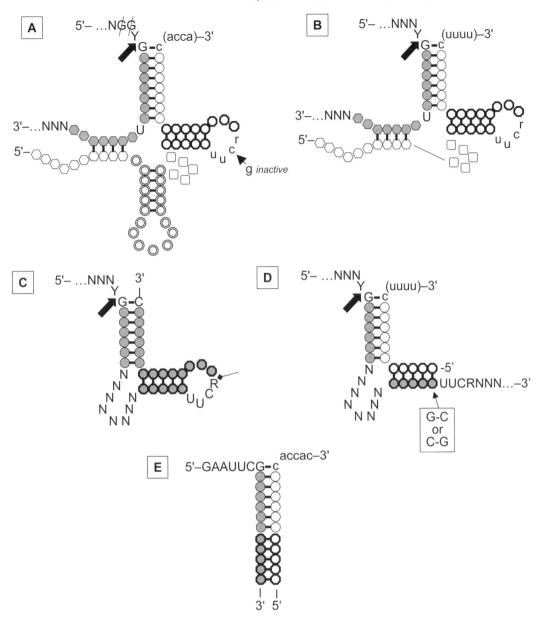

Figure 14.3 RNase P EGS designs, shown as complexes with target sequences, depicted diagrammatically with respect to tRNA secondary structure in Figure 14.2B. Target RNA is denoted by filled polygons and capital font; EGS sequence is denoted by open polygons and small case font. (A) ¾ EGS, so called because the EGS is derived from approximately ¾ of the native tRNA structure. The g to c substitution in the uucr sequence abrogates RNase P activity. The 3′ accac sequence is absolutely required when any RNase P EGS is used in *E. coli*; however, it is unnecessary in higher eukaryotes, and omitted from subsequent Figures. If, however, the accac sequence is included at the 3′ end of the EGS, the −2 and −3 nucleotides of the target should not be Gs to avoid base pairing between the sequences analogous to the 5′ and 3′ trailers. Cleavage (filled arrow) occurs between −1 and +1 nucleotide of the target, which must be a pyrimidine and G, respectively. Cleavage site denoted by a solid block arrow (see Fig. 14.2B). (B) ¾ EGS with deleted acceptor stem (¾ EGS ΔAC). Inclusion of the variable loop is optional. All applicable full ¾ EGS sequence constraints also apply. EGS is shown with a run of uridines present when EGS is expressed by a RNA polymerase III promoter. (C) Minimal synthetic unimolecular model substrate of RNase P. Diamond-headed arrow indicates position where TψC loop backbone can be opened up to create a bimolecular substrate (see D) and maintain RNase P activity. (D) Short EGS (sEGS), derived from C). Typically used for synthetic, rather than expressed EGS. (E) Minimal EGS (mEGS), cleaved by *E. coli* RNAse P, but not recognized by human RNAse P. See text for full details.

referred to as an M1 guide sequence (M1GS) ribozyme.

M1GS behave similarly to conventional *trans* ribozymes in eukaryotic cells, in the sense that they cleave the target mRNA using their own intrinsic activity, not that of the host cell RNase P complex, although protein components of the human enzyme may enhance or be required for activity *in vivo* (Liu and Altman, 1995; Sharin *et al.*, 2005). The targeting sequences of M1GS ribozymes are not involved in catalysis, unlike like minimal *trans*–hammerhead and –hairpin ribozymes, where the sequences of the hybridizing arms are responsible for part of their unpredictable variation in activity. Also, like other *trans* ribozymes, the mRNA cleavage site must be accessible for binding with N-terminal guide sequence. However, the *E. coli* M1 RNA is relatively large – 377 bases. Stable expression requires an active promoter that confers nuclear localization, typically H1 (~100 bp) or U6 (~300 bp), leading to an expression cassette that is ~500–700 bases. This size can become problematic in combinatorial strategies using vectors with size limitations, and is far too large to be synthesized for use in applications requiring exogenous delivery. Nonetheless, this approach has been used to silence a variety of cellular oncogenic and viral messages in human cells (Cobaleda and Sanchez-Garcia, 2000; Guo *et al.*, 2003; Kim and Liu, 2004; Kim *et al.*, 2004a; Kim *et al.*, 2004b; Liu and Altman, 1995; Trang *et al.*, 2000a; Trang *et al.*, 2003; Trang *et al.*, 2001; Trang *et al.*, 2000b; Yu *et al.*, 2005; Zou *et al.*, 2004b; Zou *et al.*, 2003) and is also being explored for applications in bacterial systems (Guerrier-Takada *et al.*, 1997; Li *et al.*, 1992; McKinney *et al.*, 2001; McKinney *et al.*, 2004; Soler Bistue *et al.*, 2007). Further details of M1GS applications in eukaryotic cells can be found in several recent reviews (Gopalan *et al.*, 2002; Guerrier-Takada and Altman, 2000; Kim and Liu, 2004; Raj and Liu, 2003; Trang *et al.*, 2004; Zou *et al.*, 2004a).

While simple EGS:target complexes (Fig. 14.3E) are not substrates for human RNase P, an EGS that reconstitutes additional critical elements of the tRNA structure when complexed with the target can recruit eukaryotic RNase P; several EGS variants have been described.

The '¾ EGS' (Fig. 14.3A) contains structural analogues of a TψC-stem/loop, variable loop, and anti-codon stem/loop and form the 5′ leader, acceptor stem- and D stem-like structures upon annealing to the target (Yuan *et al.*, 1992). The UUC trinucleotide, identical to the TψC invariant nucleotides prior to post-transcriptional modification, is required for RNase P recognition and cleavage activity mediated by the ¾ EGS (Yang *et al.*, 2006; Yuan and Altman, 1994; Zhou *et al.*, 2002; Zhu *et al.*, 2004). Subsequent studies showed that the anti-codon-like stem/loop was not only dispensable for ¾ EGS activity, but also that its deletion to form a ¾ ΔAC EGS (Fig. 14.3B) could actually improve cleavage efficiency (Kawa *et al.*, 1998; Yuan and Altman, 1994; Yuan and Altman, 1995). The variable loop is sometimes also reduced in size (with respect to the parent tRNA) (Barnor *et al.*, 2004; Hnatyszyn *et al.*, 2001; Ma *et al.*, 2000) although deletion of both the acceptor stem and portions of the variable loop led to severe loss of activity using single molecule tRNA model substrates (Yuan and Altman, 1995).

Human RNase P determines the substrate cleavage site by a 'measurement' of the both the combined length of the acceptor and TψC stems and the measurement of the acceptor stem determined by placement of the structural discontinuity (a bulge in small EGS; see below) between the two helices. Insertion of two nucleotides in the acceptor stem causes a two-base shift downstream of the cleavage site; the same insertion into the T-stem causes only a single base shift (Yuan and Altman, 1995). The requirement for a discontinuity between the two helices is consistent with the inability of human eukaryotic RNase P to cleave stem EGS complexes which are substrates for the *E. coli* enzyme (Fig. 14.3E; see discussion above). In small EGS model substrates (see below), extension of either helix by more than one base also reduced RNase P cleavage efficiency (Werner *et al.*, 1998), although efficiency of cleavage is not compromised with all small/EGS model substrates (Yuan and Altman, 1995). The effect of extending the base-pairing of the acceptor stem-like helix in ¾ EGS:substrate complexes has not been addressed systematically in the literature to our knowledge, although there are reports of an active ¾ ΔAC

EGS that forms a 10 base-pair acceptor stem helix with its substrate (Hnatyszyn et al., 2001; Kraus et al., 2002).

However, even smaller substrates consisting of a structure resembling an acceptor stem, bulge, and T-stem/loop (Fig. 14.3C) can be cleaved by human RNAse P in vitro (Carrara et al., 1995; Yuan and Altman, 1995). The 3′ RCCA required by E. coli RNase P is not necessary for human RNase P cleavage of the eukaryotic minimal EGS in vitro (Carrara et al., 1995) or for human RNAse P cleavage in general (Altman and Kirsebom, 1999); this may reflect in H1 RNA the absence of the known RCCA binding loop in M1 RNA (Chen and Pace, 1997). The structural bulge in the minimal human substrate is critical for RNAse P cleavage and both size and sequence affect cleavage activity (Carrara et al., 1995; Yuan and Altman, 1995). Subsequently, Werner et al. demonstrated that a bimolecular version of the minimal substrate, equivalent to opening the TψC loop after the UUCR sequence (Fig. 14.3C, D), could be cleaved by human RNase P (Werner et al., 1998). Thus, an EGS as short as 13 nucleotides can form a complex with an RNA target (Fig. 14.3D) and direct RNAse P cleavage even with chemical modifications that reduce sensitivity to nuclease degradation, leading to the possibility of introducing synthetic EGS against cellular targets (Werner et al., 1998).

Expressed EGS have been designed to target cellular (Dreyfus et al., 2004; Kovrigina et al., 2003) and viral transcripts based on these pioneering studies characterizing human RNase P activity. Long-term inhibition of HIV replication was reported in Molt 4 CD4+ T (M4C8) cells transduced by a MoMLV retroviral construct bearing a tRNA[Val]– ¾ ΔAC EGS expression construct (Hnatyszyn et al., 2001), designed to target a highly conserved site (Leavitt et al., 1994) in the U5 region of HIV. In addition, increased cell viability, inhibition of cytopathological effects, and reduced HIV replication (as monitored by p24 gag antigen production) was observed when tRNA[Val]– ¾ ΔAC EGS U5-expressing M4C8 cells were challenged with clinical HIV-1 isolates representing four different clades in separate experiments (Kraus et al., 2002). Two recent studies from the Liu lab (Li et al., 2006a; Yang et al., 2006) document the ability of ¾

mEGS to effectively inhibit human cytomegalovirus (HCMV) growth in human cells, a human herpes virus that causes disease in newborns and complications in immunocompromised patients. In one report (Li et al., 2006a), a ¾ mEGS derived from tRNase[Tyr] was designed to target a sequence present in the coding region of two capsid proteins: capsid scaffolding protein (CSP) and assemblin. In U373MG cells transduced with a LXSN retroviral vector expressing the ¾ mEGS from a U6 promoter, CSP and assemblin mRNAs and protein levels were reduced 75–80%, and HCMV growth was suppressed 800-fold. The substrate specificity of the ¾ mEGS is evident in that it blocks HCMV capsid formation, but permits earlier events in HCMV growth cycle such as genome replication. In the second report (Yang et al., 2006), a ¾ mEGS of the same type was designed to target sequence was in the N-terminal region common to transcripts encoding the essential immediate early genes 1 and 2 (IE1, IE2), regulatory proteins that are major activators of HCMV transcription rather than structural genes. An important aspect of RNase P recognition of endogenous pre-tRNA is tertiary structure but the acceptor stem and D stem/loop sequences of ¾ EGS will of necessity differ from the tRNA on which they are based, since they must be complementary to the target; consequently, the ability of different ¾ EGS/target complexes to recruit RNase P is expected to vary as well. Therefore, after confirming the basic efficacy of the initial ¾ mEGS, the authors employed an in vitro selection strategy (Zhou et al., 2002) where sequences corresponding to the tRNase[Tyr] TψC loop (excepting the UUC) and variable regions were randomized and the IE1/IE2-specific acceptor stem and D stem/loop were held constant. Using the same experimental design as Li et al, the original tRNase[Tyr]-based ¾ EGS reduced IE1/IE2 expression 80% and viral growth 150-fold. In contrast, the ¾ mEGS variant reduced IE1/IE2 expression 93% and viral growth 3000-fold.

Since reviews of EGS design are published elsewhere (Gopalan et al., 2002; Guerrier-Takada and Altman, 2000; Raj and Liu, 2004; Trang et al., 2004; Zou et al., 2004a) we will primarily consider the parameters that are primarily relevant to the decision to select the RNAse P EGS method for mRNA knockdown. EGS require an accessible target site for efficient RNA cleavage. While *in silico* secondary structure prediction methods provide a starting point, the cellular environment profoundly affects mRNA structure and the availability of potential target sequences. Unstructured mRNA sequences can be identified in a biologically relevant context by probing with complementary oligodeoxynucleotides (ODN) in cell extracts (Scherr *et al.*, 2001; Scherr *et al.*, 2000; Scherr and Rossi, 1998). Only accessible sites will permit annealing of complementary ODNs and become sensitive to RNase H degradation, detectable by quantitative PCR. Alternatively, whole cells can be incubated in dimethyl sulphate (DMS) which methylates A, C and G residues (Liu and Altman, 1995; Yang *et al.*, 2006; Zaug and Cech, 1995) and accessible sites mapped by primer extension assays.

There are some target sequence constraints to be considered when selecting RNase P EGS target sequences, which differ based on the type of EGS to be used. For all EGS types, bases 5′ and 3′ to the cleavage site on the target mRNA must be a pyrimidine and guanosine, respectively (Liu and Altman, 1996; Werner *et al.*, 1998; Yen *et al.*, 2001); these correspond to the −1 and +1 positions of a natural pre-tRNA substrate using the standard tRNA numbering system (compare Figs 14.2B and 14.3A, B, D). When using either a ¾ EGS or a ¾ ΔAC EGS (Fig. 14.3A, B), the target +1 G and the next 6 bases pair with the 3′ terminus of the EGS to form an acceptor stem-like helix, but the following base (+8) of the target is usually, but not always, a U and unpaired (Li *et al.*, 2006a; Yang *et al.*, 2006; Yuan and Altman, 1994); base +9 may also be unpaired; and the next 4–7 bases anneal with the 5′ region of the ¾ mEGS to form the D stem-like structure. The D, rather than the acceptor stem is usually extended to increase the total length of target complementarity; this avoids potential problems associated with altering the site and efficiency of cleavage, as already discussed. Extension of the D-stem hybridization arm by more than a few bases, however, could conceivably effect folding and recognition of the EGS:target complex by RNase P.

The ¾– and ¾ ΔAC EGS guide sequences also have some constraints. The 5′ and 3′ ends

are the same: typically a string of 6 unpaired nucleotides at the 5′ end (which may be derived from natural tRNA D loop nucleotides) and an unpaired 3′ base analogous to the tRNA discriminator nucleotide. Some EGS also contain an additional 3′ CCA sequence; though unnecessary for cleavage by human RNase P, inclusion does not abrogate cleavage. The remainder of the ¾ EGS encodes the bases for the TψC stem/loop, variable loop and acceptor stem, derived from a natural tRNA such as tRNA^Ser or tRNA^Tyr, although theoretically, any tRNA isotype could be used. However, the only sequence in the TψC-like loop that is absolutely required is the UUC trinucleotide; changes introduced by *in vitro* SELEX of the remaining loop nucleotides can improve activity (Yang *et al.*, 2006; Yuan and Altman, 1994; Zhou *et al.*, 2002). *In vitro* SELEX of variable loop and anti-codon stem/loop sequences can significantly improve target knockdown mediated by both the ¾ EGS and ¾ ΔAC EGS *in vitro* and *in vivo* (Yang *et al.*, 2006; Yuan and Altman, 1994; Zhou *et al.*, 2002). *In vitro*-selected substitutions generate non-canonical interactions that affect tertiary structure of the hybrids (Yuan and Altman, 1994). These novel tertiary interactions may function to correct the adverse effects due to the non-natural bimolecular substrate and sequence. Alternatively, SELEX optimization may allow the elimination of tertiary interactions that are suboptimal for RNase P processing alone, but which are present in the natural pre-tRNA structure as a structural compromise to permit the multiple enzymatic steps required to make a mature tRNA (Yuan and Altman, 1994). From a practical standpoint, it may be most efficient in the long run to design the EGS taking into account the requirements for specificity and knockdown for a particular target, and after initial testing, optimize cleavage by SELEX whenever maximum target reduction is essential.

All *in vivo* testing of ¾– and ¾ ΔAC EGS constructs require controls for discriminating between mere antisense and true RNase P-mediated effects. This is achieved by alterations in the TψC loop known to abrogate endogenous RNase P cleavage. A variety of alternatives have been used. The entire stem/loop has been deleted (Hnatyszyn *et al.*, 2001) or nearly to completely replaced (Altman and Kirsebom, 1999; Frank and Pace, 1998; Guerrier-Takada and Altman, 2000; Zhu *et al.*, 2004). The conserved UUC can be replaced by an alternate trinucleotide such as AAG (Dunn and Liu, 2004; Dunn *et al.*, 2001; Li *et al.*, 2006a; Yang *et al.*, 2006) or deleted altogether(Barnor *et al.*, 2004; Endo *et al.*, 2001; Ikeda *et al.*, 2006). The minimal required change to eliminate activity is UUC to UUG (Fig. 14.3A) (Guerrier-Takada and Altman, 2000). In addition, an irrelevant EGS control may be important in some systems, particularly when using transient transfection assays.

The short EGS (Fig. 14.3D), while designed primarily to be used as a synthetic, exogenously introduced molecule(Ma *et al.*, 2000; Ma *et al.*, 1998), can be effective when expressed (Barnor *et al.*, 2004; Ikeda *et al.*, 2006). There are more target sequence constraints for the short EGS than the ¾– and ¾ ΔAC EGS. As in all the EGS, the 1 and +1 bases of the EGS:target complex must be a pyrimidine and G. However, the last base pair in the T-like stem should be a G:C (to stabilize the TψC-like stem) and the following bases must be UUCR. Studies of the uni-molecular model substrate (Fig. 14.3C) have demonstrated that the length and sequence of the bulge is crucial to activity. The bulge is derived from the target in the EGS:target complex, so the sequence is fixed; only the connector length can potentially be adjusted by the positioning of the hybridizing arms. Studies of the uni-molecular model substrate suggest that maximum activity is obtained with a bulge length of 9 nucleotides (Carrara *et al.*, 1995; Yuan and Altman, 1995) although seven bases was sufficient in a study of the bimolecular system (Werner *et al.*, 1998). Extending either the acceptor stem– and TψC–stem-like helix by more than 1 base pair to increase binding specificity may have adverse effects on RNase P cleavage (Werner *et al.*, 1998), limiting the total number of complementary nucleotides to 13. To summarize: the target sequence should begin (from the 5′ end) with a pyrimidine, followed by 7–8 bases, beginning with a G, that anneal to the 3′ end of the EGS; then 7–9 (or slightly longer) unpaired nucleotides; followed by 5–6 bases ending with a G or C that complement the 5′ end of the EGS; and finally, an (unpaired) UUCR sequence.

Target binding is one of the rate-limiting factors for EGS cleavage; therefore, it is important that EGS be expressed at high levels. In addition, RNase P is highly localized in the nucleolus and is present in the nucleoplasm, and possibly in other cellular compartments as well (Jarrous, 2002). For these reasons EGS sequences are typically expressed *in vivo* using the U6 or H1 promoter; inducible U6 (Kovrigina *et al.*, 2005) and tRNA^Met (Barnor *et al.*, 2004; Endo *et al.*, 2001) promoters have also been used. In some cases, it may be advantageous to express an EGS using an RNA polymerase II (RNA pol II) promoter, although a considerable amount of flanking sequence will be appended to the EGS. This difficulty can be solved, however (Guerrier-Takada and Altman, 2000); in one example an EGS against glutamate receptor subunit NR1 was expressed using a recombinant Herpes simplex virus (HSV) vector. The EGS sequence was bracketed by autocatalytic hammerhead ribozymes, which released the EGS with specific 5′ and 3′ ends by *cis* cleavage from the primary transcript (Yen *et al.*, 2001).

The RNase P EGS methodology shares many of same advantages and disadvantages of RNAu. The robust cellular processing system and nuclear targeting are in the former category. The method can also be applied in non-mammalian species (Pei *et al.*, 2007). Disadvantages include 1) the necessity for accessible target sequences 2) limitations on the length and sequence constraints of the target, and 3) the requirement for high levels of expression. In addition, *in vitro* SELEX may be required to optimize activity for some EGS/target combinations in applications where maximum target cleavage is desirable.

There is little information regarding relative efficacies by direct comparisons of RNase P- and other RNA-based technologies, but two reports suggest that RNAi-mediated target knockdown is generally higher than ¾ EGS (Gibbons *et al.*, 2004; Zhang and Altman, 2004), although in one case, maximal levels of inhibition were achieved more rapidly with EGS than RNAi-mediated knockdown (Zhang and Altman, 2004). For the time being, the issue of the comparative efficiency of these two approaches awaits a larger body of data. RNase-P EGS may also be valuable in combinatorial knockdown strategies (see below).

tRNase Z^L^-based small guide RNAs (sgRNAs)

The tRNA 3′ processing enzyme, tRNase Z, occurs in long and short forms, tRNase Z^L and tRNase Z^S, respectively; all eukaryotes have the long form, and some have both (for nomenclature see Vogel et al., 2005). The ELAC1 and ELAC2 genes encode the human homologues HutRNase Z^S and HutRNase Z^L, respectively (Schiffer et al., 2002; Takaku et al., 2003; Tavtigian et al., 2001); however, HutRNase Z^L is the predominant form that processes nuclear-encoded pre-tRNA (Yan et al., 2006) and probably mitochondrial pre-tRNAs as well, as it contains a mitochondrial import signal (Dubrovsky et al., 2004; Levinger et al., 2004b; Yan et al., 2006). A number of in vitro studies defined the structural characteristics of mammalian tRNase Z^L substrates which were subsequently used as models for sgRNAs (Levinger et al., 2004b; Mohan et al., 1999; Nashimoto, 1995; Nashimoto, 1996; Nashimoto, 1997; Nashimoto, 2000; Nashimoto et al., 1998; Nashimoto and Kaspar, 1997; Nashimoto et al., 1999a; Nashimoto et al., 1999b; Nashimoto et al., 1999c; Shibata et al., 2005; Takaku et al., 2004).

For our purposes we will discuss four related types of sgRNAs (Fig. 14.4A–D): the 5′-half tRNA (5′½ -sgRNA) (Nashimoto, 1996); 14-nucleotide linear RNA (lin-sgRNA) (Shibata et al., 2005); RNA heptamer (hep-sgRNA) (Nashimoto et al., 1998) (Nashimoto, 2000; Shibata et al., 2005); and hook-sgRNA (hk-sgRNA) (Shibata et al., 2005; Takaku et al., 2004). The sequence of a 5′½ -sgRNA is based, as its name implies, on the first half of a tRNA (Fig. 14.4A). The first seven nucleotides are complementary to the target, the next ~20 nucleotides are derived from an endogenous tRNA D stem/loop, and the remaining five nucleotides also complementary to the target; thus the annealed sgRNA/target regions form structures analogous to the wildtype tRNA acceptor and anti-codon stems. The target sequence must contain a stem/loop structure between the two discontinuous annealed sequences to provide a structure corresponding to a TψC-stem/loop.

A tRNase Z^L minimal model substrate (Fig. 14.4E) consists of a unimolecular 12 base-pair stem/loop designed to mimic the continuous helix formed by the acceptor and TψC helix stems composing one arm of the canonical tRNA L-form tertiary structure (Fig. 14.1B), with the addition of a 3′ trailer (Nashimoto et al., 1998). The remaining sgRNAs are various bimolecular sgRNA/target configurations that reconstitute a structure similar to a micro-pre-tRNA. The simplest derivative is the lin-sgRNA (Fig. 14.4B), which is identical to the micro-pre-tRNA without the loop; the addition of two extra base pairs increases specificity without seriously compromising cleavage activity (Shibata et al., 2005). The heptamer-sgRNAs (Fig. 14.4C) require a 5 base pair stem/loop in the target, which forms the base of the 12-base pair helix; the hep-sgRNA anneals to the next 7 target nucleotides immediately 3′ to the mRNA stem/loop, completing the base-pairing of the upper part of the stem. The hook sgRNA/target complexes (Fig. 14.4D) have the inverse structural relationship; the hk-sgRNA forms the stem/loop, and single-stranded 5′ leader in the hk-sgRNA complexes with the target to form the complete 12 base pair stem/loop. The ability of these complexes to act as substrates is consistent with the observations that the most important structural features for tRNase Z^L recognition and cleavage are an approximately 12 base pair helix similar to that formed by alignment of the acceptor stem and TψC loop, plus a single-stranded 3′ trailer (Nashimoto et al., 1999b); cleavage occurs after the first 3′ unpaired base in the complex which corresponds to the discriminator nucleotide in wild-type pre-tRNAs. In contrast to RNase P substrates, a bulge between the seventh and eighth base pairs is not required and an ACCA trailer sequence actually inhibits RNase Z^L activity.

While the results of in vitro studies provided strong support for the hypothesis that tRNase Z^L is involved in sgRNA-directed mRNA knockdown, direct evidence was published only recently (Nakashima et al., 2007). The authors derived independent 293 cell lines that specifically expressed tRNase Z^L levels that were 2-fold higher and 4-fold lower than the parent cell line by overexpression and shRNA-mediated reduction, respectively, while tRNase Z^S levels were unaffected. Similar control tRNase Z^S over- and underexpressing 293 cell lines were also derived. 2′-O-methyl synthetic derivatives of 5′½ – and

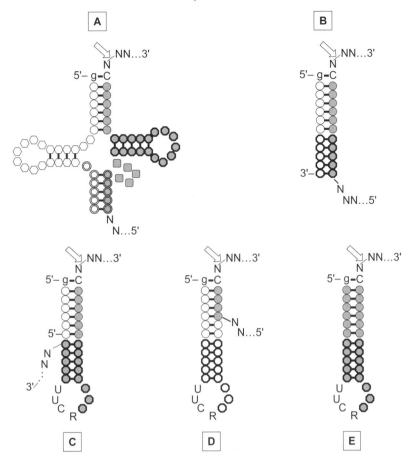

Figure 14.4 RNase Z^L sgRNA designs, shown as complexes with target sequences, depicted diagrammatically with respect to tRNA secondary structure in Fig. 14.2B. Target RNA is denoted by filled polygons and capital font; sgRNA sequence is denoted by open polygons and small case font. Cleavage site, denoted by open block arrow, is typically after the analogue of the discriminator base (Fig. 14.2B), the first unpaired 3′ nucleotide after the acceptor stem-like helix. (A) 5′-half tRNA (5′½ -sgRNA). (B) 14-nucleotide linear RNA (lin-sgRNA). (C) RNA heptamer (hep-sgRNA). (D) hook-sgRNA (hk-sgRNA). (E) tRNase Z^L minimal unimolecular model substrate. See text for full details.

lin-sgRNAs targeting the same sequence in a firefly luciferase (Fl-luc) reporter or a 2′-O-methyl hep-sgRNA targeting chloramphenicol acetyltransferase (CAT) sequence were transiently co-transfected with plasmids expressing their respective target mRNAs in each of the cell lines. The degree of reporter downregulation was directly proportional to the level of tRNase Z^L expression in all cases and was unaffected in cell lines with manipulated tRNAs Z^S levels. In addition, they demonstrated that stable expression of 5′½- and lin-sgRNAs (against the same target sequence) mediated knockdown of endogenous Bcl-2 and GSK-3B (a serine threonine kinase) mRNAs, although the final protein levels were not reported. Curiously, the level of GSK-3B

mRNA was only reduced to 43%, although protein levels were much lower, indicating that a non-cleavage repression mechanism was involved. In addition, a synthetic hep-sgRNA mediated successful knockdown of reporter expression in mice, providing proof of principle for the possibility of *in vivo* applications of this technology.

5′½-sgRNAs have been designed that take advantage of well-known structural features of HIV-1 genomic RNA (Habu *et al.*, 2005). The HIV-1 5′ leader region of contains a region known as the HIV packaging signal (HIVψ signal, not to be confused with psuedouridine, ψ, of the tRNA TψC loop, above). HIVψ contains a cluster of four stem–loops, designated SL1 to SL4, whose structures are highly conserved, bind Gag with high affinity, and are required in *cis* for efficient viral assembly. Habu *et al.* designed 5′½ -sgRNAs based on tRNA^{Arg} that were complementary to sequences surrounding SL3 (also known as the ψ stem/loop) and SL4, which is found immediately downstream of the *gag* AUG, as well as target in the *gag* coding region. These sgRNAs were stably expressed in transduced Jurkat cells from a human tRNA^{Met} promoter in a Moloney murine leukaemia virus MoMLV-based retroviral vector backbone. The SL4 5′½ -sgRNA significantly inhibited HIV-1 replication in infectious challenge assays.

Some design generalizations can be drawn from these experiments, although more data is required before they can be considered absolute. 1) 5′½ -sgRNAs work better *in vivo* than their corresponding lin-sgRNAs (although see below) (Nakashima *et al.*, 2007). 2) For 5′½ -sgRNAs, targets that contain a stem loops that more closely mimic a TψC stem/loop, particularly the loop invariant nucleotides, are better substrates (Habu *et al.*, 2005; Nakashima *et al.*, 2007; Nashimoto, 1996). Structural parameters (such as stem base/pairing) and loop sequence are both factors, but structure is probably the more important of the two. Note that this is in contrast to sequence requirements for RNase P EGS (see above), reflecting different sequence/structural requirements for substrate recognition and enzymatic activity. 3) Nuclear localization of the sgRNAs is highly preferable and possibly essential, since 3′ end processing of native nuclear pre-tRNAs occurs in the nuclear compartment. 4) Efficacy is proportional to the expression level.

While the experiments to date indicate the potential of tRNase Z^L-recruiting sgRNAs for *in vivo* knockdown of therapeutic targets, work remains to be done before rational design is a reality. The application of RNase Z^L-sgRNAs has the advantage of utilizing an efficient cellular processing system in the nucleus. However, target accessibility may be important, which could explain in part why lin-sgRNAs are less effective against sequences with the appropriate structure for targeting by 5′½-sgRNAs (Nakashima *et al.*, 2007). Target sequence length is constrained to 12–14 nucleotides, as cleavage efficiency decreases with longer sequences, raising concerns about specificity and off-target effects, although

5′½-sgRNAs targets must also contain a stem/loop. Most of the work on efficiency of substrate cleavage has been done *in vitro*, but systematic *in vivo* studies on the effects of target mismatches, specific sequences, and promoters, in addition to the factors already mentioned, are lacking. For instance, in work already discussed (Habu *et al.*, 2005), an anti-HIV-1 *gag* 5′½ -sgRNA mediated poorer target knockdown *in vivo* than a SL4 5′½ -sgRNA, which paralleled their relative efficiency of substrate cleavage by tRNase Z^L *in vitro*. The lower efficiency of the *gag* 5′½ -sgRNA was attributed to an unstable stem in the target site, an interpretation consistent with other observations (Levinger *et al.*, 2003). However, this interpretation was not verified even *in vitro* by assays of the same substrate with mutations that increased stem base-pairing (Habu *et al.*, 2005); this type of information has crucial implications both for target site selection and the evolution of escape mutants. Likewise, there are no side-by-side comparisons of the efficiency of the same sgRNA expressed from the commonly used pol III promoters, U6, H1 and various tRNAs (which can be used in different configurations). And while one report claims that sgRNA and siRNA technologies can have comparable efficiencies, equivalent effects were only observed using synthetic molecules at very different concentrations (Nakashima *et al.*, 2007).

In summary, RNAu, RNase P- and RNase Z-recruiting technologies show promise as additional methods of therapeutic mRNA knockdown. Direct comparisons of these technologies, especially with RNAi in long-term expression assays would be informative. In addition, more thorough assessments of cellular immunological responses and toxicity are required before these methods can be considered for therapeutic applications.

RNAi

RNA interference (RNAi) was initially described in the nematode *Caenorhabditis elegans* (Fire et al., 1998). These studies relied on long double-stranded RNA (dsRNA) to trigger RNAi. However, in mammalian cells, dsRNAs longer than 30 base pairs can activate the interferon pathway, leading to global downregulation of translation and RNA degradation. The discovery that synthetic, exogenously delivered short interfering 21-mer RNA duplexes (siRNAs) mediate RNAi in mammalian cells (Elbashir *et al.*, 2001) by accessing the conserved endogenous microRNA pathway, opened the door to applications of RNAi in higher eukaryotes. The demonstration that expressed siRNAs and short hairpin RNAs (shRNAs) could also trigger RNAi in mammalian cells soon followed (Brummelkamp *et al.*, 2002; Lee *et al.*, 2002; Miyagishi and Taira, 2002; Paddison *et al.*, 2002; Paul *et al.*, 2002). Subsequently, RNAi has become the method of choice for knockdown of mRNA transcripts for a variety of genetic and therapeutic applications due to its potency and specificity. Moreover, small duplex RNAs have been shown to mediate transcriptional regulation in mammalian cells using components of the RNAi machinery. Once the mechanism is better understood, it may be possible to adapt it as another mode of mRNA inhibition at the level of transcription (Janowski *et al.*, 2007) (Kehayova and Liu, 2007; Li *et al.*, 2006b; Morris, 2006; Rossi, 2007; Rusk, 2007). A comprehensive treatment of this subject is beyond the scope of this manuscript, but the details of siRNA and shRNA design are covered in many excellent reviews; we cite a selection of recent ones that provide an introduction to the topic in general and therapeutic applications in particular (Cullen, 2006; Kim and Rossi, 2007; Morris and Rossi, 2006; Pei and Tuschl, 2006; Snove and Rossi, 2006; Wiznerowicz *et al.*, 2006). We focus on aspects of RNAi applications that pertain to design strategies and therapeutic applications.

While the antisense (or guide) and sense (or passenger) strand can be expressed separately from individual promoters and anneal to form siRNA intracellularly, it is less cumbersome to express a single shRNA. Target choice is crucial; the thermodynamic properties (Khvorova *et al.*, 2003; Schwarz *et al.*, 2003), base composition, location of specific nucleotides and other characteristics are associated with 'good' si- and shRNAs (reviewed in Pei and Tuschl, 2006). Most shRNAs are expressed using U6 or H1 promoters, since transcription of the shRNA can start with the first base of the stem and the sequence terminates with a few uridines; the 5′ leader and 3′ trailer sequences that are part of the typical

pol II transcript can adversely affect the siRNA thermodynamic profile. Although U6 transcripts are primarily nuclear, shRNAs are exported to the cytoplasm by exportin 5 through recognition of a mini-helix motif where they enter RISC. While shRNAs with only 19 base-pair stems can be effective, shRNAs with base-paired stems long enough to be efficiently cleaved by Dicer may mediate more efficient target knockdown, based on studies of synthetic siRNAs, (Kim *et al.*, 2005; Rose *et al.*, 2005; Siolas *et al.*, 2005). The most common hairpin design uses the sense-loop-antisense orientation. Dicer enters the shRNA from the base of the hairpin (when the stem is of sufficient length) (Siolas *et al.*, 2005; Vermeulen *et al.*, 2005); therefore, in the sense-loop-antisense orientation the base of the stem forms what should be the more stable end of the final siRNA. The sense strand may contain all canonical Watson-Crick base pairs, or G:U wobble bases that aid construct sequencing, reduce unwanted immunostimulation, and potentially aid strand selectivity. The size and sequence of the loop can also affect activity.

Toxicity of expressed shRNAs can commonly arise from three sources: first, activation of innate cellular immunity; second, off-target affects; third, saturation of the endogenous microRNA pathway. Double-stranded RNA (dsRNA) can lead to a type 1 interferon response through activation of dsRNA-dependent protein kinase (PKR) and retinoic-acid-inducible gene-I (RIG-I) (Garcia-Sastre and Biron, 2006), which has implications for RNAi applications (de Veer *et al.*, 2005; Karpala *et al.*, 2005). Indeed, synthetic siRNAs that contain certain sequence motifs 5′-GUCCUUCAA-3′ or 5′-UGUGU-3′ stimulate the innate immune response (Hornung *et al.*, 2005; Judge *et al.*, 2005), although siRNAs without these motifs have been reported to induce inflammatory cytokine production in some cell types, depending on the transfection method (Yoo *et al.*, 2006). However, the immunostimulatory properties of synthetic siRNAs can be eliminated by incorporation of modified nucleotides into one strand (Judge *et al.*, 2006). However, expressed shRNAs (Robbins *et al.*, 2006) and long hairpin RNAs (lhRNAs) with stems longer than 30 base pairs that contain G:U mismatches also do not induce interferon responses in some cell types (M. Sano, personal communication).

Off target effects are potentially severe due to the extraordinary efficiency of RNAi. The thermodynamic asymmetry of the duplex determines which strand is incorporated into the RNA-induced silencing complex (RISC). When both strands of an siRNA are incorporated into RISC due to poor strand selectivity, the potential off-target sequence space is doubled. In addition, the passenger strand competes with the guide strand for access to RISC and reduces knockdown of the desired target mRNA. Relative strand selectivity should be assessed, to a first approximation, in reporter assays such as psiCHECK (Promega) that measure the targeting ability of each strand independently, before moving to a more biologically relevant assay. While protein reduction is the desired endpoint, it should occur in concert with knockdown of the cognate mRNA; the absence of mRNA degradation can be an indication of an indirect, off-target effect. While some target:RNAi mismatches, such as those near the mRNA cleavage site, are known to be deleterious, the effect of other target:RNAi mismatches is less understood, which also impacts the ability to accurately predict off-target effects, particularly since mismatched shRNAs have the potential to act as microRNAs (Doench et al., 2003). Seven bases of complementarity at the 5′ end of the microRNA (bases 2–8, referred to as the seed sequence) can be sufficient to confer regulation of a target (Brennecke et al., 2005), although perfect seed pairing is not required (Didiano and Hobert, 2006; Miranda et al., 2006). MicroRNA down-regulation is much more efficient when multiple binding sites are present in an mRNA 3′ UTR with optimal spacing, (Saetrom et al., 2007), but it is not yet possible to predict microRNA target sites with high accuracy, particularly since most mRNAs are thought to be targeted by multiple, different microRNAs, which may interact cooperatively. Consequently, RNAi off-target effects can be difficult to predict in any specific instance. However, in general, RNAi is extremely specific, particularly when intracellular siRNA concentrations (expressed or exogenously introduced) are minimized.

Ectopically expressed shRNAs can also cause cellular toxicity *in vivo* by saturating the RNAi pathway and denying endogenous microRNAs access to crucial processing and transport components (An et al., 2006; Grimm et al., 2006). Replacing the U6 promoter with the weaker H1 promoter eliminated the toxic effect of one shRNA(An et al., 2006). We have recently described a chimeric tRNA^{Lys3}–shRNA expression system that produces varying levels of mature shRNA (Scherer et al., 2007), which may be useful in avoiding this type of toxicity. Alternatively, inducible promoters may be appropriate in some settings; our lab has described a system in which an shRNA targeting *rev* was expressed using a TAT-inducible HIV-1 LTR/*Drosophila* hsp70 minimal heat shock promoter. The promoter is only active in HIV-1 infected cells and upon induction produced an siRNA that inhibited HIV replication in cultured T-lymphocytes and monocytes (Unwalla et al., 2006; Unwalla et al., 2004). Inducible H1 and U6 promoters have also been described for RNAi gene silencing in plasmid (Henriksen et al., 2007) and lentiviral vector backbones (Aagaard et al., 2007).

In light of these issues, one practical scheme for choosing and testing shRNAs might be to: 1) select a the set of sequences within the target mRNA to consider, based on the biology of the system; 2) interrogate those sequences for potent RNAi trigger sequences using an on-line program (Pei and Tuschl, 2006); 3) run a BLAST search and possibly a microRNA target screen for obvious off-target hits; 4) design the expression constructs and perform reporter assays; and 5) test at least 2 shRNAs against the same target for phenotypic convergence in a biologically relevant assay. For clinical applications it is also important to 6) check for indicators of interferon responses; and 7) test directly for off-target effects using array analysis and/or animal models.

Combinatorial approaches and gene therapy of HIV

Although effective shRNAs can mediate very potent and specific knockdown, a single base change within the target sequence may eliminate target knockdown. HIV is extremely prone to generating escape mutants in response to shRNAs, even in highly conserved sequences (Boden et al., 2003; Das et al., 2004; Sabariegos et al., 2006), which poses problems for long-term viral inhibition. Moreover, mutations outside the target sequence can allow viral escape due to secondary structure changes in target, although this problem is likely to be less severe for RNAi relative to some of the other RNA knockdown methods already discussed (Westerhout *et al.*, 2005). A combinatorial approach targeting multiple conserved sequences in HIV is likely to be more successful.

Combinatorial therapy can be divided into two classes: multiple targeting employing the same method (simple multiplexing), or different methods of knockdown (compound multiplexing). The combination of two modified U1 snRNAs reduced HIV-1 mRNA levels to a greater degree than either one alone (Sajic *et al.*, 2007). The use of two RNase P EGS against a single target did not improve knockdown compared to single EGS in transient assays (Zhang and Altman, 2004); however, greater efficiency was observed with stable expression of double EGS (Plehn-Dujowich and Altman, 1998). This data, albeit limited, indicates that the use of two RNAu or RNase P EGS effectors does not tax the cellular processing machinery. A number of strategies are currently being developed to express multiple shRNAs including: long hairpin RNAs (lsRNAs) that generate multiple siRNAs; polycistronic shRNAs and microRNA mimics expressed from a single promoter; and triple promoter combinations (reviewed in Kim and Rossi, 2007; Scherer and Rossi, 2007). However, simple RNAi multiplexing strategies require extra care to avoid difficulties attendant upon saturation effects than RNAu, RNase P EGS and tRNAse Z^L sgRNA approaches, particularly since all RNAi toxic effects are concentration dependent.

Compound multiplexing strategies have a great deal of potential in HIV gene therapeutic applications, due to their different modes of action. RISC-mediated RNAi is cytoplasmic, while RNAu, RNase P EGS and tRNase Z^L sgRNA are based in the nucleus; aptamers and ribozymes can function potentially in any cellular compartment. While RNAi is extremely potent, the cellular pathways employed by RNAu, RNase P EGS and tRNase Z^L sgRNA are more robust. RNAi is the most specific and least susceptible to target structure interference of the trigger methods discussed; RNAu, RNase

P EGS and tRNase Z^L sgRNA can use shorter target sequences, which while less specific may be advantageous when conserved sequences are shorter than 19 nucleotides. Various double combinations have been explored, and in general perform better than the single effectors (reviewed in Scherer and Rossi, 2007), but more data is required before generalizations concerning the best tactics can be made.

However, we have described a construct that illustrates compound multiplexing principles. Triple–R is a lentiviral construct containing three anti-HIV expression cassettes: a TAR RNA decoy embedded in a U16 snoRNA directing nucleolar localization, expressed from a U6 promoter; a U6-driven shRNA targeting a sequence in the common exon of *tat* and *rev* transcripts, coding for essential regulatory proteins; and a chimeric VA1- anti-CCR5 *trans*-cleaving hammerhead ribozymes expressed from the VA1 promoter. The three individual inhibitory cassettes exploit a broad range of possibilities: two autonomous effectors and an RNA trigger; a viral protein, viral mRNA, and cellular mRNA target, (coding for the HIV-1 macrophage co-receptor); that are important in both pre-entry and post-entry viral replication stages. T-cell lines and CD34+ stem cell-derived macrophages transduced *in vitro* with Triple-R were more resistant to HIV-1 replication than cells transduced with vector carrying one or two effector constructs (Li *et al.*, 2005). Moreover, haematopoietic stem cells (HSCs) transduced with Triple-R and transplanted in a SCID-hu mouse xenograft model gave rise to phenotypically normal transgenic T cells that were resistant to *ex vivo* HIV infection (Anderson *et al.*, 2007). Human clinical trials with the Triple-R construct in AIDS/lymphoma patients are likely to begin sometime in 2007. In the meantime, new combinations of anti-HIV RNA gene therapeutics will continue to be developed.

References

Aagaard, L., Amarzguioui, M., Sun, G., Santos, L.C., Ehsani, A., Prydz, H., and Rossi, J.J. (2007). A facile lentiviral vector system for expression of doxycycline-inducible shRNAs: knockdown of the pre-miRNA processing enzyme Drosha. Mol. Ther. *15*, 938–945.

Altman, S., and Kirsebom, L.A. (1999). Ribonuclease P. The RNA world. (Cold Spring Harbor, NY: Cold Spring Harbor Laboratory Press), pp. 351–380.

An, D.S., Qin, F.X., Auyeung, V.C., Mao, S.H., Kung, S.K., Baltimore, D., and Chen, I.S. (2006). Optimization and functional effects of stable short hairpin RNA expression in primary human lymphocytes via lentiviral vectors. Mol. Ther. *14*, 494–504.

Anderson, J., Li, M.J., Palmer, B., Remling, L., Li, S., Yam, P., Yee, J.K., Rossi, J., Zaia, J., and Akkina, R. (2007). Safety and Efficacy of a Lentiviral Vector Containing Three Anti-HIV Genes-CCR5 Ribozyme, Tat-rev siRNA, and TAR Decoy-in SCID-hu Mouse-Derived T Cells. Mol. Ther. *15*, 1182–1188.

Bagheri, S., and Kashani-Sabet, M. (2004). Ribozymes in the age of molecular therapeutics. Curr Mol Med *4*, 489–506.

Barnor, J.S., Endo, Y., Habu, Y., Miyano-Kurosaki, N., Kitano, M., Yamamoto, H., and Takaku, H. (2004). Effective inhibition of HIV-1 replication in cultured cells by external guide sequences and ribonuclease P. Bioorg. Med. Chem. Lett *14*, 4941–4944.

Bartolome, J., Castillo, I., and Carreno, V. (2004). Ribozymes as antiviral agents. Minerva Med. *95*, 11–24.

Beckley, S.A., Liu, P., Stover, M.L., Gunderson, S.I., Lichtler, A.C., and Rowe, D.W. (2001). Reduction of target gene expression by a modified U1 snRNA. Mol. Cell. Biol. *21*, 2815–2825.

Boden, D., Pusch, O., Lee, F., Tucker, L., and Ramratnam, B. (2003). Human immunodeficiency virus type 1 escape from RNA interference. J. Virol. *77*, 11531–11535.

Boelens, W.C., Jansen, E.J., van Venrooij, W.J., Stripecke, R., Mattaj, I.W., and Gunderson, S.I. (1993). The human U1 snRNP-specific U1A protein inhibits polyadenylation of its own pre-mRNA. Cell *72*, 881–892.

Brennecke, J., Stark, A., Russell, R.B., and Cohen, S.M. (2005). Principles of microRNA-target recognition. PLoS Biol. *3*, e85.

Brummelkamp, T.R., Bernards, R., and Agami, R. (2002). Stable suppression of tumorigenicity by virus-mediated RNA interference. Cancer Cell *2*, 243–247.

Carrara, G., Calandra, P., Fruscoloni, P., and Tocchini-Valentini, G.P. (1995). Two helices plus a linker: a small model substrate for eukaryotic RNase P. Proc. Natl. Acad. Sci. USA *92*, 2627–2631.

Chaloin, L., Lehmann, M.J., Sczakiel, G., and Restle, T. (2002). Endogenous expression of a high-affinity pseudoknot RNA aptamer suppresses replication of HIV-1. Nucleic Acids Res. *30*, 4001–4008.

Chen, J.L., and Pace, N.R. (1997). Identification of the universally conserved core of ribonuclease P RNA. RNA *3*, 557–560.

Cobaleda, C., and Sanchez-Garcia, I. (2000). *In vivo* inhibition by a site-specific catalytic RNA subunit of RNase P designed against the BCR-ABL oncogenic products: a novel approach for cancer treatment. Blood *95*, 731–737.

Cullen, B.R. (2006). Enhancing and confirming the specificity of RNAi experiments. Nat. Method. *3*, 677–681.

Dahlberg, J.E., and Lund, E. (1987). Structure and expression of U-snRNA genes. Mol. Biol. Rep. *12*, 139–143.

Das, A.T., Brummelkamp, T.R., Westerhout, E.M., Vink, M., Madiredjo, M., Bernards, R., and Berkhout, B. (2004). Human immunodeficiency virus type 1 escapes from RNA interference-mediated inhibition. J. Virol. *78*, 2601–2605.

de Veer, M.J., Sledz, C.A., and Williams, B.R. (2005). Detection of foreign RNA: implications for RNAi. Immunol. Cell Biol. *83*, 224–228.

Didiano, D., and Hobert, O. (2006). Perfect seed pairing is not a generally reliable predictor for miRNA-target interactions. Nat. Struct. Mol. Biol. *13*, 849–851.

Doench, J.G., Petersen, C.P., and Sharp, P.A. (2003). siRNAs can function as miRNAs. Genes Dev. *17*, 438–442.

Dreyfus, D.H., Matczuk, A., and Fuleihan, R. (2004). An RNA external guide sequence ribozyme targeting human interleukin-4 receptor alpha mRNA. Int. Immunopharmacol. *4*, 1015–1027.

Dubrovsky, E.B., Dubrovskaya, V.A., Levinger, L., Schiffer, S., and Marchfelder, A. (2004). *Drosophila* RNase Z processes mitochondrial and nuclear pre-tRNA 3′ ends *in vivo*. Nucleic Acids Res. *32*, 255–262.

Dunn, W., and Liu, F. (2004). RNase P-mediated inhibition of viral growth by exogenous administration of short oligonucleotide external guide sequence. Method. Mol. Biol. *252*, 425–436.

Dunn, W., Trang, P., Khan, U., Zhu, J., and Liu, F. (2001). RNase P-mediated inhibition of cytomegalovirus protease expression and viral DNA encapsidation by oligonucleotide external guide sequences. Proc. Natl. Acad. Sci. USA *98*, 14831–14836.

Elbashir, S.M., Harborth, J., Lendeckel, W., Yalcin, A., Weber, K., and Tuschl, T. (2001). Duplexes of 21-nucleotide RNAs mediate RNA interference in cultured mammalian cells. Nature *411*, 494–498.

Endo, Y., Miyano-Kurosaki, N., Kitano, M., Habu, Y., and Takaku, H. (2001). Effective inhibition of HIV-1 replication in cultured cells by external guide sequences and ribonuclease. Nucleic Acids Res. Suppl. 213–214.

Engelke, D.R., and Hopper, A.K. (2006). Modified view of tRNA: stability amid sequence diversity. Mol. Cell *21*, 144–145.

Evans, D., Marquez, S.M., and Pace, N.R. (2006). RNase P: interface of the RNA and protein worlds. Trends Biochem. Sci. *31*, 333–341.

Famulok, M., and Mayer, G. (2005). Intramers and aptamers: applications in protein-function analyses and potential for drug screening. Chem. Biochem. *6*, 19–26.

Fire, A., Xu, S., Montgomery, M.K., Kostas, S.A., Driver, S.E., and Mello, C.C. (1998). Potent and specific genetic interference by double-stranded RNA in *Caenorhabditis elegans*. Nature *391*, 806–811.

Forster, A.C., and Altman, S. (1990). External guide sequences for an RNA enzyme. Science *249*, 783–786.

Fortes, P., Cuevas, Y., Guan, F., Liu, P., Pentlicky, S., Jung, S.P., Martinez-Chantar, M.L., Prieto, J., Rowe, D., and Gunderson, S.I. (2003). Inhibiting expression of specific genes in mammalian cells with 5′ end-mutated U1 small nuclear RNAs targeted to terminal exons of pre-mRNA. Proc. Natl. Acad. Sci. USA *100*, 8264–8269.

Frank, D.N., and Pace, N.R. (1998). Ribonuclease P: unity and diversity in a tRNA processing ribozyme. Annu. Rev. Biochem. *67*, 153–180.

Furth, P.A., Choe, W.T., Rex, J.H., Byrne, J.C., and Baker, C.C. (1994). Sequences homologous to 5′ splice sites are required for the inhibitory activity of papillomavirus late 3′ untranslated regions. Mol. Cell. Biol. *14*, 5278–5289.

Garcia-Sastre, A., and Biron, C.A. (2006). Type 1 interferons and the virus-host relationship: a lesson in detente. Science *312*, 879–882.

Geslain, R., and Ribas de Pouplana, L. (2004). Regulation of RNA function by aminoacylation and editing? Trends Genet. *20*, 604–610.

Gibbons, D.L., Shashikant, C., and Hayday, A.C. (2004). A comparative analysis of RNA targeting strategies in the thymosin beta 4 gene. J. Mol. Biol. *342*, 1069–1076.

Gilmartin, G.M. (2005). Eukaryotic mRNA 3′ processing: a common means to different ends. Genes Dev. *19*, 2517–2521.

Good, P.D., Krikos, A.J., Li, S.X., Bertrand, E., Lee, N.S., Giver, L., Ellington, A., Zaia, J.A., Rossi, J.J., and Engelke, D.R. (1997). Expression of small, therapeutic RNAs in human cell nuclei. Gene Ther. *4*, 45–54.

Gopalan, V., Vioque, A., and Altman, S. (2002). RNase P: variations and uses. J. Biol. Chem. *277*, 6759–6762.

Gorman, L., Mercatante, D.R., and Kole, R. (2000). Restoration of correct splicing of thalassemic beta-globin pre-mRNA by modified U1 snRNAs. J. Biol. Chem. *275*, 35914–35919.

Grimm, D., Streetz, K.L., Jopling, C.L., Storm, T.A., Pandey, K., Davis, C.R., Marion, P., Salazar, F., and Kay, M.A. (2006). Fatality in mice due to oversaturation of cellular microRNA/short hairpin RNA pathways. Nature *441*, 537–541.

Guerrier-Takada, C., and Altman, S. (2000). Inactivation of gene expression using ribonuclease P and external guide sequences. Method. Enzymol. *313*, 442–456.

Guerrier-Takada, C., Gardiner, K., Marsh, T., Pace, N., and Altman, S. (1983). The RNA moiety of ribonuclease P is the catalytic subunit of the enzyme. Cell *35*, 849–857.

Guerrier-Takada, C., Salavati, R., and Altman, S. (1997). Phenotypic conversion of drug-resistant bacteria to drug sensitivity. Proc. Natl. Acad. Sci. USA *94*, 8468–8472.

Gunderson, S.I., Beyer, K., Martin, G., Keller, W., Boelens, W.C., and Mattaj, L.W. (1994). The human U1A snRNP protein regulates polyadenylation via a direct interaction with poly(A) polymerase. Cell *76*, 531–541.

Gunderson, S.I., Polycarpou-Schwarz, M., and Mattaj, I.W. (1998). U1 snRNP inhibits pre-mRNA polyadenylation through a direct interaction between U1 70K and poly(A) polymerase. Mol. Cell *1*, 255–264.

Guo, R., Zou, P., Fan, H.H., Gao, F., Shang, Q.X., Cao, Y.L., and Lu, H.Z. (2003). Repression of allo-cell transplant rejection through CIITA ribonuclease P+ hepatocyte. World J. Gastroenterol. *9*, 1077–1081.

Habu, Y., Miyano-Kurosaki, N., Kitano, M., Endo, Y., Yukita, M., Ohira, S., Takaku, H., and Nashimoto, M. (2005). Inhibition of HIV-1 gene expression by retroviral vector-mediated small-guide RNAs that direct specific RNA cleavage by tRNase ZL. Nucleic Acids Res. *33*, 235–243.

Harris, M.E., Kazantsev, A.V., Chen, J.L., and Pace, N.R. (1997). Analysis of the tertiary structure of the ribonuclease P ribozyme-substrate complex by site-specific photoaffinity crosslinking. RNA *3*, 561–576.

Held, D.M., Kissel, J.D., Patterson, J.T., Nickens, D.G., and Burke, D.H. (2006). HIV-1 inactivation by nucleic acid aptamers. Front. Biosci. *11*, 89–112.

Helm, M. (2006). Post-transcriptional nucleotide modification and alternative folding of RNA. Nucleic Acids Res. *34*, 721–733.

Henriksen, J.R., Lokke, C., Hammero, M., Geerts, D., Versteeg, R., Flaegstad, T., and Einvik, C. (2007). Comparison of RNAi efficiency mediated by tetracycline-responsive H1 and U6 promoter variants in mammalian cell lines. Nucleic Acids Res. *35*, e67.

Hnatyszyn, H., Spruill, G., Young, A., Seivright, R., and Kraus, G. (2001). Long-term RNase P-mediated inhibition of HIV-1 replication and pathogenesis. Gene Ther. *8*, 1863–1871.

Hopper, A.K., and Phizicky, E.M. (2003). tRNA transfers to the limelight. Genes Dev. *17*, 162–180.

Hornung, V., Guenthner-Biller, M., Bourquin, C., Ablasser, A., Schlee, M., Uematsu, S., Noronha, A., Manoharan, M., Akira, S., de Fougerolles, A., *et al.* (2005). Sequence-specific potent induction of IFN-alpha by short interfering RNA in plasmacytoid dendritic cells through TLR7. Nat. Med. *11*, 263–270.

Hu, J., Lutz, C.S., Wilusz, J., and Tian, B. (2005). Bioinformatic identification of candidate cis-regulatory elements involved in human mRNA polyadenylation. RNA *11*, 1485–1493.

Ikeda, M., Habu, Y., Miyano-Kurosaki, N., and Takaku, H. (2006). Suppression of HIV-1 replication by a combination of endonucleolytic ribozymes (RNase P and tRNnase ZL). Nucleos. Nucleot. Nucl. Acids *25*, 427–437.

Ishizaki, J., Nevins, J.R., and Sullenger, B.A. (1996). Inhibition of cell proliferation by an RNA ligand that selectively blocks E2F function. Nat. Med. *2*, 1386–1389.

Jambhekar, A., and Derisi, J.L. (2007). Cis-acting determinants of asymmetric, cytoplasmic RNA transport. RNA *13*, 625–642.

James, W. (2007). Aptamers in the virologists' toolkit. J. Gen. Virol. *88*, 351–364.

Janowski, B.A., Younger, S.T., Hardy, D.B., Ram, R., Huffman, K.E., and Corey, D.R. (2007). Activating gene expression in mammalian cells with promoter-targeted duplex RNAs. Nat. Chem. Biol. *3*, 166–173.

Jarrous, N. (2002). Human ribonuclease P: subunits, function, and intranuclear localization. RNA *8*, 1–7.

Joshi, P.J., North, T.W., and Prasad, V.R. (2005). Aptamers directed to HIV-1 reverse transcriptase display greater efficacy over small hairpin RNAs targeted to viral RNA in blocking HIV-1 replication. Mol. Ther. *11*, 677–686.

Judge, A.D., Bola, G., Lee, A.C., and MacLachlan, I. (2006). Design of noninflammatory synthetic siRNA mediating potent gene silencing *in vivo*. Mol. Ther. *13*, 494–505.

Judge, A.D., Sood, V., Shaw, J.R., Fang, D., McClintock, K., and MacLachlan, I. (2005). Sequence-dependent stimulation of the mammalian innate immune response by synthetic siRNA. Nat. Biotechnol. *23*, 457–462.

Karpala, A.J., Doran, T.J., and Bean, A.G. (2005). Immune responses to dsRNA: implications for gene silencing technologies. Immunol Cell. Biol. *83*, 211–216.

Kawa, D., Wang, J., Yuan, Y., and Liu, F. (1998). Inhibition of viral gene expression by human ribonuclease P. RNA *4*, 1397–1406.

Kehayova, P.D., and Liu, D.R. (2007). In vivo evolution of an RNA-based transcriptional silencing domain in S. cerevisiae. Chem. Biol. *14*, 65–74.

Khan, A.U., and Lal, S.K. (2003). Ribozymes: a modern tool in medicine. J. Biomed. Sci. *10*, 457–467.

Khvorova, A., Reynolds, A., and Jayasena, S.D. (2003). Functional siRNAs and miRNAs exhibit strand bias. Cell *115*, 209–216.

Kikovska, E., Svard, S.G., and Kirsebom, L.A. (2007). Eukaryotic RNase P RNA mediates cleavage in the absence of protein. Proc. Natl. Acad. Sci. USA *104*, 2062–2067.

Kim, D.H., Behlke, M.A., Rose, S.D., Chang, M.S., Choi, S., and Rossi, J.J. (2005). Synthetic dsRNA Dicer substrates enhance RNAi potency and efficacy. Nat. Biotechnol. *23*, 222–226.

Kim, D.H., and Rossi, J.J. (2007). Strategies for silencing human disease using RNA interference. Nat. Rev. Genet. *8*, 173–184.

Kim, K., and Liu, F. (2004). In vitro selection of RNase P ribozymes that efficiently cleave a target mRNA. Method. Mol. Biol. *252*, 399–412.

Kim, K., Trang, P., Umamoto, S., Hai, R., and Liu, F. (2004a). RNase P ribozyme inhibits cytomegalovirus replication by blocking the expression of viral capsid proteins. Nucleic Acids Res. *32*, 3427–3434.

Kim, K., Umamoto, S., Trang, P., Hai, R., and Liu, F. (2004b). Intracellular expression of engineered RNase P ribozymes effectively blocks gene expression and replication of human cytomegalovirus. RNA *10*, 438–447.

Kirsebom, L.A. (2002). RNase P RNA-mediated catalysis. Biochem. Soc. *Trans*. *30*, 1153–1158.

Kolb, G., Reigadas, S., Castanotto, D., Faure, A., Ventura, M., Rossi, J.J., and Toulme, J.J. (2006). Endogenous expression of an anti-TAR aptamer reduces HIV-1 replication. RNA Biol. *3*, 150–156.

Komatsu, Y. (2004). Regulation of ribozyme activity with short oligonucleotides. Biol. Pharm. Bull. *27*, 457–462.

Kovrigina, E., Wesolowski, D., and Altman, S. (2003). Coordinate inhibition of expression of several genes for protein subunits of human nuclear RNase P. Proc. Natl. Acad. Sci. USA *100*, 1598–1602.

Kovrigina, E., Yang, L., Pfund, E., and Altman, S. (2005). Regulated expression of functional external guide sequences in mammalian cells using a U6 RNA polymerase III promoter. RNA *11*, 1588–1595.

Kraus, G., Geffin, R., Spruill, G., Young, A.K., Seivright, R., Cardona, D., Burzawa, J., and Hnatyszyn, H.J. (2002). Cross-clade inhibition of HIV-1 replication and cytopathology by using RNase P-associated external guide sequences. Proc. Natl. Acad. Sci. USA *99*, 3406–3411.

Kyriakopoulou, C., Larsson, P., Liu, L., Schuster, J., Soderbom, F., Kirsebom, L.A., and Virtanen, A. (2006). U1-like snRNAs lacking complementarity to canonical 5′ splice sites. RNA *12*, 1603–1611.

Leavitt, M.C., Yu, M., Yamada, O., Kraus, G., Looney, D., Poeschla, E., and Wong-Staal, F. (1994). Transfer of an anti-HIV-1 ribozyme gene into primary human lymphocytes. Hum. Gene Ther. *5*, 1115–1120.

Lee, N.S., Dohjima, T., Bauer, G., Li, H., Li, M.J., Ehsani, A., Salvaterra, P., and Rossi, J. (2002). Expression of small interfering RNAs targeted against HIV-1 rev transcripts in human cells. Nat. Biotechnol. *20*, 500–505.

Levinger, L., Giege, R., and Florentz, C. (2003). Pathology-related substitutions in human mitochondrial tRNA(Ile) reduce precursor 3′ end processing efficiency *in vitro*. Nucleic Acids Res. *31*, 1904–1912.

Levinger, L., Morl, M., and Florentz, C. (2004a). Mitochondrial tRNA 3′ end metabolism and human disease. Nucleic Acids Res. *32*, 5430–5441.

Levinger, L., Oestreich, I., Florentz, C., and Morl, M. (2004b). A pathogenesis-associated mutation in human mitochondrial tRNALeu(UUR) leads to reduced 3′-end processing and CCA addition. J. Mol. Biol. *337*, 535–544.

Li, H., Trang, P., Kim, K., Zhou, T., Umamoto, S., and Liu, F. (2006a). Effective inhibition of human cytomegalovirus gene expression and growth by intracellular expression of external guide sequence RNA. RNA *12*, 63–72.

Li, L.C., Okino, S.T., Zhao, H., Pookot, D., Place, R.F., Urakami, S., Enokida, H., and Dahiya, R. (2006b). Small dsRNAs induce transcriptional activation in human cells. Proc. Natl. Acad. Sci. USA *103*, 17337–17342.

Li, M.J., Kim, J., Li, S., Zaia, J., Yee, J.K., Anderson, J., Akkina, R., and Rossi, J.J. (2005). Long-term inhibition of HIV-1 infection in primary hematopoietic cells by lentiviral vector delivery of a triple combination of anti-HIV shRNA, anti-CCR5 ribozyme, and a nucleolar-localizing TAR decoy. Mol. Ther. *12*, 900–909.

Li, Y., Guerrier-Takada, C., and Altman, S. (1992). Targeted cleavage of mRNA *in vitro* by RNase P from Escherichia coli. Proc. Natl. Acad. Sci. USA *89*, 3185–3189.

Liu, D., Donegan, J., Nuovo, G., Mitra, D., and Laurence, J. (1997). Stable human immunodeficiency virus type 1 (HIV-1) resistance in transformed CD4+ monocytic cells treated with multitargeting HIV-1 antisense sequences incorporated into U1 snRNA. J. Virol. *71*, 4079–4085.

Liu, F., and Altman, S. (1995). Inhibition of viral gene expression by the catalytic RNA subunit of RNase P from Escherichia coli. Genes Dev. *9*, 471–480.

Liu, F., and Altman, S. (1996). Requirements for cleavage by a modified RNase P of a small model substrate. Nucleic Acids Res. *24*, 2690–2696.

Liu, P., Gucwa, A., Stover, M.L., Buck, E., Lichtler, A., and Rowe, D. (2002). Analysis of inhibitory action of modified U1 snRNAs on target gene expression: discrimination of two RNA targets differing by a 1 bp mismatch. Nucleic Acids Res. *30*, 2329–2339.

Liu, P., Kronenberg, M., Jiang, X., and Rowe, D. (2004). Modified U1 snRNA suppresses expression of a targeted endogenous RNA by inhibiting polyadenylation of the transcript. Nucleic Acids Res. *32*, 1512–1517.

Liu, P., Stover, M.L., Lichtler, A., and Rowe, D.W. (2005). Modification of human U1 snRNA for inhibition of gene expression at the level of pre-mRNA. Method. Mol. Biol. *309*, 321–332.

Ma, M., Benimetskaya, L., Lebedeva, I., Dignam, J., Takle, G., and Stein, C.A. (2000). Intracellular mRNA cleavage induced through activation of RNase P by nuclease-resistant external guide sequences. Nat. Biotechnol. *18*, 58–61.

Ma, M.Y., Jacob-Samuel, B., Dignam, J.C., Pace, U., Goldberg, A.R., and George, S.T. (1998). Nuclease-resistant external guide sequence-induced cleavage of target RNA by human ribonuclease P. Antisense Nucleic Acid Drug Dev. *8*, 415–426.

Mandel, C.R., Kaneko, S., Zhang, H., Gebauer, D., Vethantham, V., Manley, J.L., and Tong, L. (2006). Polyadenylation factor CPSF-73 is the pre-mRNA 3′-end-processing endonuclease. Nature *444*, 953–956.

Martell, R.E., Nevins, J.R., and Sullenger, B.A. (2002). Optimizing aptamer activity for gene therapy applications using expression cassette SELEX. Mol. Ther. *6*, 30–34.

McClain, W.H., Guerrier-Takada, C., and Altman, S. (1987). Model substrates for an RNA enzyme. Science *238*, 527–530.

McKinney, J., Guerrier-Takada, C., Wesolowski, D., and Altman, S. (2001). Inhibition of Escherichia coli viability by external guide sequences complementary to two essential genes. Proc. Natl. Acad. Sci. USA *98*, 6605–6610.

McKinney, J.S., Zhang, H., Kubori, T., Galan, J.E., and Altman, S. (2004). Disruption of type III secretion in Salmonella enterica serovar Typhimurium by external guide sequences. Nucleic Acids Res. *32*, 848–854.

Mi, J., Zhang, X., Rabbani, Z.N., Liu, Y., Su, Z., Vujaskovic, Z., Kontos, C.D., Sullenger, B.A., and Clary, B.M. (2006). H1 RNA polymerase III promoter-driven expression of an RNA aptamer leads to high-level inhibition of intracellular protein activity. Nucleic Acids Res. *34*, 3577–3584.

Michienzi, A., Li, S., Zaia, J.A., and Rossi, J.J. (2002). A nucleolar TAR decoy inhibitor of HIV-1 replication. Proc. Natl. Acad. Sci. USA *99*, 14047–14052.

Miranda, K.C., Huynh, T., Tay, Y., Ang, Y.S., Tam, W.L., Thomson, A.M., Lim, B., and Rigoutsos, I. (2006). A pattern-based method for the identification of MicroRNA binding sites and their corresponding heteroduplexes. Cell *126*, 1203–1217.

Miyagishi, M., and Taira, K. (2002). U6 promoter-driven siRNAs with four uridine 3′ overhangs efficiently suppress targeted gene expression in mammalian cells. Nat. Biotechnol. *20*, 497–500.

Mohan, A., Whyte, S., Wang, X., Nashimoto, M., and Levinger, L. (1999). The 3′ end CCA of mature tRNA is an antideterminant for eukaryotic 3′-tRNase. RNA *5*, 245–256.

Morl, M., and Marchfelder, A. (2001). The final cut. The importance of tRNA 3′-processing. EMBO Rep. *2*, 17–20.

Morris, K.V. (2006). Therapeutic potential of siRNA-mediated transcriptional gene silencing. Biotechniques Suppl. 7–13.

Morris, K.V., and Rossi, J.J. (2006). Lentiviral-mediated delivery of siRNAs for antiviral therapy. Gene Ther. *13*, 553–558.

Muller, S., Strohbach, D., and Wolf, J. (2006). Sensors made of RNA: tailored ribozymes for detection of small organic molecules, metals, nucleic acids and proteins. IEE Proc NanoBiotechnol. *153*, 31–40.

Nakanishi, K., and Nureki, O. (2005). Recent progress of structural biology of tRNA processing and modification. Mol. Cell *19*, 157–166.

Nakashima, A., Takaku, H., Shibata, H.S., Negishi, Y., Takagi, M., Tamura, M., and Nashimoto, M. (2007). Gene silencing by the tRNA maturase tRNase ZL under the direction of small-guide RNA. Gene Ther. *14*, 78–85.

Nashimoto, M. (1995). Conversion of mammalian tRNA 3′ processing endoribonuclease to four-base-recognizing RNA cutters. Nucleic Acids Res. *23*, 3642–3647.

Nashimoto, M. (1996). Specific cleavage of target RNAs from HIV-1 with 5′ half tRNA by mammalian tRNA 3′ processing endoribonuclease. RNA *2*, 523–524.

Nashimoto, M. (1997). Distribution of both lengths and 5′ terminal nucleotides of mammalian pre-tRNA 3′ trailers reflects properties of 3′ processing endoribonuclease. Nucleic Acids Res. *25*, 1148–1154.

Nashimoto, M. (2000). Anomalous RNA substrates for mammalian tRNA 3′ processing endoribonuclease. FEBS Lett. *472*, 179–186.

Nashimoto, M., Geary, S., Tamura, M., and Kaspar, R. (1998). RNA heptamers that direct RNA cleavage by mammalian tRNA 3′ processing endoribonuclease. Nucleic Acids Res. *26*, 2565–2572.

Nashimoto, M., and Kaspar, R. (1997). Specific cleavage of a target RNA from HIV-1 by mammalian tRNA 3′ processing endoribonuclease directed by an RNA heptamer. Nucleic Acids Symp. Ser. 22–25.

Nashimoto, M., Tamura, M., and Kaspar, R.L. (1999a). Minimum requirements for substrates of mammalian tRNA 3′ processing endoribonuclease. Biochemistry *38*, 12089–12096.

Nashimoto, M., Tamura, M., and Kaspar, R.L. (1999b). Selection of cleavage site by mammalian tRNA

3' processing endoribonuclease. J. Mol. Biol. *287*, 727–740.

Nashimoto, M., Wesemann, D.R., Geary, S., Tamura, M., and Kaspar, R.L. (1999c). Long 5' leaders inhibit removal of a 3' trailer from a precursor tRNA by mammalian tRNA 3' processing endoribonuclease. Nucleic Acids Res. 27, 2770–2776.

Niwa, M., MacDonald, C.C., and Berget, S.M. (1992). Are vertebrate exons scanned during splice-site selection? Nature *360*, 277–280.

Oh, B.K., Frank, D.N., and Pace, N.R. (1998). Participation of the 3'-CCA of tRNA in the binding of catalytic Mg2+ ions by ribonuclease P. Biochemistry 37, 7277–7283.

Oh, B.K., and Pace, N.R. (1994). Interaction of the 3'-end of tRNA with ribonuclease P RNA. Nucleic Acids Res. *22*, 4087–4094.

Paddison, P.J., Caudy, A.A., Bernstein, E., Hannon, G.J., and Conklin, D.S. (2002). Short hairpin RNAs (shRNAs) induce sequence-specific silencing in mammalian cells. Genes Dev. *16*, 948–958.

Paul, C.P., Good, P.D., Winer, I., and Engelke, D.R. (2002). Effective expression of small interfering RNA in human cells. Nat. Biotechnol. *20*, 505–508.

Pei, D.S., Sun, Y.H., Long, Y., and Zhu, Z.Y. (2007). Inhibition of no tail (ntl) gene expression in zebrafish by external guide sequence (EGS) technique. Mol. Biol. Rep. doi:10.1007/s11033-007-9063-9.

Pei, Y., and Tuschl, T. (2006). On the art of identifying effective and specific siRNAs. Nat. Method. 3, 670–676.

Peracchi, A. (2004). Prospects for antiviral ribozymes and deoxyribozymes. Rev. Med. Virol. *14*, 47–64.

Plehn-Dujowich, D., and Altman, S. (1998). Effective inhibition of influenza virus production in cultured cells by external guide sequences and ribonuclease P. Proc. Natl. Acad. Sci. USA 95, 7327–7332.

Que-Gewirth, N.S., and Sullenger, B.A. (2007). Gene therapy progress and prospects: RNA aptamers. Gene Ther. *14*, 283–291.

Raj, S., and Liu, F. (2004). In vitro selection of external guide sequences for directing human RNase P to cleave a target mRNA. Method. Mol. Biol. *252*, 413–424.

Raj, S.M., and Liu, F. (2003). Engineering of RNase P ribozyme for gene-targeting applications. Gene *313*, 59–69.

Reich, C., Olsen, G.J., Pace, B., and Pace, N.R. (1988). Role of the protein moiety of ribonuclease P, a ribonucleoprotein enzyme. Science 239, 178–181.

Reiner, R., Ben-Asouli, Y., Krilovetzky, I., and Jarrous, N. (2006). A role for the catalytic ribonucleoprotein RNase P in RNA polymerase III transcription. Genes Dev. *20*, 1621–1635.

Robbins, M.A., Li, M., Leung, I., Li, H., Boyer, D.V., Song, Y., Behlke, M.A., and Rossi, J.J. (2006). Stable expression of shRNAs in human CD34(+) progenitor cells can avoid induction of interferon responses to siRNAs *in vitro*. Nat. Biotechnol. *24*, 566–571.

Rose, S.D., Kim, D.H., Amarzguioui, M., Heidel, J.D., Collingwood, M.A., Davis, M.E., Rossi, J.J., and Behlke, M.A. (2005). Functional polarity is introduced by Dicer processing of short substrate RNAs. Nucleic Acids Res. 33, 4140–4156.

Rossi, J.J. (1999). The application of ribozymes to HIV infection. Curr. Opin. Mol. Ther. *1*, 316–322.

Rossi, J.J. (2007). Transcriptional activation by small RNA duplexes. Nat. Chem. Biol. 3, 136–137.

Rusk, N. (2007). Expanding the RNA tool box. Nat. Method. *4*, 297.

Sabariegos, R., Gimenez-Barcons, M., Tapia, N., Clotet, B., and Martinez, M.A. (2006). Sequence homology required by human immunodeficiency virus type 1 to escape from short interfering RNAs. J. Virol. *80*, 571–577.

Saetrom, P., Heale, B.S., Snove, O., Jr., Aagaard, L., Alluin, J., and Rossi, J.J. (2007). Distance constraints between microRNA target sites dictate efficacy and cooperativity. Nucleic Acids Res. 35, 2333–2342.

Sajic, R., Lee, K., Asai, K., Sakac, D., Branch, D.R., Upton, C., and Cochrane, A. (2007). Use of modified U1 snRNAs to inhibit HIV-1 replication. Nucleic Acids Res. 35, 247–255.

Scherer, L.J., Frank, R., and Rossi, J.J. (2007). Optimization and characterization of tRNA-shRNA expression constructs. Nucleic Acids Res 35, 2620–2628.

Scherer, L.J., and Rossi, J.J. (2007). Progress and prospects: RNA-based therapies for treatment of HIV infection. Gene Ther 14, 1057–1064.

Scherr, M., LeBon, J., Castanotto, D., Cunliffe, H.E., Meltzer, P.S., Ganser, A., Riggs, A.D., and Rossi, J.J. (2001). Detection of antisense and ribozyme accessible sites on native mRNAs: application to NCOA3 mRNA. Mol. Ther. *4*, 454–460.

Scherr, M., Reed, M., Huang, C.F., Riggs, A.D., and Rossi, J.J. (2000). Oligonucleotide scanning of native mRNAs in extracts predicts intracellular ribozyme efficiency: ribozyme-mediated reduction of the murine DNA methyltransferase. Mol. Ther. *2*, 26–38.

Scherr, M., and Rossi, J.J. (1998). Rapid determination and quantitation of the accessibility to native RNAs by antisense oligodeoxynucleotides in murine cell extracts. Nucleic Acids Res. *26*, 5079–5085.

Schiffer, S., Rosch, S., and Marchfelder, A. (2002). Assigning a function to a conserved group of proteins: the tRNA 3'-processing enzymes. EMBO J. *21*, 2769–2777.

Schwarz, D.S., Hutvagner, G., Du, T., Xu, Z., Aronin, N., and Zamore, P.D. (2003). Asymmetry in the assembly of the RNAi enzyme complex. Cell *115*, 199–208.

Sharin, E., Schein, A., Mann, H., Ben-Asouli, Y., and Jarrous, N. (2005). RNase P: role of distinct protein cofactors in tRNA substrate recognition and RNA-based catalysis. Nucleic Acids Res. 33, 5120–5132.

Shibata, H.S., Takaku, H., Takagi, M., and Nashimoto, M. (2005). The T loop structure is dispensable for substrate recognition by tRNase ZL. J. Biol. Chem. *280*, 22326–22334.

Siolas, D., Lerner, C., Burchard, J., Ge, W., Linsley, P.S., Paddison, P.J., Hannon, G.J., and Cleary, M.A. (2005). Synthetic shRNAs as potent RNAi triggers. Nat. Biotechnol. *23*, 227–231.

Snove, O., Jr., and Rossi, J.J. (2006). Expressing short hairpin RNAs *in vivo*. Nat. Method. 3, 689–695.

Soler Bistue, A.J., Ha, H., Sarno, R., Don, M., Zorreguieta, A., and Tolmasky, M.E. (2007). External Guide Sequences Targeting the aac(6')-Ib mRNA Induce Inhibition of Amikacin Resistance. Antimicrob Agents Chemother.

Sprinzl, M. (2006). Chemistry of aminoacylation and peptide bond formation on the 3'terminus of tRNA. J. Biosci 31, 489–496.

Sullenger, B.A., Gallardo, H.F., Ungers, G.E., and Gilboa, E. (1990). Overexpression of TAR sequences renders cells resistant to human immunodeficiency virus replication. Cell 63, 601–608.

Takaku, H., Minagawa, A., Takagi, M., and Nashimoto, M. (2003). A candidate prostate cancer susceptibility gene encodes tRNA 3' processing endoribonuclease. Nucleic Acids Res. 31, 2272–2278.

Takaku, H., Minagawa, A., Takagi, M., and Nashimoto, M. (2004). A novel 4-base-recognizing RNA cutter that can remove the single 3' terminal nucleotides from RNA molecules. Nucleic Acids Res. 32, e91.

Tavtigian, S.V., Simard, J., Teng, D.H., Abtin, V., Baumgard, M., Beck, A., Camp, N.J., Carillo, A.R., Chen, Y., Dayananth, P., *et al.* (2001). A candidate prostate cancer susceptibility gene at chromosome 17p. Nat. Genet. 27, 172–180.

Trang, P., Kilani, A., Kim, J., and Liu, F. (2000a). A ribozyme derived from the catalytic subunit of RNase P from Escherichia coli is highly effective in inhibiting replication of herpes simplex virus 1. J. Mol. Biol. 301, 817–826.

Trang, P., Kim, K., and Liu, F. (2004). Developing RNase P ribozymes for gene-targeting and antiviral therapy. Cell MicroBiol. 6, 499–508.

Trang, P., Kim, K., Zhu, J., and Liu, F. (2003). Expression of an RNase P ribozyme against the mRNA encoding human cytomegalovirus protease inhibits viral capsid protein processing and growth. J. Mol. Biol. 328, 1123–1135.

Trang, P., Lee, J., Kilani, A.F., Kim, J., and Liu, F. (2001). Effective inhibition of herpes simplex virus 1 gene expression and growth by engineered RNase P ribozyme. Nucleic Acids Res. 29, 5071–5078.

Trang, P., Lee, M., Nepomuceno, E., Kim, J., Zhu, H., and Liu, F. (2000b). Effective inhibition of human cytomegalovirus gene expression and replication by a ribozyme derived from the catalytic RNA subunit of RNase P from Escherichia coli. Proc. Natl. Acad. Sci. USA 97, 5812–5817.

Unwalla, H.J., Li, H.T., Bahner, I., Li, M.J., Kohn, D., and Rossi, J.J. (2006). Novel Pol II fusion promoter directs human immunodeficiency virus type 1-inducible coexpression of a short hairpin RNA and protein. J. Virol. 80, 1863–1873.

Unwalla, H.J., Li, M.J., Kim, J.D., Li, H.T., Ehsani, A., Alluin, J., and Rossi, J.J. (2004). Negative feedback inhibition of HIV-1 by TAT-inducible expression of siRNA. Nat. Biotechnol. 22, 1573–1578.

Vermeulen, A., Behlen, L., Reynolds, A., Wolfson, A., Marshall, W.S., Karpilow, J., and Khvorova, A. (2005). The contributions of dsRNA structure to Dicer specificity and efficiency. RNA 11, 674–682.

Vogel, A., Schilling, O., Spath, B., and Marchfelder, A. (2005). The tRNase Z family of proteins: physiological functions, substrate specificity and structural properties. Biol. Chem. 386, 1253–1264.

Wegscheid, B., and Hartmann, R.K. (2006). The precursor tRNA 3'-CCA interaction with Escherichia coli RNase P RNA is essential for catalysis by RNase P *in vivo*. RNA 12, 2135–2148.

Weiner, A.M. (2004). tRNA maturation: RNA polymerization without a nucleic acid template. Curr. Biol. 14, R883–885.

Weiner, A.M. (2005). E Pluribus Unum: 3' end formation of polyadenylated mRNAs, histone mRNAs, and U snRNAs. Mol. Cell 20, 168–170.

Werner, M., Rosa, E., Nordstrom, J.L., Goldberg, A.R., and George, S.T. (1998). Short oligonucleotides as external guide sequences for site-specific cleavage of RNA molecules with human RNase P. RNA 4, 847–855.

Westerhout, E.M., Ooms, M., Vink, M., Das, A.T., and Berkhout, B. (2005). HIV-1 can escape from RNA interference by evolving an alternative structure in its RNA genome. Nucleic Acids Res. 33, 796–804.

Will, C.L., and Lührmann, R. (2006). Spliceosome structure and function. The RNA World Vol 3. (Cold Spring Harbor, NY: Cold Spring Harbor Laboratory Press), pp. 369–400.

Willingham, A.T., and Gingeras, T.R. (2006). TUF love for 'junk' DNA. Cell 125, 1215–1220.

Wiznerowicz, M., Szulc, J., and Trono, D. (2006). Tuning silence: conditional systems for RNA interference. Nat. Method. 3, 682–688.

Wolin, S.L., and Matera, A.G. (1999). The trials and travels of tRNA. Genes Dev. 13, 1–10.

Xiao, S., Scott, F., Fierke, C.A., and Engelke, D.R. (2002). Eukaryotic ribonuclease P: a plurality of ribonucleoprotein enzymes. Annu. Rev. Biochem. 71, 165–189.

Xiong, Y., and Steitz, T.A. (2006). A story with a good ending: tRNA 3'-end maturation by CCA-adding enzymes. Curr. Opin. Struct. Biol. 16, 12–17.

Yan, H., Zareen, N., and Levinger, L. (2006). Naturally occurring mutations in human mitochondrial pre-tRNASer(UCN) can affect the transfer ribonuclease Z cleavage site, processing kinetics, and substrate secondary structure. J. Biol. Chem. 281, 3926–3935.

Yang, Y.H., Li, H., Zhou, T., Kim, K., and Liu, F. (2006). Engineered external guide sequences are highly effective in inducing RNase P for inhibition of gene expression and replication of human cytomegalovirus. Nucleic Acids Res. 34, 575–583.

Yen, L., Gonzalez-Zulueta, M., Feldman, A., Yuan, Y., Fryer, H., Dawson, T., Dawson, V., and Kalb, R.G. (2001). Reduction of functional N-methyl-D-aspartate receptors in neurons by RNase P-mediated cleavage of the NR1 mRNA. J. Neurochem. 76, 1386–1394.

Yoo, J.W., Hong, S.W., Kim, S., and Lee, D.K. (2006). Inflammatory cytokine induction by siRNAs is cell type- and transfection reagent-specific. Biochem. Biophys. Res. Commun. 347, 1053–1058.

Yu, X., Trang, P., Shah, S., Atanasov, I., Kim, Y.H., Bai, Y., Zhou, Z.H., and Liu, F. (2005). Dissecting human cytomegalovirus gene function and capsid maturation by ribozyme targeting and electron cryomicroscopy. Proc. Natl. Acad. Sci. USA *102*, 7103–7108.

Yuan, Y., and Altman, S. (1994). Selection of guide sequences that direct efficient cleavage of mRNA by human ribonuclease P. Science *263*, 1269–1273.

Yuan, Y., and Altman, S. (1995). Substrate recognition by human RNase P: identification of small, model substrates for the enzyme. EMBO J. *14*, 159–168.

Yuan, Y., Hwang, E.S., and Altman, S. (1992). Targeted cleavage of mRNA by human RNase P. Proc. Natl. Acad. Sci. USA *89*, 8006–8010.

Zareen, N., Hopkinson, A., and Levinger, L. (2006). Residues in two homology blocks on the amino side of the tRNase Z His domain contribute unexpectedly to pre-tRNA 3′ end processing. RNA *12*, 1104–1115.

Zaug, A.J., and Cech, T.R. (1995). Analysis of the structure of Tetrahymena nuclear RNAs *in vivo*: telomerase RNA, the self-splicing rRNA intron, and U2 snRNA. RNA *1*, 363–374.

Zhang, H., and Altman, S. (2004). Inhibition of the expression of the human RNase P protein subunits Rpp21, Rpp25, Rpp29 by external guide sequences (EGSs) and siRNA. J. Mol. Biol. *342*, 1077–1083.

Zhao, J., Hyman, L., and Moore, C. (1999). Formation of mRNA 3′ ends in eukaryotes: mechanism, regulation, and interrelationships with other steps in mRNA synthesis. MicroBiol. Mol. Biol. Rev *63*, 405–445.

Zheng, Z.M., and Baker, C.C. (2006). Papillomavirus genome structure, expression, and post-transcriptional regulation. Front Biosci *11*, 2286–2302.

Zhou, T., Kim, J., Kilani, A.F., Kim, K., Dunn, W., Jo, S., Nepomuceno, E., and Liu, F. (2002). In vitro selection of external guide sequences for directing RNase P-mediated inhibition of viral gene expression. J. Biol. Chem. *277*, 30112–30120.

Zhu, J., Trang, P., Kim, K., Zhou, T., Deng, H., and Liu, F. (2004). Effective inhibition of Rta expression and lytic replication of Kaposi's sarcoma-associated herpesvirus by human RNase P. Proc. Natl. Acad. Sci. USA *101*, 9073–9078.

Zou, H., Chan, K., Trang, P., and Liu, F. (2004a). General design and construction of RNase P ribozymes for gene-targeting applications. Method. Mol. Biol. *252*, 385–398.

Zou, H., Lee, J., Kilani, A.F., Kim, K., Trang, P., Kim, J., and Liu, F. (2004b). Engineered RNase P ribozymes increase their cleavage activities and efficacies in inhibiting viral gene expression in cells by enhancing the rate of cleavage and binding of the target mRNA. J. Biol. Chem. *279*, 32063–32070.

Zou, H., Lee, J., Umamoto, S., Kilani, A.F., Kim, J., Trang, P., Zhou, T., and Liu, F. (2003). Engineered RNase P ribozymes are efficient in cleaving a human cytomegalovirus mRNA *in vitro* and are effective in inhibiting viral gene expression and growth in human cells. J. Biol. Chem. *278*, 37265–37274.

Index

A

Adenosine deaminase (ADAR1) 67
Antigen PNA 166–167; *see also* peptide nucleic acids (PNAs)
Antisense Igf2r RNA 143–144
Aptamers 203–204
Arabidopsis 2, 38, 45, 68, 76–77, 93, 111, 120, 175, 177
Argonaute 31–32, 35, 50, 54–55, 60–62, 65, 75, 80–82, 86, 110, 120–122, 159, 168, 171, 174–175, 178, 182; *see also* RISC; RNAi

B

B2 RNA 138–139; *see also* ncRNAs
Borna disease virus 9–10

C

Caenorhabditis elegans 73–74, 79–80, 91, 112, 150, 173, 175–177, 216
CCR5 35, 181
Cis cleaving 6–17
Chromatin 19–25, 53, 65, 68, 143, 168, 178–179, 182–185, 191; *see also* epigenetic; histone; nucleosome
Cyclin D1 181

D

DHFR; *see* Dihydrofolate reductase
Dicer 119–121, 125, 174–175; *see also* RISC; RNAi
Dihydrofolate reductase (DHFR) 135, 139–140
DNA methylation 14–15, 19, 24–25, 34–35, 45, 102, 115–116, 124, 144, 168, 173, 175, 177–178, 191–192, 195
DNMT3a 24, 34–35, 177–179, 182, 184; *see also* epigenetic
DNA transposons 112–113, 118–119, 195
Drosha 61, 74, 82, 119; *see also* RISC; RNAi
Drosophila 29, 45, 59, 61–67, 75, 80, 81, 86, 112, 120, 150, 173, 175–177
Duplex DNA 166, 170

E

Enhancer of zeste homologue (Ezh2) 24, 35, 178–179, 182
Encephalomyocarditis virus 9–10

ENSEMBL 154,
Epigenetic 19–25, 177; *see also* chromatin
Evf-2 RNA 138–139; *see also* ncRNA

H

Hammerhead ribozyme 2–12, 203, 219
HCMV 211
HCV 101, 201
Hepatitis B virus 10
Histone acetyltransferase 22–23, 191, 196
Histone deacetylase 22–23, 46, 55, 137, 177–178, 182, 196
Histone 19–24, 31–39, 45–50, 66, 168, 174–178, 194
Histone code 19
Histone methylation 31–39, 174–178, 182, 194, 196
HIV-1 7–10, 25, 75, 91–105, 114, 141–142, 174, 201–203, 218–220
HOX 29, 31, 35–40, 78; *see also* Polycomb
HP1 23, 24, 31, 53, 66–68, 178, 180
HSR-1 136–137; *see also* ncRNA

J

Junk DNA 150, 160

L

Let-7 73–74, 78–79, 83–84
Lin-4 73–74, 78
Lin-14 74, 78,
LINE 35, 112, 114–115, 117, 119
L1 35, 112, 115–116, 118, 122–124

M

M1 guide sequence (M1GS) 210
microRNA (miRNA) 35, 73–87, 90–105, 109–110, 125–126, 149, 189, 192, 195, 217–219

N

ncRNAs 133–144, 190–191, 195
NF-κB 181
NRON 133–136; *see also* ncRNA
Nuclear factor of activated T cells (NFAT) 134, 136
Nucleosome 19–25, 182; *see also* epigenetic

P

PACT 110, 174
P-bodies 82–83; *see also* RNAi; RISC
Peptide nucleic acids (PNAs) 165–171
piRNA 110, 121–123, 158–160
PIWI 112, 121–123, 158, 175–177
PNA *see* Peptide nucleic acids
Poly A polymerase (PAP) 205–206
Polycomb 31, 34, 39, 64, 68, 143, 176
Post-transcriptional gene silencing (PTGS) 29, 30–33, 63, 173–177, 216
Progesterone receptor (PR) 166–171, 191–192
Pykons 110, 148–160

R

RACE 5' 167
RDRC 46, 51–54
Retrotransposons 114, 115–116, 118, 122–123, 138
Ribozyme 1–12, 143, 189, 203, 208–210, 219
RISC 29, 60–63, 66–67, 109, 173–174
RITS 31, 46, 49–55, 65, 175–176
RNA activation 189–196
RNA-dependent RNA polymerase (RDRP) 30, 33, 45, 54, 63; *see also* RNAi
RNA interference 19–40, 45–55, 59–61, 65, 91, 104, 109, 111, 150, 156–158, 173–174, 189, 201–202, 216
RNAse P 208–213, 216
RNAi 6–7, 9, 11, 13, 29–40, 45–55, 59–68, 82, 91–92, 98, 102–104, 109–112, 117–118, 120, 127, 134, 150, 156, 158, 165, 168, 174–180, 184, 189–194, 201, 202, 213, 216–219
RNA polymerase II (RNAPII) 8, 19, 21–22, 24, 35, 37, 39, 46, 49–50, 117, 133, 139, 174, 176–177, 181–182, 203, 207

S

Schizosaccharomyces pombe (*S. pombe*) 45, 173, 175–179
SELEX 7, 142, 203, 212,
7SK RNA 140
SINE 117, 123–125
siRNA 173; *see also* RNAi
Sleeping disease virus 9–10
Small activating RNA 189–197
SRA 136–137; *see also* ncRNA
SUP alleles 175,
Suv39H6 176–178

T

TAR RNA 141, 174, 204; *see also* HIV-1
TAT 218; *see also* HIV-1
Tetrahymena 31
Trans cleaving 5–12
Transcriptional gene silencing (TGS) 29, 33–34, 48, 63, 100, 109, 173–177, 216
TRBP 178, 182; *see also* HIV-1
tRNA 207–210, 214
Tsix 23, 143

U

U1 RNA 140
U1 snRNP 205–206

V

VEGF 165

X

X-chromosome 32–34, 143, 190–191
Xist 33, 34, 143

Plate 10.1 Combinatorial arrangements of pyknons in 3′UTRs – reproduced from Rigoutsos et al., 2006. (Copyright 2006, National Academy of Sciences U.S.A.).

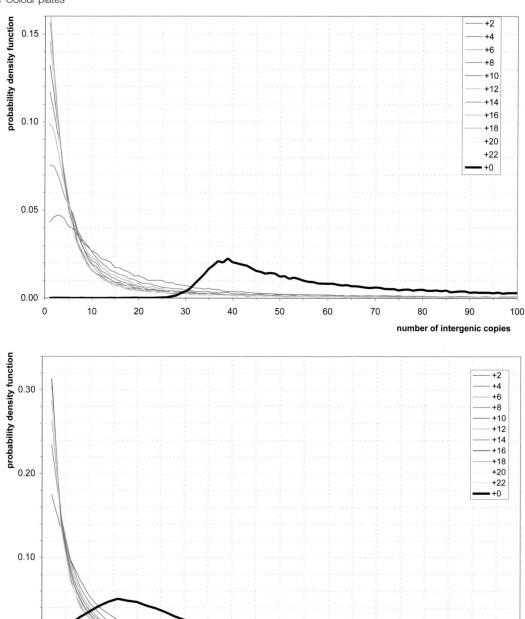

Plate 10.2 Pyknons and their natural boundaries. Probability density functions for the number of *intergenic* (top) and *intronic* (bottom) copies of the variable-length strings derived by counting the instances of 3′UTR-conserved pyknons after having shifted them to the right by *d* positions. In both cases, the black curves correspond to $d=0$, i.e. to the pyknons in $P_{3'UTR}$. (Copyright 2006, national Academy of Sciences, U.S.A.).

Plate 12.6 Mechanistic insights into RNA-directed transcriptional gene silencing in human cells. Small antisense RNAs can be generated to specifically recognize gene promoter regions ultimately resulting in transcriptional silencing of the targeted genes expression. How the small RNAs recognize the promoter has remained an enigma. Recent direct evidence shows that a low-copy promoter associated RNA (grey) is present at the targeted gene/promoter loci and becomes bound by the small antisense RNAs (Han, 2007). The small RNAs (red) are thought to recruit or be in complex with a transcriptional silencing complex which can then direct specific epigenetic modifications to the local chromatin leading at the targeted loci leading ultimately to transcriptional silencing of the antisense RNA targeted gene.

Other Books of Interest

Plasmids: Current Research and Future Trends   2008

Pasteurellaceae: Biology, Genomics and Molecular Aspects   2008

Vibrio cholerae: Genomics and Molecular Biology   2008

Pathogenic Fungi: Insights in Molecular Biology   2008

Helicobacter pylori: Molecular Genetics and Cellular Biology   2008

Corynebacteria: Genomics and Molecular Biology   2008

Staphylococcus: Molecular Genetics   2008

Leishmania: After The Genome   2008

Archaea: New Models for Prokaryotic Biology   2008

RNA and the Regulation of Gene Expression   2008

Legionella Molecular Microbiology   2008

Molecular Oral Microbiology   2008

Epigenetics   2008

Animal Viruses: Molecular Biology   2008

Segmented Double-Stranded RNA Viruses   2008

Acinetobacter Molecular Biology   2008

Pseudomonas: Genomics and Molecular Biology   2008

Microbial Biodegradation: Genomics and Molecular Biology   2008

The Cyanobacteria: Molecular Biology, Genomics and Evolution   2008

Coronaviruses: Molecular and Cellular Biology   2007

Real-Time PCR in Microbiology: From Diagnosis to Characterisation   2007

Bacteriophage: Genetics and Molecular Biology   2007

Candida: Comparative and Functional Genomics   2007

Bacillus: Cellular and Molecular Biology   2007

AIDS Vaccine Development: Challenges and Opportunities   2007

Alpha Herpesviruses: Molecular and Cellular Biology   2007

Pathogenic *Treponema*: Molecular and Cellular Biology   2007

PCR Troubleshooting: The Essential Guide   2006

Influenza Virology: Current Topics   2006

Microbial Subversion of Immunity: Current Topics   2006

Cytomegaloviruses: Molecular Biology and Immunology   2006

Papillomavirus Research: From Natural History To Vaccines and Beyond   2006

Epstein Barr Virus   2005

HIV Chemotherapy: A Critical Review   2005

Probiotics and Prebiotics: Scientific Aspects   2005

Foodborne Pathogens: Microbiology and Molecular Biology   2005

Caister Academic Press   www.caister.com